Pelagic Snails

The Biology of Holoplanktonic Gastropod Mollusks

Pelagic Snails

The Biology of Holoplanktonic Gastropod Mollusks

Carol M. Lalli and Ronald W. Gilmer

Stanford University Press
Stanford, California 1989

Stanford University Press, Stanford, California

Printed in the United States of America

CIP data appear at the end of the book

Published with the assistance of the
Harbor Branch Oceanographic Institute

We dedicate this book to
Lyle E. Koester,
who nurtured an appreciation
and love of nature in his daughter,
and to
Ralph A. and Lorna B. Gilmer,
whose delights in science and nature
carried over to their son.

Preface

The purpose of this book is to draw attention to some unusual and poorly known gastropods that are highly specialized for life in the open ocean. These mollusks, unlike most species in their phylum, live out their entire life cycles in an environment without solid substrates. Consequently, they have become adept at swimming or floating or attaching to drifting objects and, as members of the planktonic community, they have developed life styles quite different from those of their benthic relatives.

Although these animals have been known and studied since the late seventeenth century, much of the information has remained scattered in scientific journals, expedition reports, or specialized monographs. Because this is particularly true of information relating to the biology and ecology of planktonic gastropods, the book concentrates on these topics. Less attention is given to taxonomy and distribution, since these areas have been considered fully in other works, but each chapter does conclude with a list of described species and references to published identification keys. Details of internal anatomy have also been neglected here, except where these are directly relevant to biological topics.

We have often noticed, in working with students new to a scientific field, that many become overwhelmed by the sheer amount of literature that has accumulated and assume that there is nothing further to be discovered. To forestall this erroneous impression, we have tried to point out where published observations are contradictory and where more research is needed.

We have tried to use only photographs or drawings that accurately portray living animals, to give the reader a more palpable sense of these unusual mollusks. Many previously published figures have represented pelagic mollusks as they appear after collection by nets and rapid preservation in formaldehyde or alcohol; these often present distorted creatures that bear little or no resemblance to the live animals. Information on the biology of planktonic gastropods often has been based upon interpretations of anatomical studies of preserved specimens. To correct or expand upon that earlier work, we have relied extensively upon recent and previously unpublished *in situ* biological observations of live animals.

We hope that this book will prove useful to invertebrate zoologists, marine biologists, biological and geological oceanographers, zooplanktologists, and professional and amateur malacologists—and of interest to all those who appreciate the beauty and diversity of the adaptations of animals to their environments.

We wish to acknowledge the generosity of Dr. T. R. Parsons and Dr. G. R. Harbison for providing space, technical facilities, and encouragement in this project. Some financial support for R.W.G. was provided by National Science Foundation grants OCE 82-09341 and OCE 85-16083 to G. R. Harbison. We also acknowledge the support of the Woods Hole Oceanographic Institution, and we especially thank the Harbor Branch Oceanographic Institution for providing publication expenses for the color figures.

We thank Drs. M. Arai, P. Cornelius, and F. Boero for their helpful advice about the hydroids that grow on shelled pteropods, Dr. F. G. Hochberg for sharing his knowledge of the parasites that infect planktonic gastropods, and Dr. T. Pafort for discussions regarding her findings about reproduction in species of *Clio*. Dr. Catherine Thiriot very kindly supplied the scanning electron micrographs of radulae and larval shells of heteropods, and G. R. Harbison allowed us to use his unpublished observations and photographs of living *Hydromyles*.

P. Linley provided technical assistance in preparing scanning electron micrographs of radulae. B. Pratt, B. Rokeby, and E. Carefoot generously provided illustrative assistance, and T. Smoyer assisted in photographic reproductions. Certain photographs were kindly provided by M. Jones, J. Carlton, and R. Reichelt. We also wish to thank the anonymous reviewer of the manuscript for the publisher.

This work is contribution #615 of the Harbor Branch Oceanographic Institution.

C.M.L.
R.W.G.

Contents

1 Introduction 1
References Cited 6

2 The Janthinid Snails
Raft Builders 8
External Anatomy 9
Float Formation 11
Food and Feeding 12
Reproduction and Development 15
Epifauna and Parasites 20
Evolution 22
List of Recognized Species 23
References Cited 24

3 The Heteropods
Visual Predators 27
External Anatomy 29
Swimming and Buoyancy 34
Food and Feeding 37
Reproduction and Development 44
Epifauna and Parasites 49
Shell Mineralogy and Sediments 50
Evolution 51
List of Recognized Species 52
References Cited 53

4 The Thecosomes
Shelled Pteropods 58
External Anatomy 62
Swimming and Buoyancy 73

Food, Feeding, and Trophic Relationships 80
Reproduction and Development 100
Epifauna and Parasites 126
Pteropods as Ecological and Palaeoecological Indicators 131
Pteropods and the Carbonate Cycle 139
Evolution 145
List of Recognized Species 150
References Cited 152

5 The Gymnosomes
Shell-less Pteropods 167

Swimming, External Anatomy, and Feeding Structures 168
Food and Feeding 178
Reproduction and Development 189
Parasites 199
Suborder Gymnoptera 199
Evolution 204
List of Recognized Species 208
References Cited 209

6 The Planktonic Nudibranchs
Swimming Sea Slugs 214

Family Phylliroidae 215
Family Glaucidae 224
Family Fionidae 229
Evolution 232
List of Recognized Species 234
References Cited 234

Glossary 239
Systematic Index 249
Subject Index 253

Tables and Figures

Tables

1 Major taxonomic divisions of the Phylum Mollusca 2
2 Major taxonomic divisions of the holoplanktonic gastropods 4
3 Characteristics of egg capsules and numbers of young produced by oviparous species of *Janthina* 20
4 Respiration rates of *Cardiapoda placenta* 43
5 *In situ* swimming rates of thecosomes 75
6 Relationship between shell length and the diameter of mucous feeding webs of euthecosomes 84
7 Gut-content analyses in the Limacinidae 85
8 Gut-content analyses in the Cavoliniidae 87
9 Gut-content analyses in the Pseudothecosomata 92
10 Chemical constituents of Euthecosomata 94
11 Minimal energy requirements and assimilation efficiency of carbon for *Cavolinia longirostris* 97
12 Respiration and ammonium-excretion rates of *Corolla* sp. and *Gleba cordata* 98
13 Numbers of egg masses and eggs deposited by *Limacina helicina* and *L. retroversa* 104
14 Growth rates and times required to attain maturational stages in tropical species of *Limacina* 109
15 Size relationships of mating cavoliniids 110
16 Hydrozoan epifauna associated with euthecosomes 128
17 Infestation of euthecosomes by the parasitic copepod *Cardiodectes medusaeus* 130
18 Relative population densities of dominant euthecosomes in the plankton and in sediments off Barbados, West Indies 136
19 Sinking rates of euthecosomes under experimental conditions 143

20 Feeding structures in gymnosome genera 174
21 Comparison of numbers of *Limacina retroversa* and *L. helicina* eaten by *Clione limacina* 181
22 Growth and feeding rates for *Clione limacina* fed on either large or small *Limacina retroversa* 184
23 Prey species of gymnosomatous pteropods 188
24 Fecundity of *Paedoclione doliiformis* at different temperatures and food concentrations 193
25 Numbers of eggs produced by *Clione limacina* from the Denmark Strait and Labrador Sea 197

Figures

1 Janthinid snails 10
2 Radular teeth of *Janthina prolongata* 13
3 Reproductive anatomy of *Janthina pallida* 16
4 A spermatozeugma of *Janthina pallida* 17
5 Egg capsules of *Janthina umbilicata* 19
6 A janthinid veliger shell 21
7 Atlantid heteropods 29
8 Carinariid heteropods 31
9 Pigmentation in *Cardiapoda* species 32
10 Pterotracheid heteropods 33
11 A quiescent atlantid heteropod with attached mucous strands 36
12 Heteropod radulae 38
13 Heteropod veliger larvae 46
14 Veliger shells of heteropods 47
15 *Limacina retroversa* 62
16 *Limacina helicina* 63
17 Cavoliniid thecosomes 64
18 Mantle appendages of *Cavolinia uncinata* 66
19 *Peraclis reticulata* 67
20 Comparative morphology of *Limacina*, *Peraclis*, and *Cymbulia* 68
21 *Corolla* sp. 70
22 *Gleba cordata* 71
23 *Desmopterus papilio* 73
24 Swimming in *Limacina retroversa* 74
25 Feeding in euthecosomes 81
26 *Cavolinia uncinata* feeding with a mucous web 82
27 Radular teeth of thecosomes 83

28 Feeding in pseudothecosomes 89
29 *Gleba cordata* in the feeding position 90
30 Particle-size distribution of food captured in the feeding webs of *Gleba cordata* 91
31 Reproductive anatomy of *Limacina* 101
32 Reproduction in *Limacina helicina* 102
33 Developmental stages of *Limacina retroversa* 105
34 The veliger of *Limacina helicina* 106
35 Reproduction in *Limacina inflata* 107
36 Mating in *Cavolinia uncinata* 111
37 Reproduction in *Diacria trispinosa* 111
38 A "metamorphosed" individual of *Clio pyramidata* after net collection 113
39 Morphological changes in euthecosomes resulting from chemical preservation 114
40 Development in *Cavolinia uncinata* 115
41 Shell weight versus tissue body weight for *Clio pyramidata* 118
42 Microstructure of thecosome shells 120
43 Growth of the shell of *Cuvierina columnella* 121
44 Spawning in *Corolla* sp. 124
45 Development in *Gleba* 125
46 Growth of the pseudoconch and wingplate of *Corolla* sp. 126
47 Hydroid epifauna of euthecosomes 127
48 Vertical distribution of thecosomes in Subantarctic and Antarctic waters 133
49 The global distribution of pteropod shells in marine sediments 135
50 Gymnosomatous pteropods 169
51 Swimming in *Clione limacina* 170
52 *Spongiobranchaea australis* 171
53 *Paedoclione doliiformis* 172
54 *Crucibranchaea* sp. in the resting and feeding positions 173
55 Gymnosome radulae and hooks 176
56 The buccal mass of *Crucibranchaea* sp. 177
57 *Clione limacina* feeding on *Limacina helicina* 179
58 The buccal mass of *Clione limacina* 180
59 Size selection of prey by *Clione limacina* 182
60 Feeding rates of *Clione limacina* 183
61 *Paedoclione doliiformis* feeding on *Limacina retroversa* 185
62 Gymnosome predation on thecosomes 187
63 Reproduction in gymnosomes 191

64 Spawning in *Paedoclione doliiformis* 192
65 An egg of *Paedoclione doliiformis* 194
66 The veliger of *Paedoclione doliiformis* 194
67 Veliger shells of *Paedoclione doliiformis* 195
68 Metamorphosis in *Paedoclione doliiformis* 196
69 The veliger shell of *Clione limacina* 198
70 *Hydromyles gaudichaudii* 200
71 *Phylliroë* species 216
72 The life cycle of *Phylliroë bucephala* 218
73 *Cephalopyge trematoides* 222
74 *Glaucus atlanticus* 225
75 *Glaucilla marginata* 226
76 Radular teeth of *Glaucus atlanticus* 227

Eight pages of color photographs follow page 80

Pelagic Snails

The Biology of Holoplanktonic Gastropod Mollusks

I

Introduction

The Phylum Mollusca probably comprises at least 100,000 described and undescribed living species, making it the second largest animal phylum after arthropods. The evolutionary success of this group is evident not only in terms of numbers of species, but also in the abundance of many species, in the appearance of mollusks in almost all types of habitats, and, as Russell-Hunter (1983) has pointed out, by the fact that the biomass of lower trophic levels of many marine bottom communities (and some non-marine ecosystems) is often dominated by mollusks.

The mollusks are divided according to common anatomical traits into seven classes (Table 1). Some of the most familiar belong to the Class Gastropoda, which contains the prosobranch sea snails, each with a coiled, univalve shell; their shelled or unshelled opisthobranch relatives, such as the nudibranchs and other sea slugs; and the largely terrestrial and freshwater pulmonate snails and slugs, many of them shelled, many unshelled. Equally well known are the Class Bivalvia, encompassing the bivalved clams; the Class Cephalopoda, which includes the multi-armed squids and octopuses; and the Class Polyplacophora, comprising the dorsoventrally flattened chitons, each with a shell divided into eight plates. The remaining classes contain far fewer species and less well-known members. These are the Class Scaphopoda, containing burrowers with elongate, tusk-shaped shells; the Class Monoplacophora, which embraces deep-sea limpet-like species with a univalve shell but suggestions of segmentation in the internal anatomy; and the Class Aplacophora, composed of worm-shaped animals with no shell but with the body surface covered with calcareous spicules.

Their plasticity of anatomical form and their equally diverse ways of performing physiological functions have made the mollusks highly successful invaders of both freshwater and terrestrial environments. More than half of the mollusks, however, have remained marine inhabitants, and most of these

TABLE I
Major taxonomic divisions of the Phylum Mollusca

Taxon	Approximate number of described species[a]	Adult habitat	
		Benthic	Pelagic
Class Monoplacophora	10	all	
Class Aplacophora	250	all	
Class Polyplacophora	650	all	
Class Gastropoda	67,000		
Subclass Prosobranchia		most	some
Subclass Opisthobranchia	3,000[b]	most	some
Subclass Pulmonata		all	
Class Bivalvia	15,000 (31,000[c])	all	
Class Scaphopoda	350	all	
Class Cephalopoda	1,000	some	most
TOTAL	87,000 (103,000)	most	some

[a] From Solem, 1974.
[b] From Thompson, 1976.
[c] From Russell-Hunter, 1983.

marine species spend their adult lives associated with the sea floor, where they form part of the benthic or bottom communities. These marine benthic mollusks either attach to a hard substrate, burrow or bore into the sea bottom, or crawl over the sea floor. Representative species of mollusks can be found at all depths, from the highest intertidal reaches to the deepest parts of the ocean. Many of these benthic species do, however, spend their larval life either floating or swimming in the water column before settling to the bottom to assume their adult form and habits. These small, free-swimming, veliger larvae are part of the meroplankton, or temporary plankton. Depending on the species, the larvae may spend only a few hours in the water column or they may drift in the water currents for several weeks or months, thereby dispersing the species by being carried to different areas before settling and metamorphosing on the sea floor.

Despite the fact that the water column itself offers a much larger environmental volume than the sea floor, few mollusks have succeeded in adapting to the pelagic realm to the extent that they spend their entire lives as swimmers or drifters. The most familiar pelagic mollusks are the cephalopods, notably the squids, which rival fish in their swimming ability. Squids, fishes, and marine mammals are referred to collectively as nekton, from the Greek word for swimmers. The term is applied to those larger marine animals that are powerful enough to swim against a current and thus can control and direct their own movements in the ocean.

Cephalopods first assumed a pelagic existence by partitioning an external shell into chambers. Only the largest and newest chamber was occupied by the animal itself; the smaller and partially closed chambers were filled with gas secreted by the blood of the animal. Although fossil evidence clearly demonstrates that this buoyancy mechanism was present in many now-extinct cephalopods, only one living cephalopod, the chambered nautilus, retains this ancient system. *Nautilus* regulates its depth within the water column by adjusting the gas content of its chambers to achieve neutral buoyancy. All other modern cephalopods have either an internal shell (*Spirula*) or an internal vestige of the shell (the pen in squids), or they have completely lost this molluscan feature (*Octopus*). The reduction or loss of the shell has resulted in a streamlining of the body, which facilitates passage through a resistant, fluid environment. Cephalopod evolution has also depended on increased size, a concomitant increase in muscular ability, and the development of a mantle siphon that produces jet propulsion, a fast and powerful method of swimming. The increase in size has been extraordinary; certain cephalopods are, in fact, the largest known invertebrates. For example, the deep-sea squid (*Architeuthis*) is known to attain lengths of over 20 meters (with outstretched tentacles) and weights exceeding 270 kilograms. Increased size and an increase in swimming power necessarily lead to increased metabolic demands and greater intake of food. Thus the evolution of cephalopods also has involved the development of predatory habits and a specialized nervous system to coordinate rapid movements and the perception and capture of prey. All of the modern cephalopods are skillful hunters or scavengers of large-sized prey. They rely to a large extent upon visual perception to locate food, and upon the coordination of their sucker-bearing arms in the capture and manipulation of prey.

The only other molluscan group that includes permanently pelagic species is the Class Gastropoda. None of these open-ocean gastropods is large enough or powerful enough to be ascribed to the nekton; rather, they form part of the planktonic communities of drifting organisms. Of the approximately 40,000 marine gastropod species, only about 140 are holoplanktonic, that is, inhabitants of the water column throughout their life cycle. They include the janthinid snails, the heteropods, the shelled and unshelled pteropods, and a few nudibranchs (Table 2). The evolution of these planktonic gastropods has proceeded by very different strategies from those that mark the development of cephalopods. The chapters that follow will deal with the holoplanktonic gastropods and their anatomical, physiological, and behavioral adaptations to an unusual molluscan habitat.

It is somewhat remarkable that so few mollusks have successfully invaded the vast and easily accessible open ocean, for many mollusks have adapted

TABLE 2
Major taxonomic divisions of the holoplanktonic gastropods

Taxon	Approximate number of planktonic species
Class Gastropoda	
Subclass Prosobranchia	
Order Mesogastropoda	
Suborder Heteropoda	28–29[a]
Suborder Ptenoglossa	
Family Janthinidae	6–16
Subclass Opisthobranchia	
Order Thecosomata	48–58
Order Gymnosomata	46–51
Order Nudibranchia	6+

[a] Maximum numbers include species of uncertain validity.

to life in much more rigorous environments on land and in fresh water. Certainly one of the constraints to invasion of the pelagic realm has been the molluscan trademark, the shell. This hard skeletal structure—which provides protection from predators, or from wave action and abrasion in intertidal species, or prevents desiccation in species exposed to air—becomes a cumbersome burden to a swimming or floating mollusk. Like the modern cephalopods, many of the pelagic gastropods accordingly have a reduced shell or none at all in the adult stage. But unlike cephalopods, which became large, fast swimmers, some planktonic gastropods have lessened their specific gravity and thus increased buoyancy by a reduction in size or weight, and many of the species that will be discussed are of very small size, some measuring only a millimeter or so in length or diameter. The typical gastropod creeping foot has no function in a mollusk that never touches a solid substrate. Accordingly, the pelagic species exhibit some remarkable modifications of this structure: the janthinids can construct bubble rafts from mucus secretions of the foot; the heteropods have produced a single swimming fin; and the pteropods have evolved paired swimming wings. The types of food available in the open ocean differ from those encountered by benthic animals, and considerable changes in feeding mechanisms have evolved in both herbivorous and carnivorous species of pelagic snails. Oceanic predators, too, differ from those that prey upon benthic mollusks, and the various defensive strategies of planktonic gastropods involve changes in coloration and behavior. Alterations in reproductive methods and developmental patterns also are apparent in these holoplanktonic species.

Some of the anatomical changes exhibited by the holoplanktonic gastropods are so extreme that early workers failed to recognize the animals as mollusks, classifying certain of them with planktonic worms, ctenophores, other gelatinous zooplankton, and even fishes. Later, as more specimens were collected and careful attention was given to internal anatomy, these animals were properly placed with the prosobranch or opisthobranch gastropods.

Pelagic gastropods traditionally have been collected with fine-meshed plankton nets towed slowly through the water at various depths. This method has been effective in capturing small individuals and small-sized species, but it rarely samples larger, faster-swimming gastropods, which sense the pressure wave in advance of the net and evade capture. Routine plankton collections generally have been preserved immediately in unbuffered formaldehyde, but this treatment results in eventual dissolution of the shell and in extreme contraction and distortion of the body in both shelled and unshelled species. Both of these preservation effects make taxonomic identification of museum specimens more difficult, and biological interpretations based solely on anatomical studies of contracted specimens often have proved erroneous.

Many of the biological and behavioral observations reported in the literature have been made by researchers who either dip-netted animals from surface waters or transferred live specimens from net tows to uncrowded aquaria immediately after capture. However, such methods have often been limited by injury to net-collected specimens and the subsequent low survival rates in the laboratory. Despite many attempts at laboratory rearing, only a few hardy species have been maintained in culture for even short periods, and few have survived well enough to provide much information about their biology.

With the advent of scuba-diving techniques, it has become possible for scientists to observe animals firsthand in shallow depths of the ocean. Much of the new material included in this book is the result of divers' observations of undisturbed gastropods in their natural environments. To the uninitiated, scuba diving off a ship in the open ocean places the diver in a vast, disorienting body of blue water, but, with a little training of the eye and the correct lighting, a myriad of small, oceanic organisms can be observed (e.g. Hamner, 1974; Harbison and Madin, 1979). Many of the photographs in this book were made by scuba divers using underwater cameras, and they illustrate animals that have been minimally disturbed. By using large glass jars, divers can collect undamaged individuals along with their associated symbionts or parasites or prey items. The study of animal behavior in natural environments, known as ethology, has been recognized as an important part of terrestrial biology for a number of years, but its extension to the oceanic realm has only just begun, as observers enter the ocean (see Hamner, 1985). The recent employment of research submersibles for observations of mid-water zooplankton is also beginning to increase our knowledge of species that live at depths below those accessible by scuba diving.

Although many of the planktonic mollusks are rare and therefore make indeterminate contributions to the biomass and ecology of marine communities, they are nonetheless of interest in illustrating the evolution of unusual anatomical adaptations and unique, indeed sometimes bizarre, life styles in the pelagic environment. Still, some planktonic gastropods, especially certain shelled pteropods (thecosomes), are abundant and widespread in the surface waters of the oceans. Some species of *Limacina* certainly rank among the most abundant of any gastropods, including benthic species, and one (*L. inflata*) may well be the most abundant gastropod species in the world. These and other abundant planktonic gastropods may have significant impacts upon the ecology of epipelagic marine communities. They either graze on phytoplankton or prey on smaller zooplankton and are themselves food for larger organisms, including sea birds, whales, and commercially important species of fish such as mackerel, herring, and salmon. In some cases, the ecological association may be deleterious: some species of thecosomes, when eaten in large numbers by fish, cause a condition known as "black gut" that renders the fish unmarketable. Certain thecosomes also have been implicated in transmitting to herring the paralytic toxin of a phytoplankton species on which they feed; the toxin is harmless to the mollusks but fatal to the fish that feed on them (White, 1977). The shelled pteropods (along with a smaller number of heteropods) also make significant contributions to sediments on the sea floor; the shells of dead animals accumulate in certain areas to form a deposit known as pteropod ooze. Consequently, the global cycle of carbon dioxide is affected by the significant amounts of bicarbonate ions removed from surface waters and the subsequent deposition of calcium carbonate in the form of aragonite, which is the major inorganic constituent of pteropod and heteropod shells (Berner, 1977). The shell deposits of particular species, which have accumulated over thousands or millions of years, also can be used to detect past changes in oceanic chemistry and climate.

References Cited

Berner, R. A. 1977. Sedimentation and dissolution of pteropods in the ocean. In: *The Fate of Fossil Fuel CO_2 in the Oceans*. R. T. Andersen and A. Malahoff, eds. New York: Plenum. Pp. 243–60.

Hamner, W. M. 1974. Blue water plankton. *Natn. geogr. 146*: 530–45.

———. 1985. The importance of ethology for investigation of marine zooplankton. *Bull. mar. Sci. 37*: 414–24.

Harbison, G. R., and L. P. Madin. 1979. Diving—A new view of plankton biology. *Oceanus* 22: 18–27.

Russell-Hunter, W. D. 1983. Overview: Planetary distribution of and ecological constraints upon the Mollusca. In: *The Mollusca*, vol. 6, Ecology. W. D. Russell-Hunter, ed. New York: Academic Press. Pp. 1–27.

Solem, A. 1974. *The Shell Makers: Introducing Mollusks*. New York: Wiley. 289 pp.

Thompson, T. E. 1976. *Biology of Opisthobranch Molluscs*, vol. 1. London: Ray Society. 207 pp.

White, A. W. 1977. Dinoflagellate toxins as probable cause of an Atlantic herring (*Clupea harengus harengus*) kill, and pteropods as apparent vector. *J. Fish. Res. Bd Can.* 34: 2421–24.

2

The Janthinid Snails
Raft Builders

Class Gastropoda
  Subclass Prosobranchia
    Order Mesogastropoda
      Suborder Ptenoglossa
        Family Janthinidae

Of all the holoplanktonic gastropods, the janthinid snails have undergone the least change in external appearance, and they remain easily recognizable as prosobranch gastropods because of their external, dextrally coiled shells (Color Fig. 1; Figs. 1a, d). Instead, their adaptation for an open-ocean existence has occurred largely through behavioral change, the major modification being the ability of the animals to construct a raft of air bubbles, from which they hang suspended, upside-down, from the water surface. The animals are incapable of swimming and, in fact, are unable to construct a new float while underwater. This dependency on remaining at the air-water interface makes them part of the pleuston, a collective term for the organisms confined to the uppermost few centimeters of the ocean.

These animals are more likely than other pelagic gastropods to be encountered by amateur or professional shell collectors, since they are not uncommonly washed up on beaches. The janthinids, though inhabitants of the surface waters of tropical and subtropical areas, may be carried by winds and currents as far north as Ireland, southwestern England, and British Columbia, and as far south as New Zealand, where they are blown ashore. The moderate-sized, violet-colored shells of *Janthina* are conspicuous on sandy beaches and are treasured for their beauty.

This family comprises two genera: the well-known *Janthina* and the less familiar *Recluzia*. A species of *Janthina* was first illustrated and described in the early seventeenth century by Fabius Columna, and since that time, numerous references to this genus have appeared in the literature. Most of the earlier works dealt with the systematics of these pelagic snails, and by the 1950s, 60 species of *Janthina* had been described. The most recent monograph was published in 1953 by Laursen, who reduced the number of species to five. The genus *Recluzia* has received much less attention and probably contains only two or three valid species, although at least 11 species have been described. It is now clear that these pelagic gastropods are closely related to the benthic epitoniid snails, and the two groups are combined in the Suborder Ptenoglossa.

External Anatomy

The shell of *Janthina* (Color Fig. 1) is immediately recognizable by its violet to blue color, the color also of the exposed body of the animal. This violet pigmentation is frequently encountered in other tropical, surface-living zooplankton, and it has been suggested that it is protective against exposure to high levels of ultraviolet light. However, these pigments have an absorption maximum in the range of 630–660 nm (red light), and such wavelengths are not considered to be particularly harmful. Cheng (1975) suggests, instead, that this coloration matches that of the surrounding clear oceanic water and thus is probably effective in concealing pleustonic animals from visual predators. Countershading is evident, with those areas of the shell directed downward in the water being lighter in color than areas close to the surface. This, too, appears to be protective camouflage, since potential predators approaching from below may have difficulty in distinguishing a lightly colored shell against a light background of sky, and predatory birds would find it difficult to spot a *Janthina* from above, because it would blend into a dark background. The rather thin shells of janthinids are composed of both forms of calcium carbonate, calcite and aragonite. Shell growth is not well understood, but mineral seems to be added to the two layers in irregular increments (Bøggild, 1930). Shell shape varies from trochoid to globose, and size ranges from a maximum height of 9.0 mm in *Janthina umbilicata* (Laursen, 1953) to 39.5 mm in *J. janthina* (Bennett, 1966).

The large head of *Janthina* is of the typical gastropod form, but the snout can be everted and distended during feeding. There is a pair of tentacles, both forked, which seem to be sensory organs. There are no visible eyes, but the animals are reported to sense and withdraw from the close presence of observers or objects placed above them (Wilson and Wilson, 1956; Bayer, 1963).

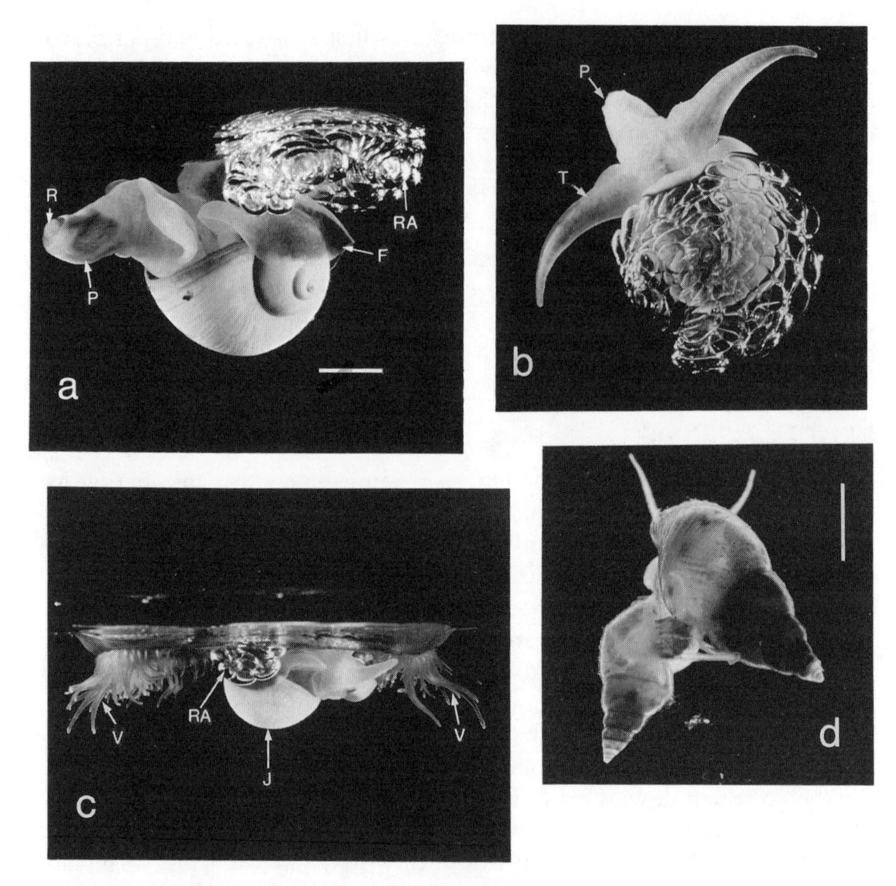

FIG. 1. Janthinid snails: *a*, *Janthina prolongata* hanging from its bubble raft at the sea surface, its proboscis extended to grasp nearby prey; *b*, the same individual viewed from above; *c*, *J. prolongata* feeding on the coelenterate *Velella velella*; *d*, mating individuals of *Recluzia* sp. F, foot; J, *Janthina* shell; P, proboscis; R, radula; RA, raft of air bubbles; T, tentacle; V, tentacles of the prey. Scale lines represent 3 mm.

The foot, a relatively broad, conspicuous feature of *Janthina*, is well supplied with glands opening onto the sole. Three foot regions are distinguished on the bases of position, external appearance, and function: the propodium, or anterior third of the foot, which is concerned with float formation; the mesopodium, or central area, which is involved in both float-building and egg-capsule attachment; and the postpodium, or the posterior third, which is fastened to the float. The adults do not have an operculum (the chitinous or calcareous structure on the foot of many gastropods that seals the shell aperture when the animal withdraws into its shell); in *Janthina*, the animal must remain attached to its extended float, and an operculum has no function.

The large gill (or ctenidium) lies within the mantle cavity, that space between the body of the animal and the largest shell whorl. The gill morphology is similar to that of other mesogastropods, and the gill has a respiratory function. An osphradium, associated with the gill, is presumably a chemosensory organ, as in other gastropods.

Another genus of janthinid snails, *Recluzia*, is very poorly known, perhaps because it is much less conspicuous in its coloration (Fig. 1d). An olive-tan periostracum covers the brown, high-spired shell, which reaches a maximum height of about 34.5 mm (Poorman, 1980). The body of *Recluzia* is yellow. Despite differences in appearance, the habits of *Recluzia*, including the formation of bubble floats, are similar to those of *Janthina*.

Float Formation

Float formation often has been observed in captive specimens of *Janthina* (Fraenkel, 1927; Laursen, 1953; Wilson and Wilson, 1956; Bayer, 1963). Only the foot of the snail is involved in the process. The sole of the anterior part of the foot (or the propodium), which normally clings to the base of the float, is stretched upward, breaking the water surface. The center of the propodium then becomes depressed and the sides of the foot fold over this hollow to encase a bubble of air. The black surface of the propodium is covered with glands that immediately coat the entrapped air with mucus. The propodium then is pulled below the water surface and is pressed against the proximal end of the existing float, where the new air bubble is cemented to the raft with the hardening mucus. The entire sequence takes as little as 10 seconds (Bayer, 1963) and can be repeated six to ten times in succession before a pause.

An animal detached from its float will sink and is unable to reconstruct a raft unless it is provided with access to air. One individual was observed to construct an entirely new float, capable of buoying the animal at the surface, in about 1 hour (Bayer, 1963). The size of the bubble raft varies, of course, with the size of the individual; the largest recorded raft (in *J. prolongata*) was 13 cm in length and 2 cm in width (Laursen, 1953). Individuals that detach from the float to crawl freely over the surface of prey presumably either reattach to the original float, some of which have been observed stuck to the margins of prey (David in Wilson and Wilson, 1956), or construct a new float after feeding.

The float of *Janthina* is firm, elastic, and dry to the touch. It varies from colorless to a pink or violet color, perhaps depending on the species or the age of the float. The 5- to 8-cm–long raft of *Recluzia rollandiana* is composed of brown air bubbles (Abbott, 1963, 1974) and presumably is formed in the manner described above.

Janthina janthina produces veliger larvae that swim freely in the water column for a period of time. These young eventually produce a long mucous stalk with a ball of air bubbles at the end (Siemroth, 1895, cited in Laursen, 1953). How this structure is formed is not known, but this early buoyancy device serves to bring the young to the water surface, where they begin to construct the definitive raft. It is not known whether the young of other *Janthina* species initiate float formation in the same way; Bayer (1963) has suggested that the young of some species metamorphose upon contact with suitable prey and make their first floats only after feeding on the adult food.

Food and Feeding

Almost all mollusks, with the exception of the bivalves and some parasites, are equipped with a unique feeding organ called the radula. This ribbon-like structure, which holds rows of variously shaped chitinous teeth, is housed in the buccal cavity. Teeth worn down by feeding activities are continuously replaced by new teeth formed at the posterior end of the radula, usually in a special radular sac.

The carnivorous janthinids lack a median radular tooth but have an indefinite number of long, hook-shaped, lateral teeth in each row (Fig. 2), and these are ideally suited for grasping prey. Because of a similarity in the arrangement, numbers, and shape of the teeth, the janthinids have been taxonomically allied with the benthic Epitoniidae (or wentletrap snails) in the Suborder Ptenoglossa. Despite the difference in habitat, species in both gastropod groups use their radular teeth for feeding on coelenterate prey. The benthic epitoniids are usually associated with sea anemones and corals, upon which they prey (Robertson, 1963); the janthinids commonly associate with pelagic coelenterates, primarily siphonophores and chondrophores, which form their principal food species (Fig. 1c).

Although distribution records remain sketchy (Laursen, 1953), all five species of *Janthina* appear to coexist in warm surface waters. Sagaidachnyi (1978) has, however, noted some separation of *J. exigua* and *J. pallida* from the other species in the Indian Ocean. Because of the co-occurrence of species, and because identifications of species often have been uncertain or neglected, it is difficult to say whether the janthinid species differ in selection of preferred prey. *Janthina janthina* appears to be the most numerous of the species, and Bayer (1963) has observed it feeding in the laboratory on two coelenterates, *Velella velella* (the by-the-wind sailor) and *Physalia physalis* (the Portuguese man-of-war), with which it is normally associated in open

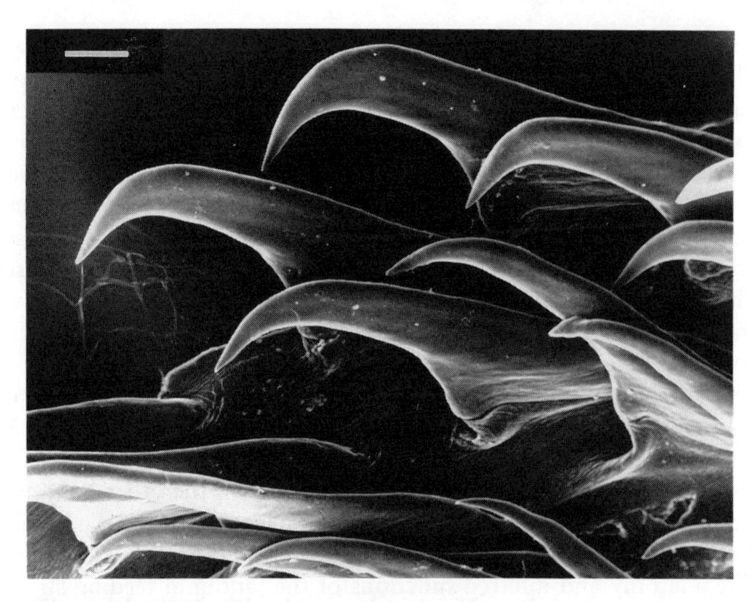

FIG. 2. The radular teeth of *Janthina prolongata.* The scale line represents 70 μm.

ocean waters. He also observed *J. janthina* feeding on *J. pallida*, and Laursen (1953) too reported evidence of cannibalism among janthinids, finding radular teeth of *J. janthina* in the gut of a specimen of the same species and janthinid embryos in the stomach contents of unidentified janthinid adults. In laboratory experiments, *J. prolongata* will eat both *Velella* and another pleustonic coelenterate, *Porpita porpita*, and the snail is capable of detecting these prey at distances of 5 to 10 cm (Bieri, 1966). *Velella* also is eaten by *J. pallida* (Bayer, 1963; Bennett, 1966). Specific feeding observations of *J. exigua* and *J. umbilicata* are not available. In addition, Laursen (1953) reported finding copepods and *Halobates* (a water-striding marine insect) in the guts of *Janthina* species. Whether or not these were incidental items contained within the guts of ingested coelenterate prey cannot be determined.

Since *Janthina* is incapable of swimming to actively locate prey, it must depend on passive chance encounters with suitable prey organisms. Bayer (1963) has reported that *J. janthina*, when coming in contact with suitable prey, extends its head and everts its proboscis to expose the radular teeth. The snail grasps the edge of the coelenterate prey with its lips and teeth and begins to chew and ingest pieces of the prey. Probably depending on the relative sizes of the prey and predator, *Janthina* may either abandon its float to crawl over the surface of the prey or it may remain attached to its raft while feeding (Bayer, 1963; Bieri, 1966).

The stinging nematocysts on the tentacles of prey appear to have no effect on the predator, if indeed they are discharged, and are ingested along with other prey tissue. Various authors (e.g. David in Wilson and Wilson, 1956; Ganapati and Rao, 1959) have observed the release of a purple dye into the water during feeding encounters, and some have suggested that this dye may have an anaesthetic action on the prey, or that it may prevent the discharge of nematocysts. However, the release of purple fluid is not always observed (Bieri, 1966), and in some cases it is not clear whether the dye is released by the hypobranchial gland of the janthinid or by the prey, many of which contain a similar colored dye that may leach out when pieces of tissue are removed in the feeding process.

The amount of prey specimen ingested and the time required to do so undoubtedly also depend on relative sizes of prey and predator. Bayer (1963) observed two snails of moderate size consume most of a *Physalia* with a 10-cm–long float in less than 24 hours. On the other hand, an average-sized *Velella* (3–4 cm) can provide sufficient nutrient for a number of very small *J. pallida* over a substantial time period (Bayer, 1963).

The anatomy and inferred functions of the janthinid feeding structures have been described in detail by Graham (1965). During feeding, the walls of the buccal cavity are everted to form a short proboscis. This action exposes the radula and pulls the radular teeth into an erect position. The teeth can then be used for grasping prey tissue. When the radula is withdrawn, the teeth collapse to hold prey tissue against the radular ribbon. In this way the food is conveyed to the buccal cavity and esophagus. A pair of quadrangular, chitinous jaws with cutting edges is present on each side of the radula, but the role of the jaws in feeding and ingestion remains unclear. The feeding structures of *Recluzia* are similar, but include, as well, paired cuticular stylets at the openings of the salivary glands; this suggests that a salivary secretion, possibly toxic, may be injected into prey (Fretter and Graham, 1962).

Little is known about diet and feeding behavior in the genus *Recluzia*. Gilmer (pers. obs.) collected small *Recluzia* (sp.?) attached to *Velella* from Gulf Stream waters and observed the feeding behavior of these snails in shipboard aquaria. *Recluzia* did not pursue any searching activity, but immediately crawled onto the surface of *Velella* upon direct contact with the coelenterate. The snails most often removed tissues from the underside of *Velella*, and an average-sized prey (3–4 cm long) provided sufficient food for six *Recluzia* (the largest, 6 mm in shell length) during three days of observation. Although *Recluzia* frequently shared the same *Velella* with *Fiona*, a nudibranch predator of the coelenterate (see Chapter 6), no interactions were observed between these mollusks. *Recluzia* would not feed on *Physalia*, however; snails would immediately detach when placed on the underside or tentacles of this siphonophore. It has also been reported (Burkenroad in Abbott, 1963) that *Recluzia rollandiana* feeds on the coelenterate *Minyas*, a floating sea anemone, and that the snail's feces contain nematocysts of this prey.

Janthina, *Recluzia*, and some pelagic nudibranchs (*Fiona*, *Glaucus*, *Glaucilla*; see Chapter 6) are among the very few predators of pleustonic coelenterates. Besides these mollusks, only sea turtles (Lane, 1960; Bingham and Albertson, 1974) and marlin (Salvini-Plawen, 1972) have been reported as major predators of *Physalia*. Coelenterates are probably not a preferred food for turtles, however, since they appear to be affected by the nematocyst toxins, exhibiting swollen eyes and facial tissue from the stings of their prey (Lane, 1960). Since *Janthina* species are known to exist in swarms covering as much as 200 nautical miles (Siemroth, 1895, cited in Laursen, 1953), they may have considerable impact in reducing populations of *Physalia*, *Velella*, and *Porpita*. Even small populations of these coelenterates usually harbor at least several janthinids. On the other hand, there are no known predators of janthinids, though it is reasonable to expect that some fish and birds may consume these snails.

Reproduction and Development

Reproductive biology and larval development have not been described for *Recluzia*. Studies of reproduction in species of *Janthina*, however, have revealed several unusual features, but there are still many unresolved problems in this area of research. Both *Janthina* and *Recluzia* are hermaphroditic, but it is unclear whether sex change takes place only once during the life cycle, whether animals can act simultaneously as males and females, and whether self-fertilization can occur. The answers to these questions depend partly on histological examination of the gonads in a large number of different sizes (and ages) of individuals in the same species. Ankel (1930) described all small individuals of *Janthina* (presumably the species *pallida*) as being males and all large individuals as females; Graham (1954) made the same observation in *J. janthina*. This suggests that *Janthina* is a protandrous hermaphrodite, that is, that young individuals are males and permanently change to the female condition at a later time in the life cycle. Laursen (1953), however, reported finding separate males and females of the same size and species in addition to sterile individuals (neither male nor female), hermaphroditic specimens (both male and female), and very large males. His findings, which seem to apply to all the species in his worldwide collections, suggest a more complex life cycle in which sex change may occur more than once or may be dictated by local environmental conditions. Obviously more research is needed to resolve this issue.

The detailed anatomy of the reproductive tract of *Janthina* (Fig. 3) has

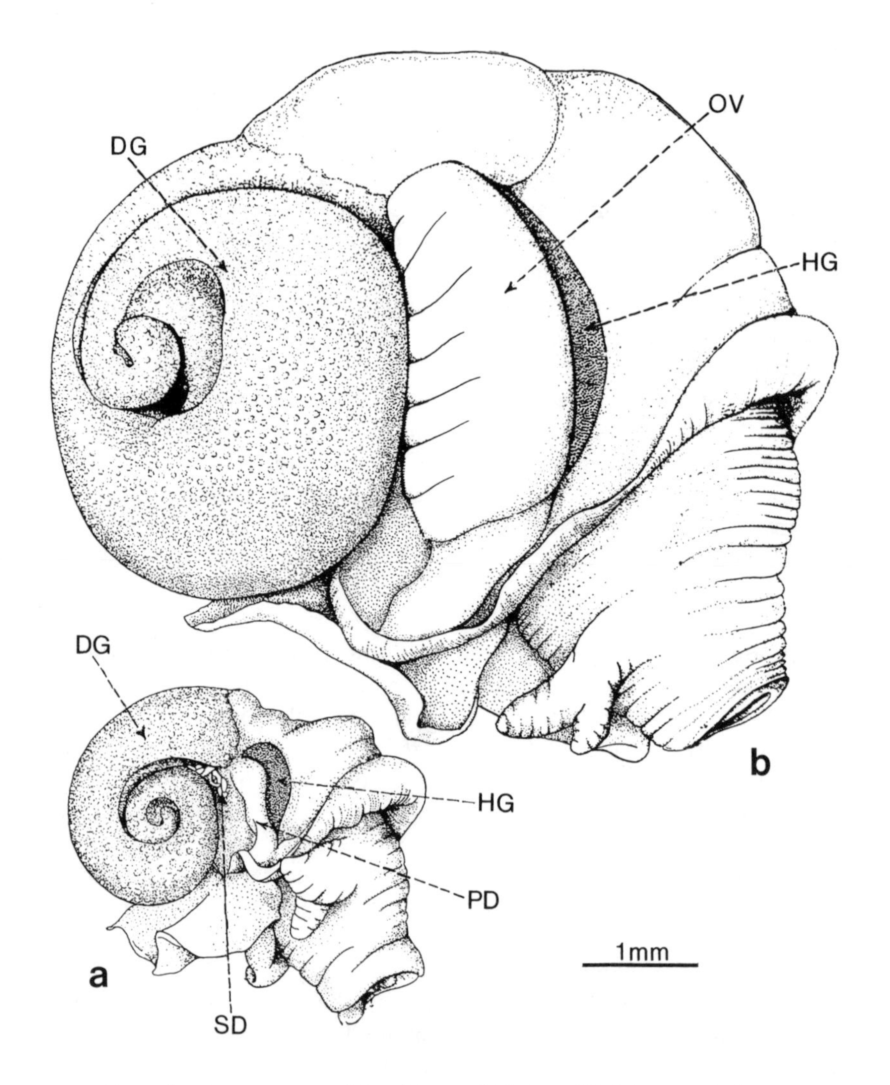

FIG. 3. Reproductive anatomy of *Janthina pallida* (modified from Ankel, 1930): *a*, a male; *b*, a female. DG, digestive gland and gonad; HG, hypobranchial (mantle) gland; OV, glandular portion of oviduct; PD, pallial glandular section of sperm duct; SD, lower section of sperm duct.

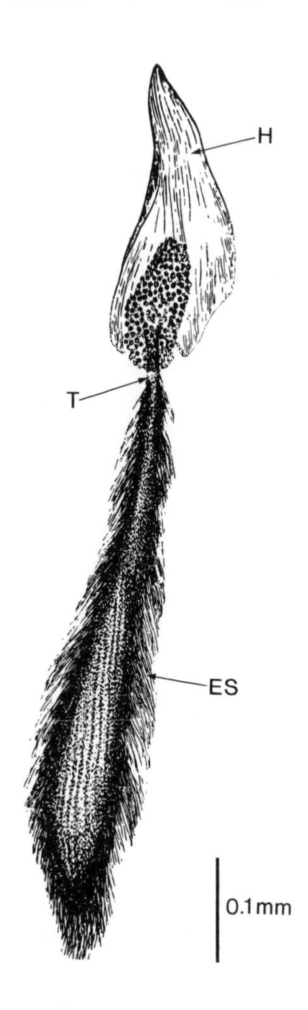

FIG. 4. A spermatozeugma of *Janthina pallida* formed by large numbers of normal (eupyrene) sperm attached to the tail of a single, large, atypical (oligopyrene) spermatozoan (modified from Ankel, 1930). ES, attached eupyrene sperm; H, fibrous, undulating plate forming the head of the oligopyrene spermatozoan; T, tail of the oligopyrene spermatozoan.

been reviewed and described most recently by Laursen (1953) and Graham (1954). The biology and ecology of these animals exhibit several unusual features: there is no copulatory organ; two types of sperm are produced; and the females have no special organs for receiving and storing sperm. As described by Ankel (1930) and others, male *Janthina* produce large numbers of typical (eupyrene) spermatozoa that contain a haploid complement of chromosomes and are used to fertilize eggs. They also produce smaller numbers of very large, oligopyrene spermatozoa; each is composed of a large flattened head covered by a membrane capable of undulating movement and a long, slender tail. The oligopyrene sperm contain only part of the haploid set of chromosomes and are not capable of fertilization. Instead, they appear to function as carriers of the normal sperm, which attach in large numbers to the tail (Fig. 4). This composite structure of a single oligopyrene spermatozoon with attached eupyrene sperm is referred to as a spermatozeugma.

Wilson and Wilson (1956) described the release of aggregates of white spermatozeugmata from a single captive *J. janthina*. These were released several times over the left side of the shell. Since there is no other means of internal fertilization, it has been suggested (e.g. Graham, 1954) that the oligopyrene spermatozoa serve to convey eupyrene sperm through the water to other individuals of the same species. However, living spermatozeugmata of *Janthina* (Ankel, 1930; Wilson and Wilson, 1956) and of species of *Epitonium* (Robertson, 1983) are not motile; in fact, they sink to the bottom of laboratory containers. In epitoniids, spermatozeugmata do not swim from a male to a female. Instead, two individuals pair side by side in "pseudocopulation," and spermatozeugmata enter the female's mantle cavity via ciliary currents (Robertson, 1983). Unless the nonmotility of spermatozeugmata is a laboratory artifact, it is reasonable to suppose that the transfer of sperm in janthinids is achieved in a similar fashion. In any case, eupyrene sperm make their way to the ovary, where fertilization occurs (Ankel, 1930; Graham, 1954). The success of this type of copulation implies that individuals in swarms of mixed species can recognize sexually mature individuals of the correct sex and species.

It is not known whether males can receive sperm from another individual and store it until the sex change occurs, although there seems to be no special provision to do so. Nor do we know whether self-fertilization is possible, although Graham (1954) found no anatomical reason why this should not be a viable alternative to cross-fertilization.

With respect to embryonic development, *Janthina janthina* differs markedly from other species of the genus. Since it releases free-swimming veliger larvae into the sea, *J. janthina* is regarded as viviparous; the other species are oviparous, laying eggs in capsules attached to the underside of the parent float.

In *J. janthina*, embryonic development takes place entirely within the ovary and genital duct of the female. The early embryos are nourished by yolk contained in the eggs, whereas later embryos ingest mucus produced by glands in the oviduct (Graham, 1954). Veliger larvae are expelled from the mantle cavity of the parent, at the left side of the shell, in dark, pellet-shaped masses measuring about 3 by 4 mm (Bayer, 1963). Wilson and Wilson (1956) observed one individual release, within a short time, at least 36 masses containing 20 or more veligers each, or a minimum of 720 young. Early veligers have tiny cap-shaped shells with a maximum width of 100 to 230 μm; late veligers develop a spirally coiled, brownish-purple shell. Veligers have a bilobed, ciliated membrane (or velum) that is used for both swimming and food collection. The length of time they continue to develop and grow within the water column, before rising to the surface to make their first bubble raft, is not known. Live veligers of this species were the first of any marine snail to be observed and described; Forsskål's notes on these larvae were published in 1775, predating other descriptions of prosobranch veligers by 60 years (Wolff, 1968).

The other four species of *Janthina*, and at least one species of *Recluzia* (*R. rollandiana*), all release eggs enclosed in capsules that are cemented to the lower surface of the float (Color Fig. 1). Anatomical observations have shown that a number of fertilized eggs are packaged, together with mucus (or albumen?) for nutrient, into capsules secreted within the genital duct. Each capsule has a compressed, pear-shaped form with a number of spines over the apical surface and a stalk for attachment (Fig. 5); the egg capsules of different species are distinguished by size and relative number and length of spines (see Laursen, 1953). The egg capsules are extruded, stalk foremost, through an opening on the left side of the mesopodium, and apparently are

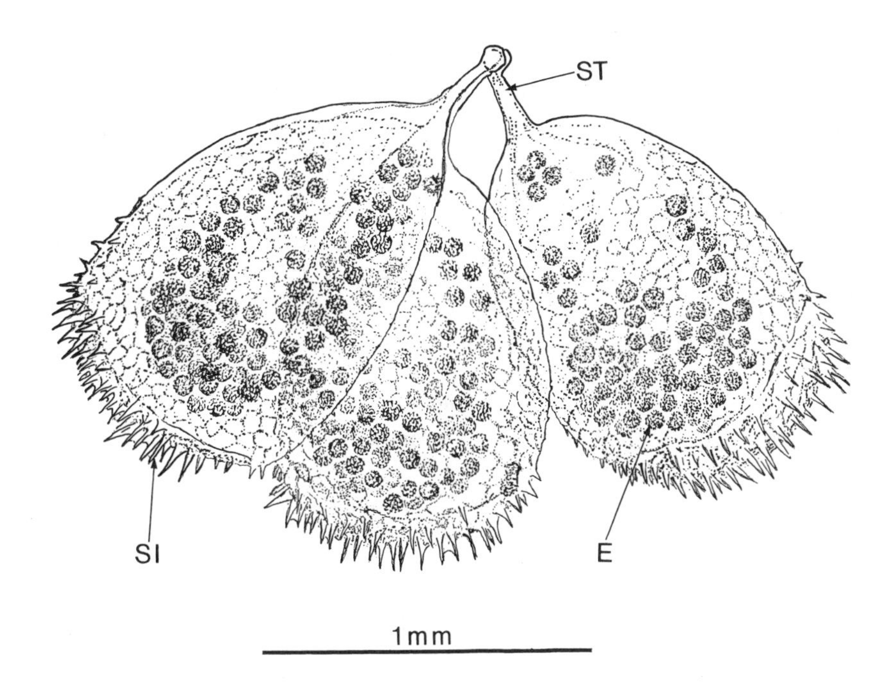

FIG. 5. Egg capsules of *Janthina umbilicata* (modified from Laursen, 1953). E, embryo; SI, spine; ST, stalk.

TABLE 3
Characteristics of egg capsules and numbers of young produced by oviparous species of Janthina

Species	Maximum adult shell height (mm)	Number of egg capsules per cm of float length	Length of capsules (mm)	Number of eggs or larvae per capsule	Maximum number of young produced per female per unit time
J. umbilicata	9	"comparatively small"	<2.0	80	?
J. exigua	17	50 (max. total, 250)	2.0	180	44,000
J. pallida	24	70 (max. total, 490)	4–7	5,500	2,700,000
J. prolongata	39	30 (max. total, 305)	7–8	7,300	2,200,000

NOTE: Data from Laursen, 1953.

cemented to the float with the same mucus used in float building (Laursen, 1953). This process is thought to take place over some period of time, since different capsules on a single float contain young in various stages of development, from early embryos to veligers.

Laursen's data (1953) on size and number of egg capsules and number of young per capsule for each oviparous species are summarized in Table 3, together with estimates of total young produced. The two largest species, *J. prolongata* and *J. pallida*, both produce over 2 million young, presumably within a short time period. The smallest species, *J. umbilicata*, undoubtedly produces the fewest young, although data are too few to permit an estimate. In all species, the estimated numbers of young produced are no doubt minimal, since we do not know how long an individual adult will continue to be reproductive.

The young leave the egg capsules through slits that develop between the spines. It is believed that the young of all species are released as free-swimming veligers (Fig. 6). Larval shells of *Janthina janthina* and *J. pallida* have been figured and described by Robertson (1971), those of an unidentified species, by Richter and Thorson (1975). The young possess eyespots, statocysts, and an operculum, all of which disappear during metamorphosis (Fraenkel, 1927). Jung (1975) has described fossilized larval shells of two species of *Janthina* from sediments off northern Venezuela.

Epifauna and Parasites

The shells of these mollusks provide a site of attachment for other animals in an environment that offers few solid substrates. Janthinid shells commonly are covered with attached clusters of small, stalked barnacles of

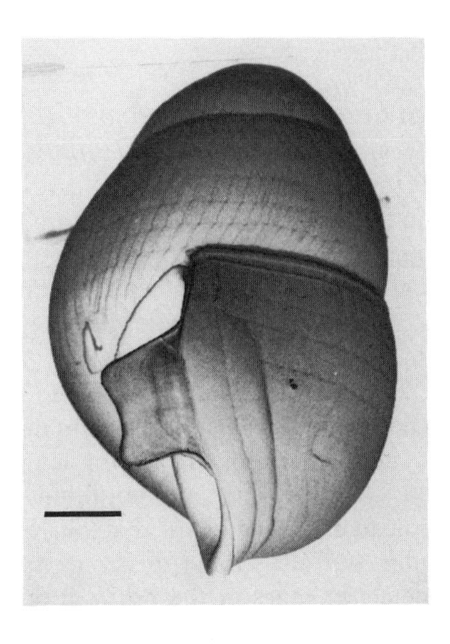

FIG. 6. The shell of a janthinid veliger collected from the plankton off Barbados, West Indies (by A. Sears). The scale line represents 25 μm.

the genus *Lepas*. Wilson and Wilson (1956) found several *Janthina* specimens bedecked with hydroid colonies of *Laomedea geniculata*. Laursen (1953) found floats of *Janthina* inhabited by a small swimming crab that apparently uses the float as a resting site, and he reported egg ribbons of the nudibranch *Glaucus atlanticus* (see Chapter 6) attached to *Janthina* shells. Probably none of these are specific associations; both *Lepas* and *Laomedea* are known to be opportunistic in settling on any drifting substrate, such as bottles, driftwood, and ships, and opportunism is probably the basis of the other associations noted.

At least three species of *Janthina* (*exigua*, *janthina*, and *prolongata*) serve as the intermediate hosts of a parasitic copepod, *Cardiodectes*, of the family Pennellidae (formerly known as the Lernaeoceridae) (Ho, 1966). Ho found copepodite stages and adults swimming freely in the mantle cavities of *Janthina prolongata* (= *globosa*), and the intermediate chalimus stages were attached to the gill lamellae of the host. The incidence of infestation by this parasite was high in *Janthina* populations collected off Barbados, West Indies; 115 snails out of 125 specimens examined contained this copepod. The definitive host or hosts of this unidentified copepod are not known, but the final hosts of other *Cardiodectes* species are normally lanternfishes (Perkins, 1983).

In addition, Reimer (1975, 1976) has found larvae of the cestode *Scolex pleuronectis*, and metacercaria of a trematode, *Lecithocladium* sp., in *Janthina prolongata*.

Recluzia does not seem to have been examined for either parasites or epifauna.

Evolution

Although ptenoglossate snails are known from the Jurassic period (Fretter and Graham, 1962), about 200 million years ago, the species of *Janthina* are much more recent in origin. They were the last addition to the calcareous plankton, appearing for the first time in the late Cenozoic (Lowenstam, 1974). Though the fossil record gives no clue to how the first appearance of raft-building snails occurred, it is not difficult to speculate from anatomical and ecological evidence on the transition from benthic ancestors, like the wentletrap snails, to the present-day pelagic janthinids.

Little anatomical change occurs between the benthic and pelagic ptenoglossans. What changes have occurred can be related to a loss of function in a pelagic habitat or to an enhancement or modification of preexisting features that came to perform new functions. Of the former, the janthinids have lost eyes and statocysts: eyes play no role in drifting animals that locate prey and mates by passive chance encounter and are unable to flee from predators; statocysts, which function to detect changes in the position of the body relative to its environment, play no essential role in an animal that spends most of its life suspended from a float. The operculum also becomes unnecessary in animals that can no longer retreat into the shell when attached to their floats, and it too has disappeared, though it is still present in the larval stage.

The foot, which served for crawling over solid substrates in benthic ancestors, became only slightly modified in appearance and can still perform a locomotory function in animals that detach from the float to crawl over the surface of larger prey to feed. The major changes in this structure have been the use of the foot to capture air from the atmosphere and an increase in glandular area that permits the secretion of sufficient mucus for raft building. It should be noted, however, that all benthic gastropods also have pedal glands, which facilitate crawling and, in some species, are concerned with egg-capsule deposition as they are in *Janthina*.

That those ancestral benthic snails, perhaps crawling over the surface of tropical *Sargassum* weed and undoubtedly feeding on attached coelenterates, should give rise to pelagic tropical janthinids that feed on floating coelenterates is well within the range of reasonable speculation. Even the assumption of a suspended position at the water surface is not unique in the molluscan phylum. For example, snails of the marine genus *Hydrobia* assume this position temporarily, also by using a small mucous raft to suspend themselves at the air-water interface (Newell, 1962). The transition to a pleustonic existence undoubtedly expanded the food supply of the early janthinids; *Physalia*, *Velella*, and *Porpita* commonly occur in large swarms and apparently have few other predators, perhaps because other animals are deterred by the nematocyst toxins of these prey. The raft of *Janthina*, which projects above the water surface, is subject to the same velocity and direction of drift by wind as are the floats and sails of its coelenterate prey. In this way, prey and predator remain together and the stability of the relationship persists.

It is tempting to suggest that the reproductive system of *Janthina*, with its unusual features, has undergone the greatest evolutionary change, but in fact there is little that is unique among gastropods. Most benthic mesogastropods are of separate sexes (i.e. male or female throughout their lives), and most have internal fertilization via a copulatory organ. In contrast, all species of *Janthina* are hermaphroditic and aphallic. Hermaphroditism is frequently linked with a pelagic existence, for it is common in other zooplankton such as chaetognaths (arrow worms), ctenophores (comb jellies), some medusae (jellyfish), salps, and larvaceans, as well as in some other pelagic gastropods (see Chapters 4, 5, and 6). Evolutionary change from a dioecious (separate sexes) state to the monoecious (hermaphroditic) condition is not without precedent in other mesogastropods; it is known in the limpet families Capulidae and Calyptraeidae and occurs in the epitoniids, which are closely related to *Janthina*. The ability of an animal to act as either a male or a female, or to self-fertilize, or to store exogenous sperm attained when it is young and male, is of utmost importance if the animal is not physically able to seek out a mate at the appropriate breeding time, or if population densities are so low as to preclude encounters with potential mates. The second condition may apply to at least some epitoniids; the first condition certainly applies to sessile mollusks, like the above-mentioned limpets, and to *Janthina*, which depends upon chance encounters with other individuals of its own species.

The absence of a copulatory organ and the production of two types of sperm are features that are also found in other ptenoglossa (the epitoniids) and in the benthic Cerithiacea (needle whelks) (Fretter, 1984). As Fretter and Graham (1962) have pointed out, these reproductive features seem more likely to be an indication of genetic affinity than to have been acquired independently. Why these groups have evolved spermatozeugmata will remain an open question, until perhaps we have more information on the biology and ecology of all the groups concerned.

List of Recognized Species

(Synonymy for *Janthina* is given in Laursen, 1953)

Janthina janthina (Linnaeus, 1758)

(The generic name is spelled with an initial *J* in the original publication; *Ianthina* is an emendation according to Abbott, 1974.)

J. exigua Lamarck, 1816
J. prolongata Blainville, 1822 (= *J. globosa* Swainson, 1823)
J. umbilicata d'Orbigny, 1840
J. pallida Thompson, 1841
Recluzia johnii Holten, 1803
(The genus *Recluzia* is in need of revision; many of the species listed are probably synonymous.)
R. lutea Bennett, 1840
R. turrita Philippi, 1848
R. jehennei Petit, 1853
R. rollandiana Petit, 1853
R. hargravesi Cox, 1870
R. palmeri (Dall, 1871)
R. montrouzieri Souverbie, 1872
R. erythraca Jickeli, 1882
R. insignis Pilsbry & Lowe, 1932
R. bensoni Adams, 1955

References Cited

Works with recommended keys or aids for the identification of species are indicated by an asterisk.

Abbott, R. T. 1963. The janthinid genus *Recluzia* in the western Atlantic. *Nautilus* 76: 151.
*———. 1974. *American Seashells.* New York: Van Nostrand Reinhold. 663 pp.
Ankel, W. E. 1930. Die atypische Spermatogenese von *Janthina* (Prosobranchia, Ptenoglossa). *Z. Zellforsch. mikr. Anat.* 11: 491–608.
Bayer, F. M. 1963. Observations on pelagic mollusks associated with the siphonophores *Velella* and *Physalia. Bull. mar. Sci. Gulf Caribb.* 13: 454–66.
Bennett, I. 1966. Some pelagic molluscs and associated animals in southeastern Australian waters. *J. malac. Soc. Aust.* 9: 40–51.
Bieri, R. 1966. Feeding preferences and rates of the snail *Ianthina prolongata*, the barnacle *Lepas anserifera*, the nudibranchs *Glaucus atlanticus* and *Fiona pinnata*, and the food web in the marine neuston. *Publs Seto mar. biol. Lab.* 14: 161–70.
Bingham, F. O., and H. D. Albertson. 1974. Observations on beach strandings of the *Physalia* (Portuguese man-of-war) community. *Veliger* 17: 220–24.

Bøggild, O. B. 1930. The shell structure of the mollusks. *K. danske Vidensk. Selsk. Skr.* 9: 231–326.

Cheng, L. 1975. Marine pleuston—Animals at the sea-air interface. *Oceanogr. mar. Biol. a. Rev.* 13: 181–212.

Fraenkel, G. 1927. Biologische Beobachtungen an *Janthina. Z. Morph. Ökol. Tiere* 7: 597–608.

Fretter, V. 1984. Prosobranchs. In: *The Mollusca*, vol. 7, Reproduction. A. S. Tompa, N. H. Verdouk, and J. A. M. van den Biggelaar, eds. New York: Academic Press. Pp. 1–45.

Fretter, V., and A. Graham. 1962. *British Prosobranch Molluscs*. London: Ray Society. 755 pp.

Ganapati, P. N., and D. V. Subba Rao. 1959. Notes on the feeding habits of *Ianthina janthina* Linnaeus. *J. mar. biol. Ass. India* 1: 251–52.

Graham, A. 1954. Some observations on the reproductive tract of *Ianthina janthina* (L.). *Proc. malac. Soc. Lond.* 31: 1–6.

———. 1965. The buccal mass of ianthinid prosobranchs. *Proc. malac. Soc. Lond.* 36: 323–38.

Ho, J.-S. 1966. Larval stages of *Cardiodectes* sp. (Caligoida: Lernaeoceriformes), a copepod parasitic on fishes. *Bull. mar. Sci.* 16: 159–99.

Jung, P. 1975. Quaternary larval gastropods from Leg 15, Site 147, Deep Sea Drilling Project. Preliminary report. *Veliger* 18: 109–26.

Lane, C. E. 1960. The Portuguese man-of-war. *Scient. Am.* 202: 158–68.

*Laursen, D. 1953. The genus *Ianthina*: A monograph. *Dana Rep.* No. 38. 40 pp.

Lowenstam, H. A. 1974. Impact of life on chemical and physical processes. In: *The Sea*, vol. 5. E. D. Goldberg, ed. New York: Wiley. Pp. 715–96.

Newell, R. 1962. Behavioural aspects of the ecology of *Peringia* (= *Hydrobia*) *ulvae* (Pennant) (Gasteropoda, Prosobranchia). *Proc. zool. Soc. Lond.* . 138: 49–75.

Perkins, P. S. 1983. The life history of *Cardiodectes medusaeus* (Wilson), a copepod parasite of lanternfishes (Myctophidae). *J. crustacean Biol.* 3: 70–87.

Poorman, L. H. 1980. New records from the tropical Eastern Pacific for *Recluzia palmeri* (Dall, 1871). *Veliger* 23: 183.

Reimer, L. W. 1975. Cestodenlarven in Wirbellosen der Küste von Madras. *Angew. Parasit.* 16: 2–16.

———. 1976. Metazerkarien in Wirbellosen der Küste von Madras. *Angew. Parasit.* 17: 33–43.

Richter, G., and G. Thorson. 1975. Pelagische Prosobranchier-Larven des Golfes von Neapel. *Ophelia* 13: 109–85.

Robertson, R. 1963. Wentletraps (Epitoniidae) feeding on sea anemones and corals. *Proc. malac. Soc. Lond.* 35: 51–63.

———. 1971. Scanning electron microscopy of planktonic larval marine gastropod shells. *Veliger 14*: 1–12.

———. 1983. Observations on the life history of the wentletrap *Epitonium albidum* in the West Indies. *Am. malac. Bull. 1*: 1–11.

Sagaidachnyi, A. Yu. 1978. [Pleustonic molluscs *Ianthina* (Gastropoda, Ianthinidae) in the Indian Ocean.] *Trudy Inst. Okeanol. 113*: 227–36. [In Russian, with English summary.]

Salvini-Plawen, L. 1972. Cnidaria as food-sources for marine invertebrates. *Cah. Biol. mar. 13*: 385–400.

Wilson, D. P., and M. A. Wilson. 1956. A contribution to the biology of *Ianthina janthina* (L.). *J. mar. biol. Ass. U.K. 35*: 291–305.

Wolff, T. 1968. The Danish Expedition to "Arabia Felix" (1761–1767). *Bull. Inst. océanogr., Monaco*, No. spécial 2 [Congr. int. Hist. Océanogr., 1]: 581–601.

3

The Heteropods

Visual Predators

Class Gastropoda
  Subclass Prosobranchia
    Order Mesogastropoda
      Superfamily Heteropoda
        Family Atlantidae
        Family Carinariidae
        Family Pterotracheidae

Apart from the janthinids (Chapter 2), heteropods are the only other prosobranchs that spend their entire lives in the water column. They comprise a larger group of approximately 28 species that are planktonic rather than pleustonic, and they exhibit more anatomical variation within the group than do the janthinids. Heteropods are highly modified for life in the open ocean, the major morphological adaptations including a reduction in the size and weight of the aragonite shell or a loss of the shell associated with an increase in body size; the development of a single swimming fin derived from the gastropod foot; and a tendency for the body and shell to be transparent. The larger, transparent species of heteropods have been of interest to anatomists and physiologists for over 150 years, since much of the internal anatomy, including the nervous system, can be traced without dissection. The unusually large and complex eyes of heteropods have also attracted the attention of anatomists.

The basic design of the heteropod eyes is not unlike that of fish eyes and, partly because of this similarity, heteropods once were thought to be a link between cephalopods and fish (Lamarck, 1812). It was not until 1836 that Cuvier properly placed the group among the gastropods. Delle Chiaie

(1841), Huxley (1853), and Souleyet (1852) provided the first accurate anatomical drawings of heteropods, and Souleyet made the first attempt to organize the group taxonomically as it occurs today. A number of systematic revisions made over the last century greatly increased the number of species, owing mainly to preservation artifacts, a lack of knowledge about the anatomy and biology of living animals, and to the erroneous belief that the radula was the ultimate taxonomic character. In 1949, Tesch published the results of his studies of heteropods captured by the *Dana* Expeditions. In this monograph, he corrected many misconceptions about the animals and reduced the number of described species from 107 to 22. Since that time, only a few new species have been described. Thiriot-Quiévreux (1973a) and van der Spoel (1976) have provided up-to-date reviews of the taxonomy and geographic distribution of heteropods.

Heteropods are found in all the tropical and subtropical oceans of the world, and most species are cosmopolitan. Normally, few species are present beyond 40° N and 40° S, and some are restricted to much narrower bands of latitude within the warmest oceanic waters. They are primarily epipelagic, living at depths from just below the surface to perhaps 500 m. Although a few specimens have been collected from below 1,000 m, the numbers of heteropods decline rapidly with depth. The paucity of adult specimens collected in plankton-sampling programs designed to detect diurnal migration has not permitted an accurate assessment of the vertical movements of these animals. Larger species, however, do seem to move into deeper water during the day (Pafort-van Iersel, 1983). Some of the heteropods are large enough and, at times, present in sufficiently dense concentrations to be detected at depth by sonar recordings. Blackburn (1956) reported a scattering layer at 20 m depth off Australia that was caused by *Firoloida desmaresti*. In some areas of the sea floor, heteropod shells make up as much as 8 percent of the coarse fraction of sediment known as pteropod ooze.

Heteropods are very mobile carnivores and, consequently, their population densities usually are relatively low compared with those of many other zooplanktonic groups. For example, Seapy (1974) normally found the densities of *Carinaria cristata* off southern California and northern Mexico to be less than one individual/1,000 m^3 of water; and only very rarely did densities exceed 150 animals/1,000 m^3. Because of these low population densities and the difficulties associated with capturing very small individuals or species as well as large, fast-swimming species, heteropods are unlikely to be encountered by other than professional oceanographers—and then infrequently. Nonetheless, they are among some of the most beautiful and interesting animals of the plankton.

External Anatomy

Because of considerable adaptive radiation within the heteropods, external features must be described separately for each family. The Family Atlantidae is considered to be the most primitive, and its 16 species all retain the dextrally coiled shell common to other prosobranchs (Color Fig. 2; Fig. 7). All the atlantids, however, are very small, with shell diameters usually measuring less than 10 mm. Atlantid shells are also characteristically thin-walled, which reduces weight; shell thickness varies between 3 and 40 μm (Richter, 1973b; Batten and Dumont, 1976). An even thinner shell area, called a keel, projects along the center of the largest whorl; this is a unique atlantid feature that acts as a stabilizer during swimming. The adult shells of all species are flattened in profile, which streamlines the shape and facilitates swimming. The chemical composition of the shell and keel differs, however, among the three atlantid genera. In *Atlanta*, both the shell and keel are calcified. In *Protatlanta*, the shell is calcified but the keel is composed of conchiolin, which gives it a cartilaginous nature. In *Oxygyrus*, only the early shell whorls are calcified, whereas the final adult whorl and the keel are cartilaginous-like. Thus further weight reduction is achieved in the two most advanced atlantid species by the replacement of heavy aragonite molecules with an organic shell matrix. The body of an atlantid can be completely withdrawn into the transparent, flattened shell, and the aperture

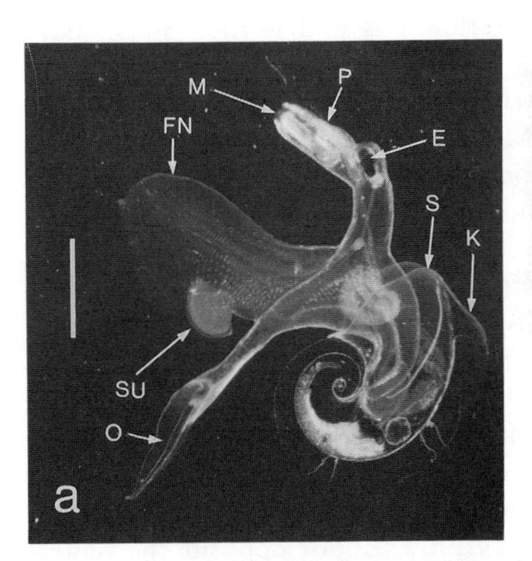

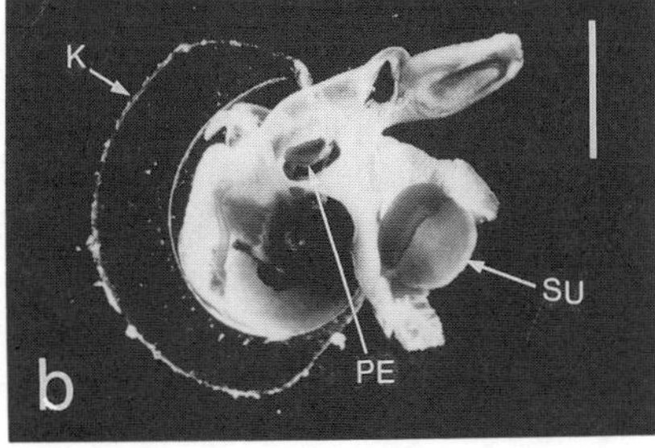

FIG. 7. Atlantid heteropods: *a*, *Atlanta inflata* in the normal swimming position; *b*, a male *Oxygyrus keraudreni*, partially retracted. E, eye; FN, swimming fin; K, keel of shell; M, mouth; O, operculum on foot; P, proboscis; PE, penis; S, shell; SU, sucker. Scale lines represent 5 mm in *a*, 3 mm in *b*.

can be sealed by the chitinous operculum on the foot. The mantle cavity is dorsally placed and houses a single gill and osphradium.

In extended animals, the foot appears as a long, relatively slender organ projecting into a single, laterally flattened, ventral fin, which is used in swimming. In all atlantid species, there is a well-developed sucker on the fin of both sexes. The head is large and elongate, with a muscular proboscis extending forward from a pair of tentacles. Radular teeth can often be seen protruding slightly from the terminal mouth. The remarkably large and complex eyes are enclosed in gelatinous capsules located at the posterior bases of the tentacles. The eyes are movable and pigmented, and contain a large crystalline lens and a retina that is unique in being shaped like a long, narrow ribbon. The retina is only a few (three to six) receptors wide and several hundred receptors long. It may be straight or curved, but in either case the field of view is restricted to a narrow strip of space a few degrees high and 80° to 180° long. To compensate for this restriction, at least one species (*Oxygyrus keraudreni*) uses systematic scanning movements of the eyes such that each retina moves through a 90° arc (Land, 1982).

In the Family Carinariidae (Color Figs. 3–5; Fig. 8), shell size has diminished relative to body size. The spired shell is external in *Carinaria* and *Pterosoma*, but is greatly reduced and embedded in tissues in the third genus, *Cardiapoda*. In all seven species, the fragile shell covers only a small part of the body; these animals are unable to retract into the shell, and consequently there is no operculum. The exposed body is drawn out into a cylindrical shape in the anterior-posterior axis and is thus much more streamlined than that of the atlantids. *Pterosoma* differs in being flattened dorsoventrally into an oblong disk, with only the tail section and proboscis remaining cylindrical (Figs. 8c, d). Maximum body length varies from 20 mm in the smallest species, *Cardiapoda richardi*, to an impressive 500 mm in *Carinaria cristata*, the largest planktonic gastropod. The published size records of soft-bodied animals, however, are often underestimates of true size; for example, *Cardiapoda placenta* often exceeds 110 mm in length in the Florida Current (Gilmer, pers. obs.), whereas the maximum size of preserved specimens is recorded as 80 mm (van der Spoel, 1976).

The transparency of the gelatinous body of the carinariids is such that the internal organs, such as the esophagus and nervous system, can clearly be seen through the body wall. Many of the internal organs—including the stomach, digestive gland, intestine, heart, kidney, and gonad—are restricted to a stalked visceral mass that projects from the body opposite the ventral swimming fin (Figs. 8a-c). The visceral mass is surrounded by the mantle and mantle cavity and is capped by the shell. The mantle cavity houses a pectinate gill that protrudes from the shell aperture into the surrounding water. The foot is now represented only by a large and muscular ventral fin,

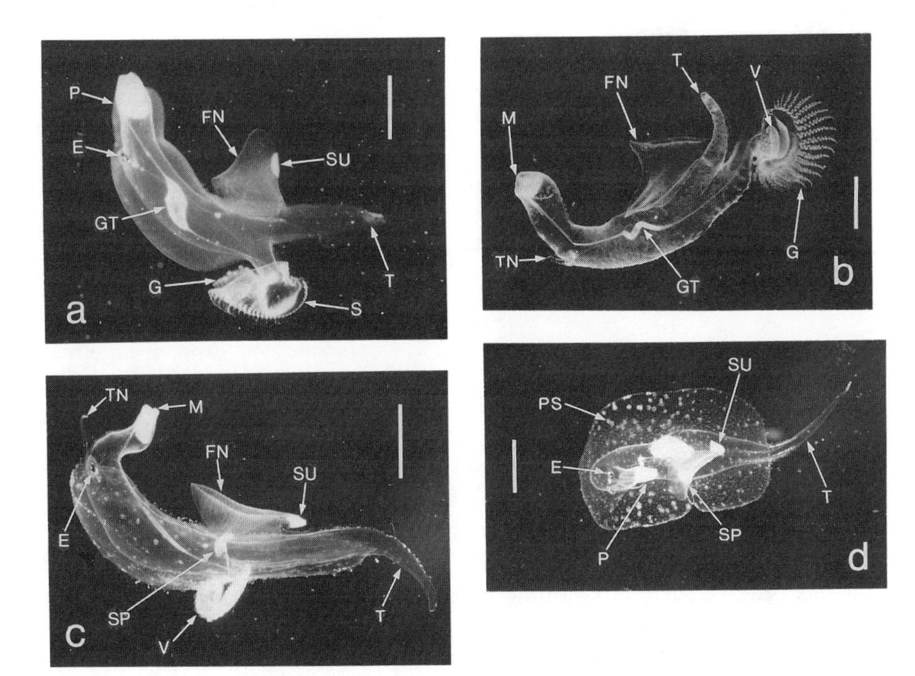

FIG. 8. Carinariid heteropods (*b*, *c*, and *d* by R. Reichelt): *a*, *Carinaria lamarcki*, lateral view; *b*, *Cardiapoda placenta*, female, lateral view; *c*, *Pterosoma planum*, female, lateral view; *d*, *P. planum*, ventral view, showing the dorsoventral flattening and expansion of the body. E, eye; FN, fin; G, gills; GT, gut; M, mouth; P, proboscis; PS, pigment spots; S, shell; SP, attached spermatophore; SU, sucker; T, tail; TN, tentacle; V, visceral nucleus. Scale lines represent 10 mm.

and a sucker is still present on the posteroventral edge of the fin in both sexes. A pair of tentacles is usually present before the eyes, but the tentacles may be unequal in size, or the right tentacle may be absent in some species. The eyes are large and, as in the atlantids, they are movable and anatomically complex (Dilly, 1969; Wolff and Schultze, 1982). The head extends forward of the tentacles into a large, muscular, contractile proboscis with the mouth and radula at its terminus.

Pigmentation in heteropods, as in most animals, is somewhat variable, and can only be studied accurately *in situ* or in freshly caught, live animals. Both sexes of *Cardiapoda placenta* are brightly pigmented, with reddish-brown spots covering the entire body, including the proboscis. Females, however, tend to have more pigment spots and are more brightly colored. The spots are highly contractile, and can change rapidly from 2 mm in diameter to less than 500 μm. Both sexes of *C. placenta* also have a tail that terminates in 12 fingerlike, reddish-brown to black extensions (Fig. 9a). The

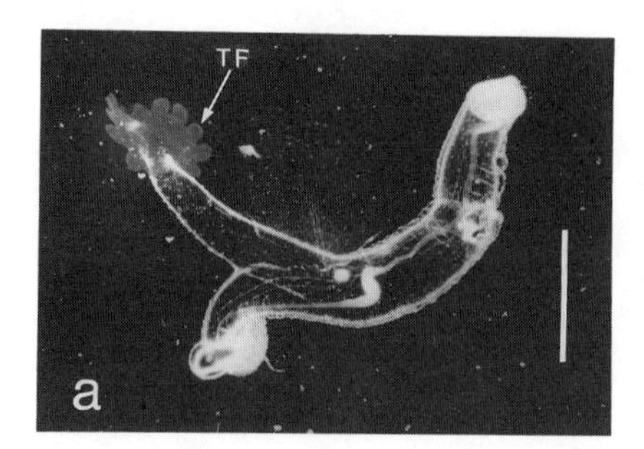

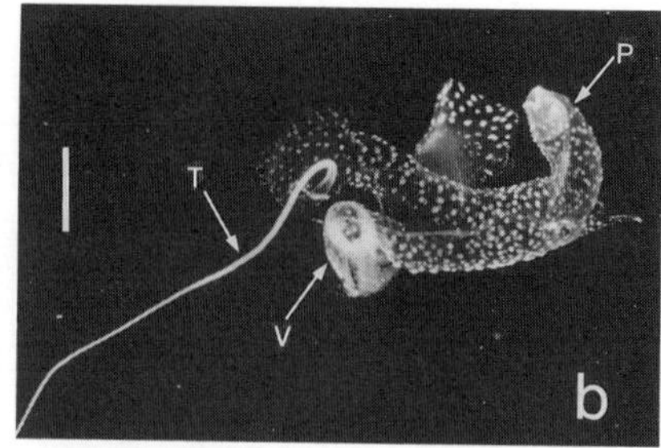

FIG. 9. Pigmentation in *Cardiapoda* species: *a*, *C. placenta* with expanded tail filaments; *b*. *C. richardi* showing pigment spots. P, proboscis; T, extension of tail; TF, tail filaments; V, visceral nucleus. Scale lines represent 5 mm.

tail filaments have often been referred to collectively as a clasper, but this function is not supported by observations of living animals. When *Cardiapoda* is disturbed, the tail filaments are rapidly contracted and then extended. They are also flashed in this manner during prey capture, when the tail slices the water in a whipping motion. It has often been noted that the tail filaments are present only in young animals of preserved material. The filaments are retained by adults, however, and in fact increase in size. They are quite fragile and can easily break off; they are also more prone to severe contraction in older individuals as the integumental musculature develops. In another species, *Cardiapoda richardi* (Color Fig. 5; Fig. 9b), the body is covered with scattered pigment spots and there is a large pigmented filament on the terminal end of the cylindrical body but anterior to the tail. In the field, this filament is spread like a fan above the animal's ventral surface. *Carinaria* and *Pterosoma* also are pigmented. Live *Carinaria* may have reddish or blue pigmentation. *Pterosoma* has large opalescent spots covering the central body in both sexes (Color Fig. 4; Fig. 8d); these spots continually change in brilliance and pattern as the animals swim.

The most advanced family, the Pterotracheidae, comprises only two genera, *Pterotrachea* and *Firoloida*, and five species (Fig. 10). There is no shell and no operculum. Tentacles are lacking in *Pterotrachea* and present in males only in *Firoloida*. In both genera, the sucker on the swimming fin is present only in males. The highly transparent body is elongate and cylindrical, and ranges in maximum length from about 40 mm (*F. desmaresti*) to 330 mm (*P. coronata*). The recorded maximal lengths of preserved animals are usually less; *Pterotrachea* species shrink by as much as 20 percent when placed in preservative (Gilmer, pers. obs.). The small visceral mass is mostly embedded in tissue within the main body axis, with gill branches projecting from the body. The dorsal visceral mass is now situated posteriorly to the ventral swimming fin, in contrast to the carinariids, in which the visceral cap is directly opposite the fin. The proboscis of pterotracheids is highly elongate and pointed terminally; consequently, the mouth opening is much smaller than in the carinariids. *Pterotrachea* is unique in having a wide gelatinous bib on the ventral body surface, extending from the anterior end of the fin to the proboscis; this structure is most highly developed in *P. scutata* (Fig. 10c).

In all species of *Pterotrachea*, there are pigment spots embedded in the transparent epidermis. Stacked layers of iridophores, which are flat platelets containing reflecting material, are present in the eyes and in the epidermal covering of the visceral nuclei of these heteropods. Seapy and Young (1986) have suggested that controlled movements of the pterotracheids can orient opaque body organs in such a way that the covering iridophores reflect light and permit concealment from visual predators. It is not clear whether the numerous red or purple spots that cover the entire body of both sexes of *P. hippocampus* are also iridophores, but these appear to change rapidly in size and color intensity (Gilmer, pers. obs.). Both sexes of this species also have a tail, which terminates in two lobes that contain a variable number of

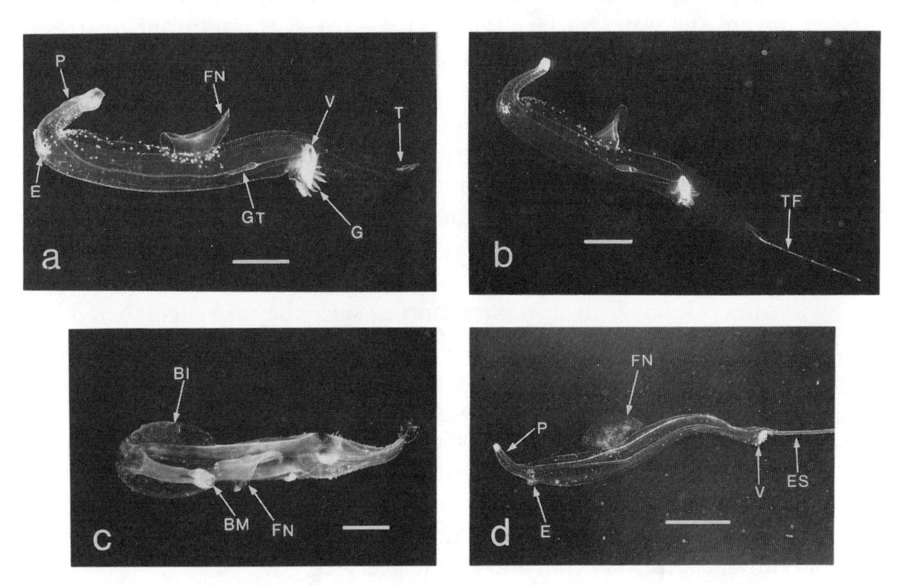

FIG. 10. Pterotracheid heteropods: *a*, lateral view of a female *Pterotrachea hippocampus*; *b*, *P. hippocampus* with expanded tail filament; *c*, ventral view of a female *P. scutata* (by L. Madin); *d*, lateral view of a female *Firoloida desmaresti* with its permanently attached egg string. BI, expanded gelatinous "bib"; BM, buccal mass at tip of proboscis; E, eye; ES, egg string; FN, fin; G, gills; GT, gut; P, proboscis; T, tail; TF, tail filament; V, visceral nucleus. Scale lines represent 10 mm.

red, fingerlike extensions similar to those described in *Cardiapoda*. An elongate tail filament (up to 60 mm long in *P. coronata*) may also be found in both sexes of species of *Pterotrachea* (Fig. 10b); the filament is composed of pigmented nodules that can contract and expand rapidly, especially if the animal is disturbed. The tail filament is not always present, however, even in the field, and further study is necessary to determine its function. In contrast to other pterotracheids and to carinariids, *Firoloida* appears to be the only genus lacking pigmentation in the integument, but males occasionally are found with a pigmented tail filament. The body of *Firoloida*, however, is so transparent that live specimens placed in a dish of water are practically invisible, except for the pigmentation in the eyes and gut.

Swimming and Buoyancy

The principal means of swimming in all heteropods is by rapid undulation of the single swimming fin, which is held upward in the water column. Thus the normal swimming position in all species is upside-down, with the viscera (and shell, if present) below and the morphologically ventral fin directed toward the water surface (Figs. 7a, 8–10). It is important to stress this point, since all heteropods have often been depicted in the literature with the dorsal surface (visceral mass) uppermost; this position is anatomically correct in schematic comparison with other mollusks, but only members of the Family Pterotracheidae are ever seen in this position in nature. Further, the designation of the heteropod fin as dorsal and the visceral mass as ventral (van der Spoel, 1976) is not only incorrect but has led to a confusion of right and left orientation, as has been pointed out by Pafort-van Iersel (1983).

Swimming is least efficient in the atlantids, since they are encumbered by a shell that is large enough to encase the entire body of a retracted animal. The thinness of the shell, however, reduces its weight relative to its size and, in *Oxygyrus* and *Protatlanta*, the relative weight is further decreased by increased proportions of organic conchiolin being incorporated into the shell matrix. Sculling motions of the single swimming fin propel the animal through the water. In all atlantid species, potentially excessive swinging of the shell and body is dampened by the outer shell whorl, which projects into a central, sharp, very thin and fragile keel (Fig. 7).

In the laboratory, flapping movements of the upwardly directed fin of *Oxygyrus keraudreni* drive the animal upward, and brief (1- to 2-second) swimming episodes alternate with longer (ca. 10-second) resting intervals in which the animal remains extended from its shell and may sink a few centimeters (Land, 1982). Scanning eye movements occur only during the sinking periods. Each eye rotates rapidly downward, then slowly returns to the normal horizontal position. In *Oxygyrus*, the two eyes may scan synchronously but are not always coordinated; in *Atlanta*, the eye movements are always conjugate. Land suggests that these eye movements not only increase an animal's field of vision but probably also enable atlantids to detect stationary objects in the water. These eye movements contrast with the eye movements of cephalopods and fishes, which are adapted to detect motion. Further, since the eyes of atlantids rotate downward, the animals are probably looking for objects (e.g. prey, predators) that reflect light from above.

In situ daytime observations (Gilmer, pers. obs) have revealed that atlantids normally swim at rates of 2 to 3 cm/second but are capable of speeds of at least three times those when disturbed. If pursued, disturbed animals withdraw into the shell, causing it to flip 90 degrees into a horizontal position in the water. This action exposes the largest surface area to the underlying water and retards sinking to measured rates of about 1.5 cm/second. Richter (1973b) has demonstrated that atlantids with flattened shells (and a corresponding increase in the horizontal surface area of the shell and keel) can slow their sinking rates when in the horizontal position by as much as 50 percent compared to species with high shell spires.

Nighttime *in situ* observations of atlantids have revealed behavioral patterns different from those seen in daylight. At night, atlantids are motionless in the water and attached to large strands of mucus, which seem to emanate from the foot (Fig. 11); similar observations have been made by Richter (1973b) on atlantids maintained in the laboratory. Gilmer (pers. obs.) has seen field specimens of *Oxygyrus keraudreni* (10 mm shell diameter) hanging from an array of mucus strings extending to almost 0.5 m above the animal. Sinking is not detectable in such animals, but if disturbed they will break free of the mucus threads and sink with the keel of the shell facing downward, cutting the water like a blade. The head and foot become slightly retracted, but the fin projects upward and appears to keep the animal in narrow profile to maximize the sinking rate. In this position, sinking may be as rapid as 10 cm/second. Few field observations have been made of atlantids, and further study may reveal other buoyancy mechanisms and escape strategies.

Swimming efficiency increases in the larger heteropod species, which are either shell-less or have a very reduced shell. In the Carinariidae and Pterotracheidae, the fin produces a screwlike undulation normally proceeding from anterior to posterior (Morton, 1964), but these animals can swim backward by reversing the direction of the fin movement (Gilmer, pers. obs.). In both families, the body is capable of rapid flexion, which facilitates both locomotion and rapid change of direction (Tesch, 1949). *Carinaria*,

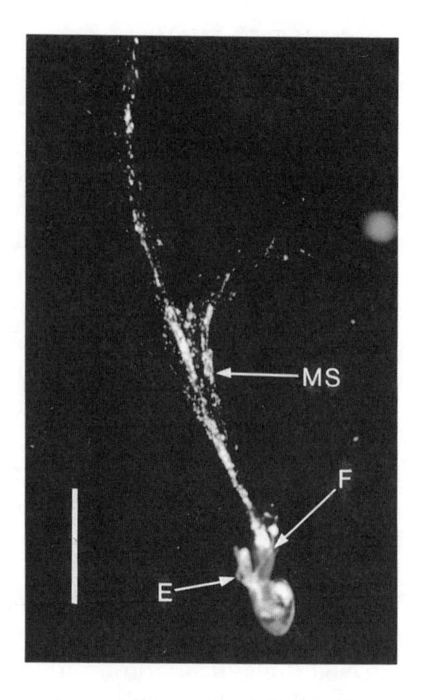

FIG. 11. A nighttime *in situ* photograph of a quiescent atlantid heteropod with attached strands of mucus. E, eye; F, foot and fin; MS, mucous strands. The scale line represents 10 mm.

Cardiapoda, *Pterotrachea*, and *Firoloida* all respond to a diver's turbulence by rapid downward swimming, which involves an undulation of the entire body in conjunction with a sculling of the swimming fin. This activity lasts for about 8 seconds before normal swimming, using only the fin, is resumed. During these few seconds, the animal can move 1 to 3 meters depending on its size. *In situ* swimming speeds vary greatly, even within a species, and depend on an animal's activity. *Cardiapoda placenta*, which has been observed most often by divers, can remain motionless or display escape swimming rates of almost 50 cm/second (Gilmer, pers. obs.).

Swimming is not a continuous activity in any of the carinariid or pterotracheid heteropods. Daylight *in situ* observations by Hamner et al. (1975) of *Cardiapoda placenta* and *Pterotrachea coronata* revealed that these animals swim actively only when pursuing prey or when avoiding capture by scuba divers (and presumably predators). At other times, they are curled into loose balls and float as if neutrally buoyant; divers could not detect any sinking. The gut appeared to be full in about 85 percent of these floating animals, indicating that they were digesting food (Gilmer, pers obs.). At night, species of *Pterotrachea* and *Firoloida desmaresti* exhibit different behavior. During darkness, the animals are positioned with the ventral surface facing down; the fin is then below the animal and may undulate slowly, but the animal does not appear to change position. In this posture, their field of vision is principally directed downward and forward, not upward as in daylight hours. Seapy (1974) has suggested that heteropods feed mostly at night, and this nighttime orientation may be advantageous in the location of bioluminescent prey.

Individuals of all of the genera of carinariid and pterotracheid heteropods have been observed floating in the field with no apparent movement. These mollusks are assisted in maintaining a particular water depth by the development of gelatinous body tissues and ionic regulation. Denton and Shaw (1961) analyzed the fluid collected from centrifuging specimens of *Pterotrachea coronata* and found it to be of lower density than sea water, principally because sulfate ions are actively replaced by lighter chloride ions through the body wall. Sulfate concentration in *Pterotrachea* body fluids is about 75 percent that of sulfate concentration in sea water. A relatively high lipid content (Ceccaldi et al., 1978) may also contribute to buoyancy control. Increased specialization directed toward the attainment of neutral buoyancy, combined with the attributes of larger body size, reduction of or loss of the shell and faster swimming, represent marked advances over the more primitive atlantids. Carinariids and pterotracheids not only can pursue a wider range of larger prey and evade predators more easily, but also are capable of conserving energy at other times through long periods of floating.

Mention should also be made of the large paired statocysts present in all heteropods. These relatively complex organs (Barber and Dilly, 1969) presumably function to sense changes in position during swimming or floating.

Food and Feeding

All heteropods are carnivores that use vision to locate zooplanktonic prey. This would seem to restrict them to living at least part of the time in lighted surface waters, where prey are visible, and to feeding during the daylight hours. At least some species, however, have been found below 200 m depth, and some species have been observed only with full guts during nighttime collections with nets or in nighttime scuba dives. This suggests that some heteropods can detect prey in very low light conditions, or that they feed upon bioluminescent prey. Indeed, Richter (1974) has noted that the heteropods with very large eyes (e.g. *Atlanta lesueuri*, *Pterotrachea coronata*, and *P. scutata*) generally are found at greater depths than species with smaller eyes. All heteropods, by virtue of their swimming abilities, are capable of actively pursuing prey. Active, visual predators like these are potentially capable of feeding on a wide range of prey species and of selecting preferred food items within this range.

The main organ of food capture is the radula, although the fin sucker of some species may also be used to manipulate prey during ingestion. The

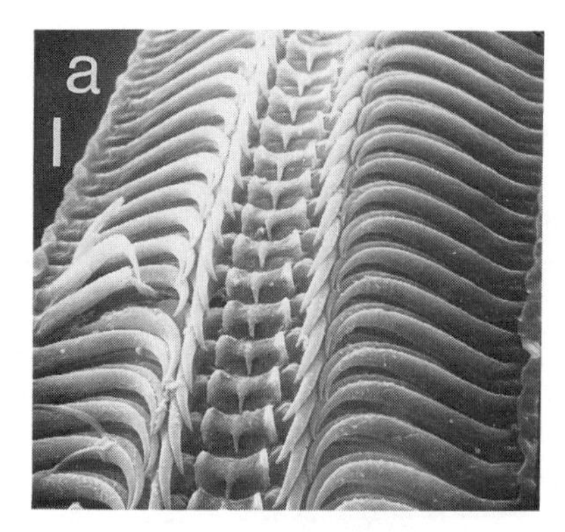

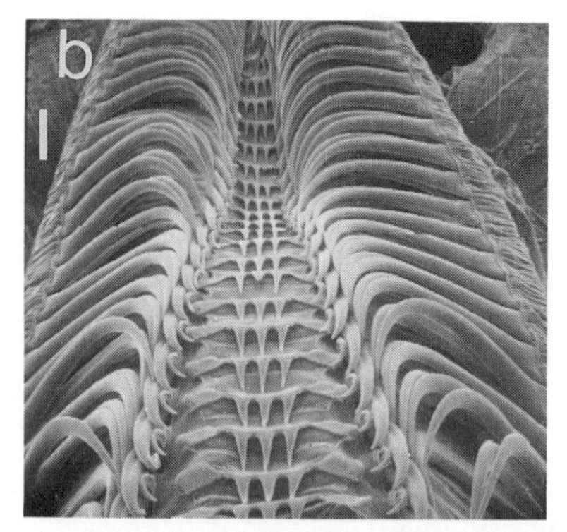

FIG. 12. Scanning electron micrographs of heteropod radulae (courtesy of C. Thiriot): *a*, *Atlanta lesueuri*; *b*, *Carinaria cristata*; *c*, *Pterotrachea minuta*. Scale lines represent 20 μm in *a* and *c*, 200 μm in *b*.

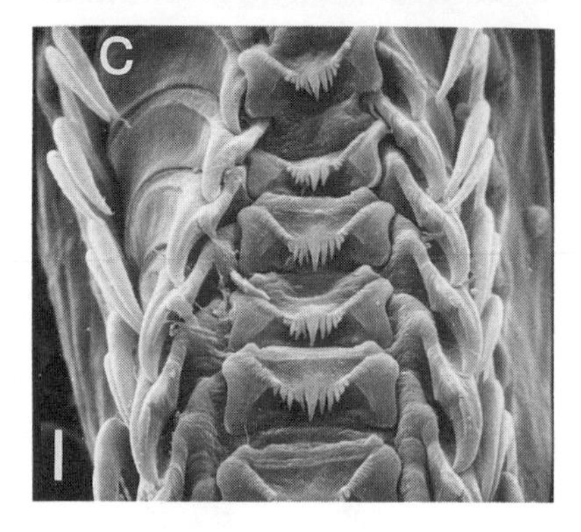

radula of heteropods (Fig. 12) is of the taenioglossate type, and is typical of the majority of mesogastropods in having seven teeth per row (radula formula = 2-1-R-1-2). There is a single central tooth (R) flanked on each side by one lateral tooth; and in the heteropods the spinelike marginal teeth (two on each outer side of a row) are specially modified for grasping prey. Species differences in the structure and function of the teeth have been described and illustrated by Thiriot-Quiévreux (1973b and Fig. 12). The large chitinous teeth can be seen clearly when they are protruded from the mouths of large specimens. Living carinariids and pterotracheids often can be induced to strike with their teeth at a moving probe—or a wiggling finger!

Unfortunately, the prey-capture and feeding mechanisms of live specimens have seldom been observed, owing to the difficulties of viewing heteropods *in situ* or of capturing undamaged specimens and maintaining them in the laboratory along with suitable living prey. Food selection has been studied mostly from examination of the stomach contents of preserved specimens. Those accounts based on prey items attached to protruded radular teeth of preserved heteropods are not included here, because such un-ingested prey may be captured incidentally and thus may not reflect food selected in nature.

Richter (1968) distinguished two types of feeding in heteropods. He characterized the atlantids as using the radula for tearing pieces of tissue from prey, and the carinariids and pterotracheids as being swallowers that use the radula to ingest prey whole. More recent work (reported below) suggests that both types of feeding behavior are exhibited by all three families.

There are few field observations of the feeding behavior of atlantids. Richter (1968), however, observed *Atlanta peroni* feeding on the shelled pteropod *Hyalocylis striata* (see Chapter 4) in the laboratory. The shell of the prey was held by the fin sucker of *Atlanta*, and the animal used its proboscis and radula to remove pieces of the prey from the shell. Thiriot-Quiévreux (1973a) observed an *Atlanta* species feeding in a similar manner on other planktonic mollusks in laboratory conditions. Prey such as shelled pteropods (*Creseis*), gastropod veliger larvae, and other *Atlanta* were moved to the mouth by the foot, the prey shell was held with the fin sucker, the proboscis was thrust into the shell aperture, and the radular teeth tore pieces from the prey. Digestion was complete in about 24 hours. These feeding observations do not necessarily indicate the preferred diet of atlantids in nature, since the prey offered in the laboratory were those available to the experimenters, but Richter (1983) has found remains of *Creseis* sp. in the gut contents of *A. inclinata*, and *A. inflata* has been observed feeding on *Creseis virgula* in nature during daylight (Gilmer, pers. obs.).

Richter also (1982) studied the natural diets of *Oxygyrus keraudreni* by examining the gut contents of 40 specimens collected by nets and 80 specimens obtained from the guts of juvenile dolphin fish (*Coryphaena*). A total of 101 zooplanktonic prey items were identified from the 58 *Oxygyrus* that contained recognizable items in their guts. Pelagic gastropods, both heteropods and thecosomatous pteropods, constituted 54 percent of the prey; 37 percent were crustaceans, mostly copepods (30 percent); and chaetognaths and foraminifera each constituted 4 percent. Some of the prey were intact and obviously had been swallowed whole.

The use of gut analyses in determining natural diet has several limitations. Species identifications of prey in gut contents are not always possible; for example, thecosome prey are identified from radular teeth or gizzard plates that are of little taxonomic value below the family level. Soft-bodied, gelatinous zooplankton would be difficult or impossible to detect in gut contents, should they be eaten. These results do clearly indicate, however, that *Oxygyrus keraudreni* takes a variety of zooplankton as food but selectively favors other gastropods over any other prey type. Further, most feeding is done during daylight hours, when 91 percent of the captured *Oxygyrus* contain food in the stomach.

The most comprehensive behavioral observations on the feeding of carinariid heteropods are those made by scuba divers in the field. Hamner et al. (1975) observed undisturbed *Cardiapoda placenta* on 38 underwater dives over the course of one year. On five of these occasions, the heteropod was feeding on *Salpa cylindrica*, and salp remains were present in the guts of seven other hand-collected specimens. The time required for prey digestion was determined by capturing feeding specimens in jars and maintaining them in the laboratory until the release of fecal material. Control specimens with empty guts, which were collected in a similar manner, released no fecal matter. Digestion times ranged from 4.5 to 6.5 hours, depending on the relative sizes of prey and predator. Live specimens of *C. placenta* were maintained in the laboratory, where they ate fish larvae; a planktonic nudibranch, *Phylliroë* (see Chapter 6); other heteropods, including *Firoloida desmaresti* and *Pterotrachea coronata*; and other individuals of their own species. The maximum digestion time for these prey organisms was 7 hours.

Hamner et al. (1975) further reported that *Cardiapoda* attacks prey located above it in the water column and at distances of up to 60 cm. This suggests not only that visual perception is acute, but also that it is adapted to detect prey against a lighter, rather than darker, background. The prey is approached at swimming speeds of up to 40 cm/second, and the heteropod either ingests smaller prey whole, by radula action combined with contractions of the proboscis, or severs pieces of larger prey by using the radular teeth. During ingestion, *Cardiapoda* may shake its body violently from side to side. Presumably this action helps to compress the prey, especially gelatinous forms like salps, into the esophagus, and aids in breaking off pieces of prey that are too large to swallow whole. Ingestion time in all observations of *Cardiapoda* was about 10 minutes.

Seapy's (1980) observations, based on prey organisms contained in the esophagus of 1,234 preserved specimens of *Carinaria cristata*, also suggest that salps are an important food for this heteropod species. At least 70 percent of the prey items consisted of small salps and related doliolids (*Thalia*, *Doliolum*), chaetognaths, and copepods. When the relative abundance of each prey species in the gut was compared with the proportion of each prey species in the plankton, it was clear that *C. cristata* feeds selectively on salps and chaetognaths. Copepods are nonpreferred prey; their relative abundance in the plankton was much higher than their relative numbers in the gut contents would suggest. Interestingly, *Carinaria* also appears to feed selectively on the young and smaller adults of its own species. Other prey found in gut contents included euphausiids, small siphonophores (*Nanomia*, *Muggiaea*), *Atlanta gaudichaudi*, planktonic worms (*Tomopteris* and polynoids), amphipods (*Hyperia*), and myctophid fish larvae. In addition to these food items, Okutani (1961) found ctenophores, scyphozoan medusae, crustacean larvae, and fish eggs in the gut contents of *Carinaria* from Japanese waters, and Woodward (1913) observed six small fish packed into the gut of a *C. lamarcki* from the Mediterranean.

The observations of Hamner et al. (1975) and Seapy (1980) show clearly that both *Cardiapoda placenta* and *Carinaria cristata* are opportunistic predators, feeding on a wide range of prey but preferring salps. *C. cristata* also selects for chaetognaths and other planktonic mollusks. There appear to be no comparable studies of food and feeding in the third genus of carinariids, *Pterosoma*.

Little information is available regarding the diets of the pterotracheids. Hamner et al. (1975) reported seeing *Pterotrachea coronata* on 27 dives. On five occasions, this species was feeding on siphonophore bracts, and digestion times in collected individuals ranged from 6 to 8 hours. When observed by scuba divers, *Pterotrachea* usually is associated with dense concentrations of siphonophores, and examination of the fecal matter released by freshly caught specimens of both *P. coronata* and *P. hippocampus* usually reveals large numbers of siphonophore nematocysts in both fired and intact states (Gilmer, pers. obs.). Hand-collected (and presumably undamaged) specimens in the laboratory would not feed on salps, ctenophores, or pteropods, but *P. hippocampus* would eat brine shrimp or live copepods when they became entrapped on the mucous surface of the heteropod's gelatinous bib area (Gilmer, pers. obs.). The proboscis swings around to the bib surface and the radular teeth pick off the snared crustaceans. This behavior also has been seen by divers in the field, and may well constitute a natural feeding mode in this species.

The only other observations on the diets of pterotracheids are those of Hirsch (1915) and Okutani (1957). Hirsch reported that *Pterotrachea* (sp.?) will eat swimming polychaetes. One heteropod 66 mm long required 10 to 20 minutes to ingest a worm 27 mm long; ingestion was accompanied by vigorous shaking of the body, as described above for *Cardiapoda*. Okutani found copepods and other crustacea in the guts of *Pterotrachea*. Hirsch also stated that *Pterotrachea* and *Firoloida desmaresti* will eat scyphozoan medusae, siphonophores, crustacea, and salps. Notwithstanding that *F. desmaresti* is a relatively common and widespread species, no other studies have been made of its feeding habits and natural prey.

Apart from cannibalism and predation by other heteropod species, fish seem to be the major predators of the group. Most of Richter's (1982) specimens of *Oxygyrus keraudreni* were obtained from the guts of juvenile dolphin fish (*Coryphaena*). Russell (1960) identified planktonic mollusks from the stomachs of three fish species. The yellowfin tuna (*Thunnus albacares*) eats *Carinaria lamarcki*, and two species of lancetfish (*Alepisaurus ferox* and *A. brevirostris*) prey on the same heteropod as well as *Atlanta peroni*, *Oxygyrus keraudreni*, *Cardiapoda placenta*, and *Pterotrachea hippocampus*. *A. peroni* has been found in the gut contents of a myctophid fish (*Centrobranchus nigroocellatus*) (Hartmann and Weikert, 1969), and Okutani (1961) reported that tuna and turtles will eat *Carinaria cristata*. The relative importance of these heteropods to the diets of the fish was not assessed, but judging from the generally low abundance of heteropods in nature and the high nutritional requirements of the fish, it is probably small. The smaller and slower-swimming heteropods, primarily atlantids, are probably also eaten by larger zooplankton, although there are few reports to substantiate this. Purcell (1981), however, reported that the siphonophore *Rosacea cymbiformis* preys on *Atlanta* (sp.?) as well as other small zooplankton. The common scyphomedusa *Pelagea noctiluca* often contains various heteropod species in gut contents and will avidly ingest heteropods offered in the laboratory (Gilmer, pers. obs.).

Predator defenses in the smaller atlantids presumably include their transparency and the shell, into which the animal can retreat while sinking. Some of the attributes that make the larger heteropods efficient carnivores may also be effective defenses against predators. Vision, for example, can be employed not only to find food, but also to sight potential predators from some distance. The larger heteropod species, which swim rapidly to approach prey quickly, probably evade predators in the same manner. The transparency of the body may aid heteropods in evading detection during their approach to prey that have some visual perception (e.g. polychaetes, larval fish, and other heteropods), but this feature is probably more important as a defense against fish predation. Carinariids and pterotracheids often are seen swimming in nature with the highly pigmented viscera or eyes bitten away and the remainder of the body intact (Tesch, 1949; Gilmer, pers. obs.). The function of the scattered, integumentary pigment spots and pigmented tail filaments of pterotracheids and carinariids is not known, although Gilmer suggests that the tail filaments may serve as lures for prey that have vision. It is also possible that scattered pigmentation on an otherwise transparent body may function as a type of camouflage, in which a large animal appears to be many smaller organisms and thus is less desirable to larger prey. Finally, large body size itself provides a refuge from the many smaller potential predators.

There are no comprehensive studies on the metabolism or energetics of any heteropod species. Biggs (1977) measured both respiration and ammonium-excretion rates of hand-collected *Pterotrachea hippocampus* weighing between 1.1 and 10.0 mg protein. The respiration rates averaged 19 ± 2.5 μl O_2/mg protein/hour; the average excretion rate was high, measuring

TABLE 4
Respiration rates of Cardiapoda placenta *determined by the Winkler technique at 26° C*

Size (mg dry weight)	Sex	Respiration Rate	
		$\mu l\ O_2$/mg/hour	ml O_2/animal/day
	Unstressed animals (n = 8)		
2.2	m	4.4	0.23
15	f	4.5	1.62
18	f	4.3	1.84
23	m	4.3	2.32
30	f	4.4	3.17
49	m	4.1	4.76
61	f	3.6	5.27
76	f	2.9	5.20
AVERAGE		4.1±0.6	
	Stressed animals[a] (n = 7)		
2.4	m	3.3	0.19
3.5	f	3.1	0.26
18	f	2.1	0.88
23	m	1.9	1.00
52	m	1.5	1.79
61	f	2.0	2.94
75	f	1.8	3.13
AVERAGE		2.2±0.7	

SOURCE: Gilmer, unpublished data.
[a] Stress was induced by pouring animals between containers, or by repeatedly prodding animals into frantic swimming for at least 1 minute prior to their being placed in respiration jars.

3.0±0.9 $\mu g\ NH_4^+$/mg protein/hour. Oxygen consumption by *Cardiapoda placenta* has been determined by Gilmer, who compared values obtained from hand-collected, unstressed specimens and specimens that had been deliberately stressed (Table 4). Healthy, unstressed animals respired at an average rate of 4.1 $\mu l\ O_2$/mg dry weight/hour; oxygen-consumption rates decreased with large size but remained approximately constant for each individual for up to 24 hours in the laboratory, regardless of preconditioning time. Stressed specimens had considerably lower respiration rates (average = 2.2 $\mu l\ O_2$/mg dry weight/hour), and the variability was much higher than in the unstressed animals. Ikeda (1974) also determined the oxygen-consumption rates of *C. placenta* (= *C. sublaevis*) collected by nets; his reported values may reflect slight temperature variation, although only very small individuals (ca. 1 mg dry weight) had respiration rates comparable to those of Gilmer's unstressed specimens. The respiration rates of Ikeda's larger specimens were approximately one-fourth the rates obtained by Gilmer on animals of similar sizes. The differences are probably due largely to collection techniques: compaction in the cod end of a plankton net damages or stresses these delicate organisms; but the animals are minimally stressed when hand-captured by scuba divers. It is therefore important to emphasize the need to carefully collect these and other fragile zooplankton for use in laboratory experiments.

Reproduction and Development

The anatomy and histology of the heteropod reproductive tract have been studied in great detail in a number of species (see Thiriot-Quiévreux, 1973a, for a review of earlier studies; also Thiriot-Quiévreux and Martoja, 1974; Martoja and Thiriot-Quiévreux, 1979). We know much less, however, about the biological and ecological aspects of reproduction.

The heteropods are dioecious and exhibit sexual dimorphism. Males are distinguished by the presence of a conspicuous penis (Fig. 7b) and, in *Pterotrachea* and *Firoloida*, by being the only sex with a sucker on the swimming fin. The sex ratios may be skewed in the genus *Atlanta*, in which it is not unusual to find nine males for every female in one collection (van der Spoel, 1976). This finding requires statistical confirmation in larger numbers of collections, but, if it is true of atlantids in general, we should seek a biological or ecological explanation for this anomaly. In pterotracheids, the situation is commonly reversed, with there being more females than males in any one collection (Aravindakshan, 1973). Dales (1952), however, found approximately equal numbers of male and female *Carinaria* in the California Current during the May breeding season. Skewed sex ratios may result from males and females inhabiting different water depths except during reproductive periods.

Though internal fertilization is known to take place, copulation has only been observed once. Scuba divers saw a female (63 mm long) and male (60 mm) *Pterotrachea hippocampus* mating in the Bismarck Sea (Indo-Pacific) at a depth of 15 meters (Harbison, pers. comm.). The animals were paired ventrally and head to head with their tails entwined. The pair moved spirally through the water, propelled by body undulations of both animals. Mating was brief, lasting somewhat over 30 seconds. The fin sucker of the male was in contact with the female during mating, but it was unclear whether this structure was used to secure a hold on the mate, as has often been claimed. The fin sucker of atlantids clearly plays a role in holding prey during ingestion, but whether it is also employed in copulation is not known. In the carinariids, each sex has a fin sucker, but it has not been observed to be used during feeding; in fact, the location of the sucker on the posterior fin margin, far from the mouth, suggests that it could not be used for holding prey. Only the males of pterotracheids have a fin sucker, and these facts suggest that both carinariids and pterotracheids use the sucker only during reproductive encounters.

Tesch (1949) was the first to observe spermatophores attached to the shells of female *Atlanta* and *Oxygyrus*, and to suggest that sperm were transferred in this manner. Spermatophore production in atlantids, *Carinaria*, *Pterosoma*, and *Firoloida* has now been confirmed (van der Spoel, 1972; Thiriot-Quiévreux and Martoja, 1974; Gilmer, pers. obs.). The bifurcated heteropod copulatory organ consists of the penis, which has a sperm groove, and a penial appendage, which has an opening to the prostate gland at its tip, at least in some species (Fretter and Graham, 1962). Presumably, spermatophores are conveyed in the sperm groove to the female; the role of the penial appendage remains uncertain. The packaging of sperm into spermatophores is unusual in benthic mollusks, but it is a common method of sperm transfer in the pelagic cephalopods and in several different groups of marine zooplankton such as copepods and euphausiids. Since the transfer of sperm via spermatophores may be achieved very rapidly, this mechanism may serve to minimize the interference of swimming motions during mating. The timing of reproduction in heteropods is not known, but Richter's data (1968) suggest that reproduction in the Mediterranean is continuous throughout the year, with early and late veliger larvae of 12 heteropod species being present in the plankton in all months.

Female heteropods lay fertilized eggs either singly or packaged within a mucoid string. Usually the eggs or egg strands are detached from the parent, but in female *Firoloida desmaresti* the egg filament is a permanent anatomical feature projecting from the posterior end of the animal (Fig. 10d) (Owre, 1964). In this species, females 17 to 30 mm long have egg filaments of approximately the same length as their body. These filaments contain young in various stages of development, from 32-cell embryos to late veligers at the distal end. Owre (1964) further reported that a female *Pterotrachea hippocampus* (46 mm long) can produce a 250-mm–long egg string. Egg-string length is apparently dependent on the species and size of the female and correlates with the numbers of young produced. Gilmer (pers. obs.) has seen a *P. coronata* (180 mm long) produce an egg string 600 mm long with approximately 11 eggs/mm, arranged in an orderly spacing. He also has seen numerous *Cardiapoda placenta* females 80 mm long release consecutive egg strings 200 to 250 mm long within a 6-hour time interval. Each string was spawned and released within 2 hours, and each contained between 5,000 and 6,000 eggs, or about 25 eggs/mm. Other female *C. placenta* as short as 32 mm can form egg strings of up to 150 mm long.

Embryonic development before hatching was described in 1876 by Fol for *Firoloida desmaresti* and for species of *Pterotrachea* and *Carinaria*.

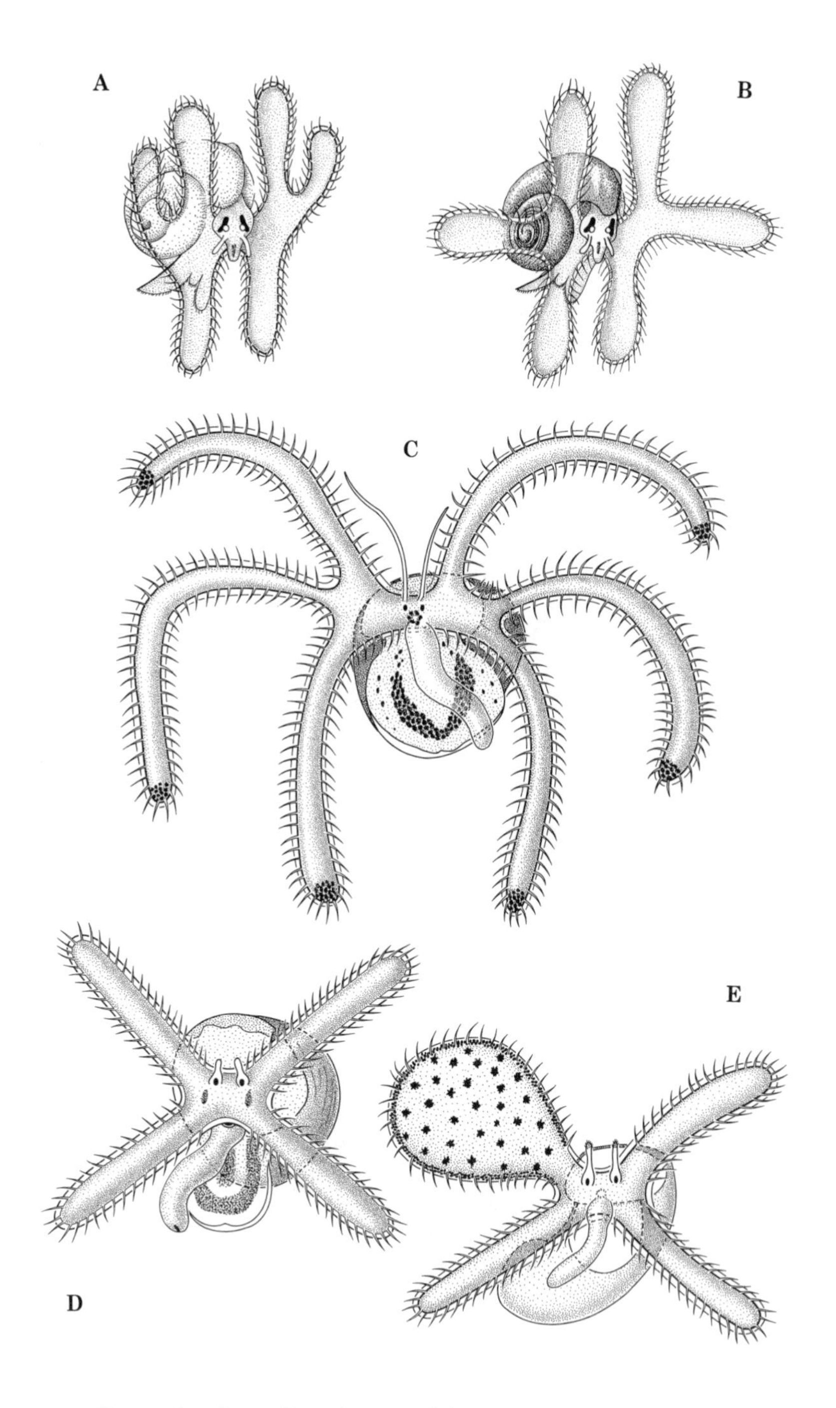

FIG. 13. Free-swimming veliger larvae of heteropods (from Thiriot-Quiévreux, 1973a): A, *Atlanta lesueuri*; B, *A. inflata*; C, *Carinaria lamarcki*; D, *Firoloida desmaresti*; E, *Pterotrachea* sp.

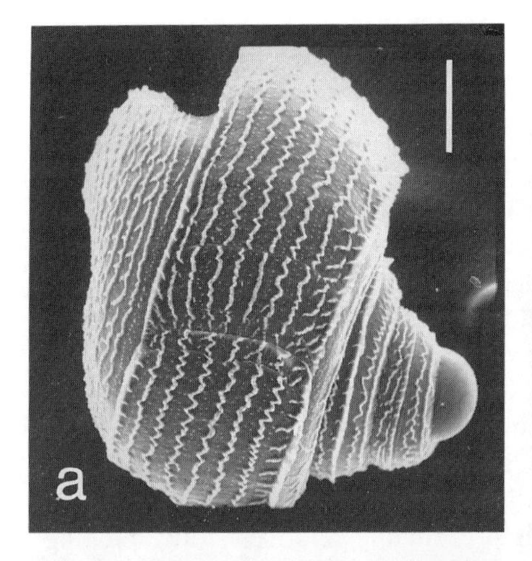

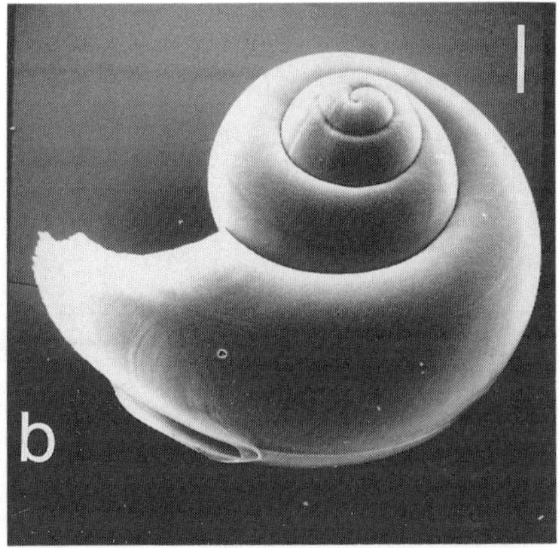

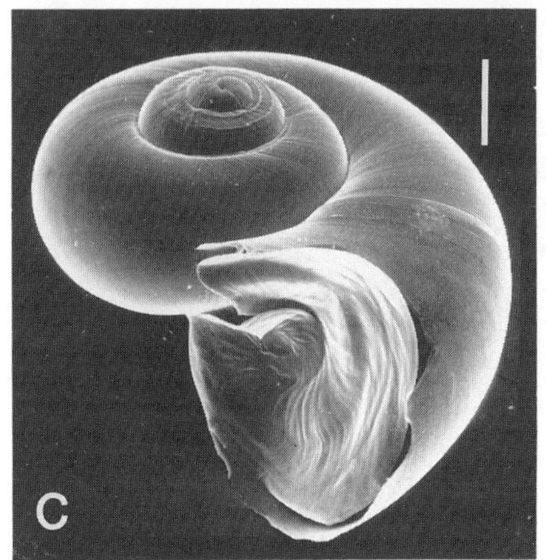

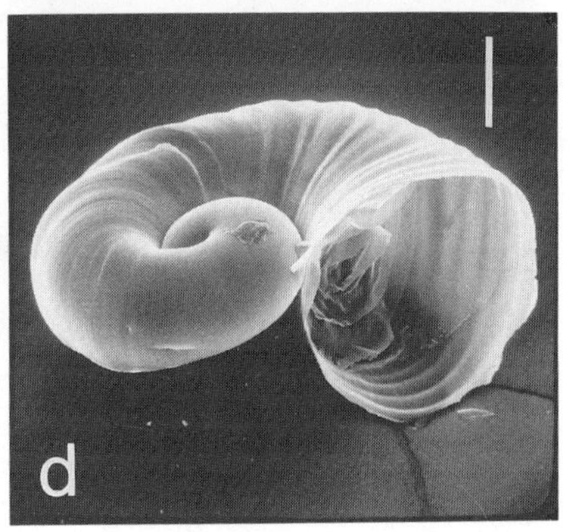

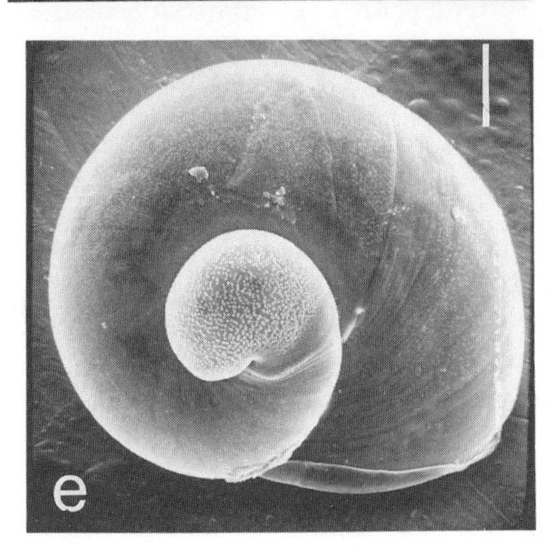

FIG. 14. Scanning electron micrographs of heteropod veliger shells: *a*, *Atlanta fusca* (from Thiriot-Quiévreux, 1973a); *b*, *A. peroni* (from Thiriot-Quiévreux, 1973a); *c*, *Carinaria lamarcki* (from Thiriot-Quiévreux, 1975); *d*, *Pterotrachea coronata* (courtesy of C. Thiriot); *e*, *Firoloida desmaresti* (from Thiriot-Quiévreux, 1972). Scale lines represent 100 μm.

Free-swimming veliger larvae (Figs. 13, 14) hatch from the eggs within a few days, and many of these have been described and identified on the bases of shell and velum characteristics (Richter, 1968; Thiriot-Quiévreux, 1973a, 1975). Jung (1975) has provided scanning electron micrographs of five species of fossil atlantid larvae recovered from sediments of Holocene to late Pleistocene age. Atlantid veligers are all distinguished from other heteropod veligers by a conspicuous slit in the center of the shell lip. Veligers of the genus *Pterosoma* apparently remain undescribed.

Heteropod veligers use the velum for both swimming and food collection, but the actual feeding mechanism and diet have never been described. The velum has two, four, or six lobes, and in at least some species the number increases with growth. The length of time from hatching to metamorphosis in nature is not known but is presumably relatively long, judging from the increase in shell and body size during this stage and from the size and complexity of the velum, which is typical of long-lived veligers. The growth of veligers coincides with the enlargement of the eyes and the differentiation of the foot. In late veligers, the propodium of the foot develops into the swimming fin, the mesopodium forms the fin sucker, and the metapodium bears the operculum. The veligers of several heteropod species have been reported to carry out more extensive diel vertical migrations than do the adults (Richter, 1973a), and these movements may be correlated with feeding cycles and the dependency of the larvae upon phytoplankton as a food source.

At metamorphosis, the veliger changes abruptly from ciliary feeding on small particles to carnivorous habits. The first meal at metamorphosis is the ingestion of its own velum, a process requiring 6 to 8 minutes in the atlantid veligers observed by Pilkington (1970). In some atlantid veligers and in a veliger of *Pterotrachea* observed by Richter (1968), the right side of the velum is ingested and the left side persists for a longer period, thus continuing to aid locomotion. Three days after metamorphosis, the young of *Pterotrachea* begin to feed on small zooplankton.

In the shell-less pterotracheids, the veliger shell and operculum are cast off at metamorphosis (Owre, 1964; Richter, 1968). In both pterotracheids and carinariids, metamorphosis involves a rapid increase in body size, owing to the development and expansion of gelatinous tissue; this increase is particularly pronounced in the first 2 to 3 days after metamorphosis (Richter, 1968). It is of interest that in newly metamorphosed female *Firoloida*, only 1.2 mm long, a small egg filament already protrudes from the tail (Owre, 1964).

Exceedingly little information exists on the growth rates of heteropods in nature. Dales (1952) reported that 40 percent of *Carinaria* collected in the

California Current during July were under 10 mm in length, and that by September the average length had increased steadily to 30 mm. These data, however, are based on relatively small numbers of collections made in an area of complex water-mass movements.

It becomes obvious from a study of the literature that significant gaps remain in our knowledge of heteropod reproduction and development. Some of these deficiencies were pointed out above, but it would also be of interest to have information on the age of sexual maturity; the location and recognition of mates, particularly in species with low population densities; the number of young produced per female in different species, including the frequency of egg-laying; the feeding method, natural diet, and growth rates of veligers and juveniles; and the mortality rates of the young. Any information on the genus *Pterosoma* would be of interest.

Epifauna and Parasites

There are few reports of symbionts or parasites occurring on heteropods, perhaps because there have been too few observations of living or well preserved specimens, or because heteropods seem to be able to clean their body or shell surfaces by removing foreign material with the radula.

Hochberg and Seapy (1985 and pers. comm.) list the following as parasites of heteropods:

Atlanta: sporozoans, nematodes
Carinaria: flagellates, ciliates, digeneans, copepods
Firoloida: sporozoans, digeneans
Pterotrachea: sporozoans, ciliates, digeneans, amphipods

Not all of the protozoans, nematodes, and digeneans listed above have been identified or described. In 1914, Collin described three new species of protozoans from heteropods: the ciliate *Syringopharynx pterotracheae* was found on the gill surface of *Pterotrachea coronata*; and a single female *Carinaria lamarcki* (= *mediterranea*) harbored both a ciliate (*Opalinopsis carinariae*) in its digestive gland and a flagellate (*Cryptobia carinariae*) in its seminal receptacle. Copepods found in two specimens of *Carinaria cristata* (= *japonica*) were postembryonic stages of the parasitic species *Cardiodectes medusaeus*; planktonic mollusks serve as intermediate hosts of this copepod, and the final hosts are myctophid fish (Perkins, 1983). Whether the hyperiid amphipods found on *Pterotrachea* were acting as parasites or as predators is not known (Harbison et al., 1977).

In addition, several specimens of *Cardiapoda placenta* have been found that were infested by a calyptoblastic hydroid (Gilmer, pers. obs.). Numerous polyps, connected by a stolon, were growing over the body surface of the heteropods. Similar hydroid infestations are sometimes present on the shells of thecosomatous pteropods.

Shell Mineralogy and Sediments

Calcareous shells are present in members of two of the three heteropod families and are lacking only in the pterotracheids. Bøggild (1930) established that the shells of atlantids and *Carinaria* are composed of aragonite and not of calcite, the more commonly encountered form of calcium carbonate. Grossman et al. (1986) have recently confirmed the total aragonite composition in several atlantid species by using X-ray diffraction. Although heteropod shells are generally small and have extremely thin walls (<50 μm), they can contain large amounts of mineral. For example, the shells of larger atlantids (e.g. *Atlanta peroni*) can weigh as much as 30 mg (dry weight) and the shells of moderate-sized (180 mm body length) specimens of *Carinaria lamarcki* may weigh 300 mg (Gilmer, pers. obs.).

The shell walls of atlantids are composed of at least three crystalline layers in which the dominant feature is always a thick internal layer of aragonite rods arranged in a complex helical pattern (Batten and Dumont, 1976). A similar helical structure is present in the shells of many thecosomatous pteropods (see Chapter 4) and it has been suggested that this crystalline arrangement may provide thin shells with flexibility. The growth patterns of heteropod shells are poorly understood, but in addition to normal apertural expansion of the whorls and also of the operculum (Tokioka, 1961), shell-wall thickness appears to increase throughout the life of an animal by irregular additions to the various shell layers (Batten and Dumont, 1976). Consequently, gravimetric techniques may provide a better indication of shell age than simple shell dimensions when dealing with adult specimens.

Heteropod shells are seldom abundant enough in marine-sediment samples (mainly pteropod oozes) to be separated from thecosomatous pteropods or other organisms. Their contribution to these sediments is usually estimated at less than 5 percent and is often considered to be less than 1 percent (e.g. Murray and Chumley, 1924). In sediment traps placed at 3,200 m off Bermuda, the contribution of atlantid shells in the coarse fraction (>0.5 mm particle size) of carbonate material (shells of foraminifera, pteropods, and heteropods) varied from 0.2 to 7.5 percent throughout a yearly cycle (Jasper and Deuser, pers. comm.).

Since heteropod shells are composed of aragonite, they are subject to the same rules of dissolution as the shells of thecosomes (see Chapter 4). Consequently, they accumulate only in relatively shallow marine areas in contrast to calcite shells, which may be present in deeper sediments. It is also worth mentioning that only juvenile shells of *Oxygyrus keraudreni* are found in sediments, since the adult shell whorl and keel are uncalcified and instead are composed of biodegradable conchiolin (Batten and Dumont, 1976).

Evolution

The time of origin of this group is uncertain, despite the fact that heteropod shells contribute to marine sediments and leave a fossil record. Heteropod remains date from at least the middle Miocene (Woodring, 1928) and possibly the early Eocene (Lowenstam, 1974). The numerous Miocene specimens representing both *Atlanta* and *Carinaria* show close similarities to living representatives. Anatomical relationships clearly place the heteropods with the taenioglossate mesogastropods, but despite the attempts of various workers to establish phylogenies, the relationships of the heteropods to other members of this large group remain obscure (Fretter and Graham, 1962). It is difficult to distinguish between anatomical similarities that are due to phylogenetic relationships and those that reflect convergent evolution.

The heteropods are clearly a specialized group that exhibits increasing adaptation to a planktonic existence. This progression is illustrated in the transition from the primitive atlantids, which have a fully developed shell, to the carinariids, with their very reduced shells, and finally to the shell-less pterotracheids. Associated with this striking progression are the development of larger body size; increases in body transparency, gelatinous body tissue, and swimming efficiency; and the achievement of apparent neutral buoyancy. These adaptations presumably equip the larger carinariids and pterotracheids for more efficient predation and more effective evasion of their own predators. At the same time, energy conservation may increase, owing to buoyancy regulation and to feeding on larger prey. The large heteropods can remain quiescent except when pursuing prey or evading predators and, further, they presumably ingest larger amounts of energy at each meal by being able to attack prey as large as or slightly larger than themselves.

But whatever the advantages of increasing specialization, the fact remains that the primitive atlantids are by far the most successful heteropod family in terms both of breadth of distribution and densities of populations. It may be that, as heteropods increase in size and swimming ability, they are forced to compete with the more successful, pelagic, visual predators, including squid and fishes. Moreover, the biology of the atlantids is the least studied and understood, especially as regards natural feeding and swimming methods. Further study of the atlantids may aid in explaining their relative success.

Certain anatomical and behavioral features remain basically unchanged across the three heteropod families. Despite some differences in the shape of the radular teeth, the feeding mechanism remains essentially the same in all heteropods. All the species studied to date are feeding generalists, preying on a wide variety of zooplanktonic species. Only the size range of prey shifts in the group, with the carinariids and pterotracheids eating larger, and therefore less abundant, species.

Present knowledge suggests that reproductive processes also are conservative within the group. The basic departure of these pelagic species from most benthic mollusks is in the production and use of spermatophores for fertilization. Some workers have used this character to attempt to establish phylogenetic relationships with the few benthic prosobranchs that share this feature; but spermatophore production seems to be a relatively common adaptation to a planktonic existence, as evidenced in many different animal phyla.

The most recent attempts to establish phylogeny within the Heteropoda have been made by Richter (1968; 1973b). His ideas are based on the larval and postlarval development of the shell and on changes in the radular teeth. In this scheme, *Atlanta* is considered the most primitive genus and *Pterotrachea* the most advanced.

List of Recognized Species

(Synonymy is given in Tesch, 1949, and van der Spoel, 1976.)

Family Atlantidae

Atlanta peroni Lesueur, 1817
A. turriculata d'Orbigny, 1836
A. fusca Souleyet, 1852
A. gaudichaudi Souleyet, 1852
A. gibbosa Souleyet, 1852
A. helicinoides Souleyet, 1852
A. inclinata Souleyet, 1852
A. inflata Souleyet, 1852
A. lesueuri Souleyet, 1852
A. pacifica Tokioka, 1955
A. peresi Frontier, 1966
A. echinogyra Richter, 1972
A. plana Richter, 1972
A. tokiokai van der Spoel & Troost, 1972

Oxygyrus keraudreni (Lesueur, 1817)
Protatlanta souleyeti (Smith, 1888)
Family Carinariidae
Carinaria cristata (Linnaeus, 1766)
C. lamarcki Péron & Lesueur, 1810
C. cithara Benson, 1835
C. galea Benson, 1835
C. challengeri Bonnevie, 1920
Pterosoma planum Lesson, 1827
Cardiapoda placenta (Lesson, 1830)
C. richardi Vayssière, 1903
Family Pterotracheidae
Pterotrachea coronata Forsskål, 1775
P. hippocampus Philippi, 1836
P. scutata Gegenbaur, 1855
?*P. minuta* Bonnevie, 1920 (a species of uncertain validity; see Seapy, 1985)
Firoloida desmaresti Lesueur, 1817

References Cited

Works with recommended keys or aids for the identification of species are indicated by an asterisk.

Aravindakshan, P. N. 1973. Distribution and ecology of *Pterotrachea coronata* (Forskål). In: *The Biology of the Indian Ocean*, vol. 3. B. Zeitschel, ed. Berlin: Springer-Verlag. Pp. 399–400.

Barber, V. C., and P. N. Dilly. 1969. Some aspects of the fine structure of the statocysts of the molluscs *Pecten* and *Pterotrachea*. *Z. Zellforsch.* 94: 462–78.

Batten, R. L., and M. P. Dumont. 1976. Shell ultrastructure of the Atlantidae (Heteropoda, Mesogastropoda) *Oxygyrus* and *Protatlanta*, with comments on *Atlanta inclinata*. *Bull. Am. Mus. nat. Hist.* 157: 263–310.

Biggs, D. C. 1977. Respiration and ammonium excretion by open ocean gelatinous zooplankton. *Limnol. Oceanogr.* 22: 108–17.

Blackburn, M. 1956. Sonic scattering layers of heteropods. *Nature, Lond.* 177: 374–75.

Bøggild, O. B. 1930. The shell structure of the mollusks. *K. danske Vidensk. Selsk. Skr.* 9: 231–326.

Ceccaldi, H. J., A. Kanazawa, and S. Teshima. 1978. Chemical composition of some Mediterranean macroplanktonic organisms. 1. Proximate analysis. *Tethys 8*: 295–98.

Collin, B. 1914. Notes protistologiques. *Archs Zool. exp. gén.* 54: 85–97.

Cuvier, G. L. C. F. D. 1836. *Le Règne animal distribué d'après son organisation pour servir de base à l'histoire naturelle des animaux et d'introduction à l'anatomie comparée*. Pteropoda. 2: 16–18. Bruxelles: Hauman.

Dales, R. P. 1952. The distribution of some heteropod molluscs off the Pacific coast of North America. *Proc. zool. Soc. Lond.* 122: 1007–15.

Delle Chiaie, S. 1841. *Descrizione e notomia degli animali invertebrati della Sicilia citeriore osservati vivi negli anni 1822–1830*, vol. 1, Molluschi cefalopedi e pteropedi. Napoli: Batelli. Pp. 84–98.

Denton, E. J., and T. I. P. Shaw. 1961. The buoyancy of gelatinous marine animals. *J. Physiol., London 161*: 14P-15P (Proceedings).

Dilly, P. N. 1969. The structure of a photoreceptor organelle in the eye of *Pterotrachea mutica. Z. Zellforsch.* 99: 420–29.

Fol, H. 1876. Étude sur le développement de mollusques. Second mémoire: Sur le développement embryonnaire et larvaire des hétéropodes. *Archs Zool. exp. gén.* 5: 104–58.

Fretter, V., and A. Graham. 1962. *British Prosobranch Molluscs*. London: Ray Society. 755 pp.

Grossman, E. L., P. R. Betzer, W. C. Dudley, and R. B. Dunbar. 1986. Stable isotopic variation in pteropods and atlantids from North Pacific sediment traps. *Mar. Micropaleont.* 10: 9–22.

Hamner, W. M., L. P. Madin, A. L. Alldredge, R. W. Gilmer, and P. P. Hamner. 1975. Underwater observations of gelatinous zooplankton: Sampling problems, feeding biology, and behavior. *Limnol. Oceanogr.* 20: 907–17.

Harbison, G. R., D. C. Biggs, and L. P. Madin. 1977. The associations of Amphipoda Hyperiidea with gelatinous zooplankton—II. Associations with Cnidaria, Ctenophora and Radiolaria. *Deep-Sea Res.* 24: 465–88.

Hartmann, J., and H. Weikert. 1969. Tagesgang eines Myctophiden (Pisces) und zweier von ihm gefressener Mollusken des Neustons. *Kiel. Meeresforsch.* 25: 328–30.

Hirsch, G. C. 1915. Die Ernährungsbiologie fleischfressender Gastropoden. *Zool. Jb.* 35: 357–504.

Hochberg, F. G., and R. R. Seapy. 1985. Parasites of holopelagic molluscs. Intl. Symp. Mar. Plankton, July, 1984, Tokai University, Shimizu, Japan. *Bull. mar. Sci.* 37: 767 (Abstract).

Huxley, T. H. 1853. On the morphology of the cephalous Mollusca as illustrated by the anatomy of certain Heteropoda and Pteropoda collected during the voyage of H.M.S. "Rattlesnake" in 1846–1850. *Phil. Trans. R. Soc. 143*: 29–65.

Ikeda, T. 1974. Nutritional ecology of marine zooplankton. *Mem. Fac. Fish. Hokkaido Univ.* 22: 1–97.

Jung, P. 1975. Quaternary larval gastropods from Leg 15, Site 147, Deep Sea Drilling Project. Preliminary report. *Veliger 18*: 109–26.

de Lamarck, J. B. P. A. 1812. *Extrait du cours de zoologie du Muséum d'Histoire Naturelle sur les animaux sans vertèbres (Mollusques, Hétéropodes)*. Paris: d'Hantel et Gabon. 127 pp.

Land, M. F. 1982. Scanning eye movements in a heteropod mollusc. *J. exp. Biol. 96*: 427–30.

Lowenstam, H. A. 1974. Impact of life on chemical and physical processes. In: *The Sea*, vol. 5. E. Goldberg, ed. New York: Wiley. Pp. 715–96.

Martoja, M., and C. Thiriot-Quiévreux. 1979. Appareil génital de *Carinaria lamarcki* (Gastropoda Heteropoda); structure et affinités. *Malacologia 19*: 63–76.

Morton, J. 1964. Locomotion. In: *Physiology of Mollusca*, vol. 1. K. M. Wilbur and C. M. Yonge, eds. New York: Academic Press. Pp. 383–423.

Murray, J., and J. Chumley. 1924. The deep-sea deposits of the Atlantic Ocean. *Trans. R. Soc. Edinb.* 54: 1–252.

Okutani, T. 1957. On pterotrachean fauna in Japanese waters. *Bull. Tokai reg. Fish. Res. Lab. 16*: 15–27.

———. 1961. Notes on the genus *Carinaria* (Heteropoda) from Japanese and adjacent waters. *Publs Seto mar. biol. Lab. 9*: 333–53.

Owre, H. B. 1964. Observations on development of the heteropod molluscs *Pterotrachea hippocampus* and *Firoloida desmaresti. Bull. mar. Sci. Gulf Caribb. 14*: 529–38.

Pafort-van Iersel, T. 1983. Distribution and variation of Carinariidae and Pterotracheidae (Heteropoda, Gastropoda) of the Amsterdam Mid North Atlantic Plankton Expedition 1980. *Beaufortia 33*: 73–96.

Perkins, P. S. 1983. The life history of *Cardiodectes medusaeus* (Wilson), a copepod parasite of lanternfishes (Myctophidae). *J. crustacean Biol. 3*: 70–87.

Pilkington, M. C. 1970. Young stages and metamorphosis in an atlantid heteropod occurring off south-eastern New Zealand. *Proc. malac. Soc. Lond. 39*: 117–24.

Purcell, J. E. 1981. Selective predation and caloric consumption by the siphonophore *Rosacea cymbiformis* in nature. *Mar. Biol. 63*: 283–94.

Richter, G. 1968. Heteropoden und Heteropodenlarven im Oberflächenplankton des Golfs von Neapel. *Pubbl. Staz. zool. Napoli.* 36: 346–400.

———. 1973a. Field and laboratory observations on the diurnal vertical migration of marine gastropod larvae. *Neth. J. Sea Res. 7*: 126–34.

———. 1973b. Zur Stammesgeschichte pelagischer Gastropoden. *Natur Mus., Frankf.* *103*: 265–75.

———. 1974. Die Heteropoden der "Meteor"-Expedition in den Indischen Ozean 1964/65. *"Meteor" Forsch.-Ergebnisse.* *17*(D): 55–78.

———. 1982. Mageninhaltsuntersuchungen an *Oxygyrus keraudreni* (Lesueur) (Atlantidae, Heteropoda). Beispiel einer Nahrungskette im tropischen Pelagial. *Senckenberg. marit.* *14*: 47–77.

———. 1983. Lebensformen und Nahrungsketten der Hochsee. Teil II. *Natur Mus., Frankf.* *113*: 166–77.

Russell, H. D. 1960. Heteropods and pteropods as food of the fish genera, *Thunnus* and *Alepisaurus*. *Nautilus* 74: 46–56.

Seapy, R. R. 1974. Distribution and abundance of the epipelagic mollusk *Carinaria japonica* in waters off southern California. *Mar. Biol.* 24: 243–50.

———. 1980. Predation by the epipelagic heteropod mollusk *Carinaria cristata* forma *japonica*. *Mar. Biol.* *60*: 137–46.

———. 1985. The pelagic genus *Pterotrachea* (Gastropoda: Heteropoda) from Hawaiian waters: A taxonomic review. *Malacologia* *26*: 125–35.

Seapy, R. R., and R. E. Young. 1986. Concealment in epipelagic pterotracheid heteropods (Gastropoda) and cranchiid squids (Cephalopoda). *J. Zool., Lond.* (A). *210*: 137–47.

Souleyet, F. L. A. 1852. *Voyage autour du monde exécuté pendant les années 1836 et 1837 sur la corvette "La Bonite"*. (F. Eydoux et F. L. A. Souleyet). Zoologie. 2: 1–664. Paris: Bertrand.

*van der Spoel, S. 1972. Notes on the identification and speciation of Heteropoda (Gastropoda). *Zoöl. Meded., Leiden.* 47: 545–60.

*———. 1976. Pseudothecosomata, Gymnosomata and Heteropoda (Gastropoda). Utrecht: Bohn, Scheltema & Holkema. 484 pp.

*Tesch, J. J. 1949. Heteropoda. *Dana Rep.* No. 34. 53 pp.

Thiriot-Quiévreux, C. 1969. Caractéristiques morphologiques des véligères planctoniques des gastéropodes de la région de Banyuls-sur-Mer. *Vie Milieu* 20(2B): 333–66.

———. 1972. Microstructures de coquilles larvaires de prosobranches au microscope électronique à balayage. *Archs Zool. exp. gén.* *113*: 553–64.

*———. 1973a. Heteropoda. *Oceanogr. mar. Biol. a. Rev.* *11*: 237–61.

———. 1973b. Observations de la radula des Hétéropodes (Mollusca Prosobranchia) au microscope électronique à balayage et interprétation fonctionnelle. *C. r. hebd. Séanc. Acad. Sci., Paris* 276(D): 761–64.

———. 1975. Observations sur les larves et les adultes de Carinariidae (Mollusca: Heteropoda) de l'Océan Atlantique Nord. *Mar. Biol.* 32: 379–88.

Thiriot-Quiévreux, C., and M. Martoja. 1974. Appareil génital femelle des Atlantidae (Mollusca Heteropoda). *Vie Milieu* 24(A): 389–412.

Tokioka, T. 1961. The structure of the operculum of the species of Atlantidae (Gastropoda: Heteropoda) as a taxonomic criterion, with records of some pelagic mollusks in the North Pacific. *Publs Seto mar. biol. Lab.* 9: 267–332.

Wolff, H. G., and A. Schultze. 1982. Strukturelle und funktionelle Aspekte der Augen der Heteropoden *Pterotrachea coronata* und *Carinaria mediterranea. Verh. Dt. zool. Ges.* 75: 291.

Woodring, W. P. 1928. Contributions to the geology and palaeontology of the West Indies. Miocene mollusks from the Bowden Jamaica. Part II: Gastropods and discussion of results. *Carnegie Inst. Wash. Publ.* No. 385. 564 pp.

Woodward, B. B. 1913. *The Life of Mollusca.* London: Methuen. 158 pp.

4

The Thecosomes
Shelled Pteropods

Class Gastropoda
  Subclass Opisthobranchia
    Order Thecosomata
      Suborder Euthecosomata
        Family Limacinidae
        Family Cavoliniidae
      Suborder Pseudothecosomata
        Family Peraclididae
        Family Cymbuliidae
        Family Desmopteridae

The thecosomes constitute one of two holoplanktonic opisthobranch orders commonly referred to as pteropods, the other group being the Order Gymnosomata (see Chapter 5). The name "pteropod" derives from the fact that portions of the molluscan foot (-poda) in both thecosomes and gymnosomes have been modified to form paired swimming wings (ptero-). At one time the two groups were placed together in a single taxonomic division, the Order Pteropoda, but it is now generally accepted that these mollusks are not that closely allied. In fact, the wings of thecosomes and gymnosomes are sited differently with respect to the head, mouth, and footlobes, and there is some doubt that the wings in the two groups have a common derivation (Boas, 1886b; Pafort-van Iersel and van der Spoel, 1979; van der Spoel, 1982). The thecosomes and gymnosomes are further separated by distinct differences in morphology, behavior, and trophic position.

The early literature contains many references to pteropods, since thecosomes with intact shells were often collected even in crude nets, and some species, including gymnosomes, were sighted at the sea surface in calm, coastal areas. The pteropods now known as *Limacina helicina* (a thecosome) and *Clione limacina* (a gymnosome) were first mentioned in a publication by F. Martens in 1675 describing voyages to Spitzbergen and Greenland. Unfortunately, Marten's work was not known to the early systematists, and credit for the first descriptions of these species now goes to Captain C. J. Phipps, who made a similar journey into northern waters in 1773 for the King of England (Phipps, 1774). Only three other pteropods were described in the eighteenth century. The history of the description of two of these species, in particular, deserves mention. A Danish expedition was sent forth in 1761 with the purpose of exploring Arabia. One of the members was a student of Linnaeus, Peter Forsskål, who recorded and illustrated numerous pelagic organisms including a pseudothecosome (*Gleba cordata*) and a euthecosome that he called *anomia tridentata* and that is now known as *Cavolinia tridentata*. Forsskål died of malaria in Yemen at the age of 31. Carston Niebuhr was the only survivor of the sea voyage. He returned overland on horseback to Copenhagen and edited and published Forsskål's notes, thus gaining recognition as the namer of these pteropods (Niebuhr, 1772; Wolff, 1968).

The taxonomic position of pteropods in relation to other animals aroused considerable debate (Rang and Souleyet, 1852; Pelseneer, 1887, 1888a, 1888b). In 1804, Cuvier established a separate order of mollusks, the Pteropoda, which included both thecosomes and gymnosomes; he later (1817) elevated the order to class status, equivalent in rank with the Cephalopoda and Gastropoda. In 1810, Péron and Lesueur divided the Pteropoda into two subgroups, selecting the presence or absence of a shell as the basic difference, though they also considered the Pteropoda to include pelagic nudibranchs, heteropods, and even some ctenophores. In 1824, careful anatomical studies by de Blainville strengthened the argument for separating thecosomes and gymnosomes, and he also removed the unrelated forms that Péron and Lesueur had included. He rejected the separate class of Cuvier, placing the pteropods among the Gastropoda, and was the first to recognize the opisthobranch affinities of these animals. Souleyet (1843) provided additional evidence for including pteropods with gastropods, but it was not until the publications of Keferstein (1866) and Boas (1886b) and Pelseneer's reports (1887, 1888a, 1888b) on pteropods collected by the H.M.S. *Challenger* that the inclusion of these animals with opisthobranchs became fully accepted. The first monograph to include a review of the natural history of pteropods was published in 1852 by Rang and Souleyet. This work contains a history of the early taxonomic controversies and records all aspects of the biology of pteropods that were known at that time, many discovered by the authors themselves. Numerous scientific oceanographic cruises followed the lead of *Challenger* in 1872, and many of these resulted in publications dealing primarily with taxonomy, anatomy, zoogeography, and descriptions of new pteropod species (see Bé and Gilmer, 1977, for a brief review).

Certain pteropods also received the attention of fishermen. In the eighteenth century, French fishermen coined the name "papillon de mer" for *Cymbulia*, a pseudothecosome often sighted swimming with fluttering wing movements along the coast of Provence (Rang and Souleyet, 1852). The name "sea butterfly" now is commonly applied to all thecosomatous pteropods.

Thecosomes are uniquely modified for planktonic life, and not only in their swimming ability. They are generally of small size and have either thin, fragile, external shells or internal gelatinous conchae, all of which reduce weight. Most, if not all, species have developed various buoyancy mechanisms that assist flotation when swimming motions cease. Further, thecosomes are the only opisthobranchs that have developed a type of feeding in which free-floating mucous webs are employed to capture planktonic organisms and small particles from the surrounding water. The two suborders of thecosomes—the euthecosomes and the pseudothecosomes—differ in anatomical appearance but share common life styles.

The Suborder Euthecosomata comprises the more familiar shelled pteropods, all of which have an external calcareous shell composed of aragonite. Euthecosomes are present in all the world seas, including the Antarctic Ocean and the central Arctic Ocean, where one species (*Limacina helicina*) lives under the permanent ice cover. The majority of species, however, are restricted to the circumglobal warm-water area. As in many animal groups, the species diversity (number of species) is highest in warm waters, but population densities (numbers of individuals per species) are greatest in cold water. Most species also are encountered in near-surface water, down to depths of approximately 200 m, but a few euthecosomes are mesopelagic or bathypelagic and thus restricted to deeper waters. Most of the euthecosomes probably undergo short diurnal vertical migrations, moving closer to the sea surface at night and migrating to deeper water during daylight hours.

Shelled pteropods often are collected in plankton nets by oceanographers, but they seldom constitute a significant fraction of the total zooplankton, in terms either of numbers or of biomass. This may be a true reflection of low population densities or of biological variability in distribution, but there is also the possibility, in many cases, that there may be an artificial bias toward underrepresentation of true densities, owing to the choice of collecting equipment, a problem reviewed by Bé and Gilmer (1977). We do know, however, that euthecosomes can be sufficiently abundant in some areas, particularly in temperate or northern cold seas, to be a significant food source for commercial fish.

Occasionally, large swarms of thecosomes are brought inshore by prevailing currents (e.g. Sakthivel, 1972b). Complaints of swimmers having been "stung" by invisible animals are sometimes attributable to large swarms or beached accumulations of *Creseis* (Vayssiere, 1915; Nishimura, 1965; Moore and Valentine, 1984), a small thecosome with a sharply pointed shell. The animals become entangled in swimming clothes, and the movement of the swimmer or sunbather results in the needle-sharp shell pricking the skin. Apart from these relatively harmless and invisible encounters, few nonprofessionals have seen a pteropod, since almost all are of very small size and require special collection gear. Some euthecosomes have been studied *in situ* by scuba divers, but no species has yet been maintained successfully in laboratory conditions for more than a few days.

Euthecosomes are of interest to marine chemists and geologists, as well as to biologists, because the calcareous shells of dead individuals may dissolve in transit to deeper water or ultimately accumulate on the sea floor in shallower areas. In certain benthic environments, the shell deposits may be abundant enough to provide information on past hydrographic and climatic conditions, as well as providing a fossil history of the group. Since settling and deposited shells also reflect a transport of carbonate into deeper water, some attention has been given to the role of euthecosomes in the global carbon-dioxide cycle. These geological aspects are considered here in the context of the total ecology of the group.

The Suborder Pseudothecosomata consists of less familiar species, some of which lack external shells. The fundamental features that set these species apart from the euthecosomes, however, are a fusion of the wings into a single swimming plate; the presence of a proboscis formed by the footlobes; and the development, in advanced species, of a gelatinous internal pseudoconch that replaces the external calcareous shell.

Until very recently, pseudothecosomes were considered to be rare inhabitants of the pelagic environment, and little was known of their biology except for that inferred from anatomical studies. Shell-less species in particular are seldom collected by plankton nets, and those that are captured are often badly damaged and distorted by the usual preservation techniques. The size of some cymbuliids, however, places them among the macrozooplankton and renders them easily observable by scuba divers. We now know, through *in situ* observations (Gilmer, 1972, 1974; Hamner, 1974; Hamner et al., 1975), that certain species of pseudothecosomes are capable of detecting a slow-moving plankton net and evading capture by rapidly swimming away. Thus the population densities of these species may be very much underestimated by conventional plankton-study techniques.

In general, pseudothecosomes tend to live deeper in the water column than do most euthecosomes, and many can be considered as mesopelagic or, more rarely, bathypelagic. Pseudothecosomatous pteropods have been found throughout the Atlantic, Pacific, and Indian Oceans, but there are no representatives in the cold waters of the Arctic and Antarctic oceans.

External Anatomy

The 34 species of euthecosomes are distinctly separated from each other on the basis of shell morphology. Shell characters also have been used to distinguish subspecies and formae with somewhat discrete distributions (van der Spoel, 1967, 1976). There is no evidence, however, that the variants of a species are significantly different biologically to warrant distinction here.

The most primitive genus, *Limacina* (Color Fig. 6; Figs. 15, 16), con-

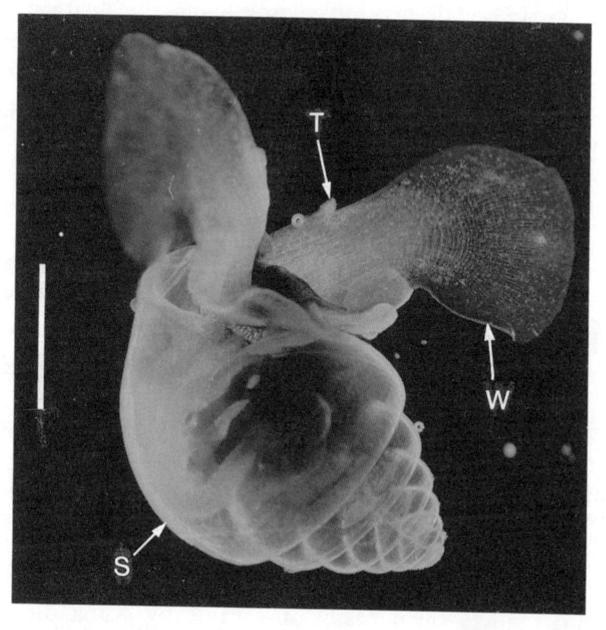

FIG. 15. *Limacina retroversa,* showing the sinistrally coiled shell that is characteristic of this euthecosome genus. S, shell; T, wing tentacle; W, wing. The scale line represents 3 mm.

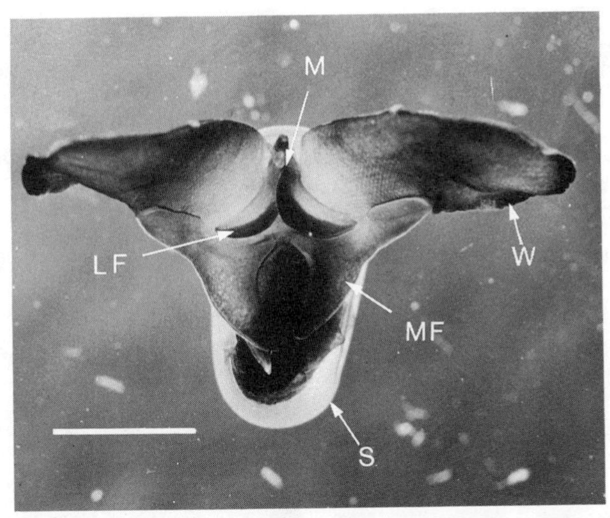

FIG. 16. A ventral view of *Limacina helicina antarctica,* showing the relationship of wings, footlobes, and head. LF, lateral footlobe; M, mouth; MF, median footlobe; S, shell; W, wing. The scale line represents 5 mm.

tains seven species with a spirally coiled shell. The shell is similar to that of prosobranch snails except that it is very thin and small (<1 mm to 15 mm in diameter) and is coiled sinistrally (counterclockwise) rather than dextrally (clockwise). The cavoliniids have lost the spiral coiling, and the internal microstructure of their shells (Fig. 42) is very different from that of limacinids (Bé et al., 1972; Rampal, 1973, 1977). Cavoliniid shells are bilaterally symmetrical, with shell shape ranging from straight and pointed (*Creseis*), bottle-shaped (*Cuvierina*), inflated or globose (*Cavolinia*), to pyramidal (*Clio*) (Color Figs. 7, 8; Fig. 17). The shells of all thecosomes are unusually thin, ranging from about 6 μm in *Limacina trochiformis* (Richter, 1973, 1976) to about 100 μm in the larger cavoliniids (Bé et al., 1972). This feature presumably reflects weight reduction for a planktonic life. During the Victorian era, shell collectors were attracted by the unique designs and subtle beauty of these tiny shells, and paid high prices for collections of the shells. The largest euthecosome is *Clio recurva*, in which shell length may measure 30 mm (Fig. 47a). The animals can withdraw into their shells for protection, but only members of the genus *Limacina* have an operculum on the foot with which to seal the opening, and this is often lost in mature adults.

In the normal swimming position, the head, footlobes, and wings are extended from the shell aperture (Figs. 16, 17). The two wings of the euthecosomes are separate and lie dorsally and laterally to the footlobes. The

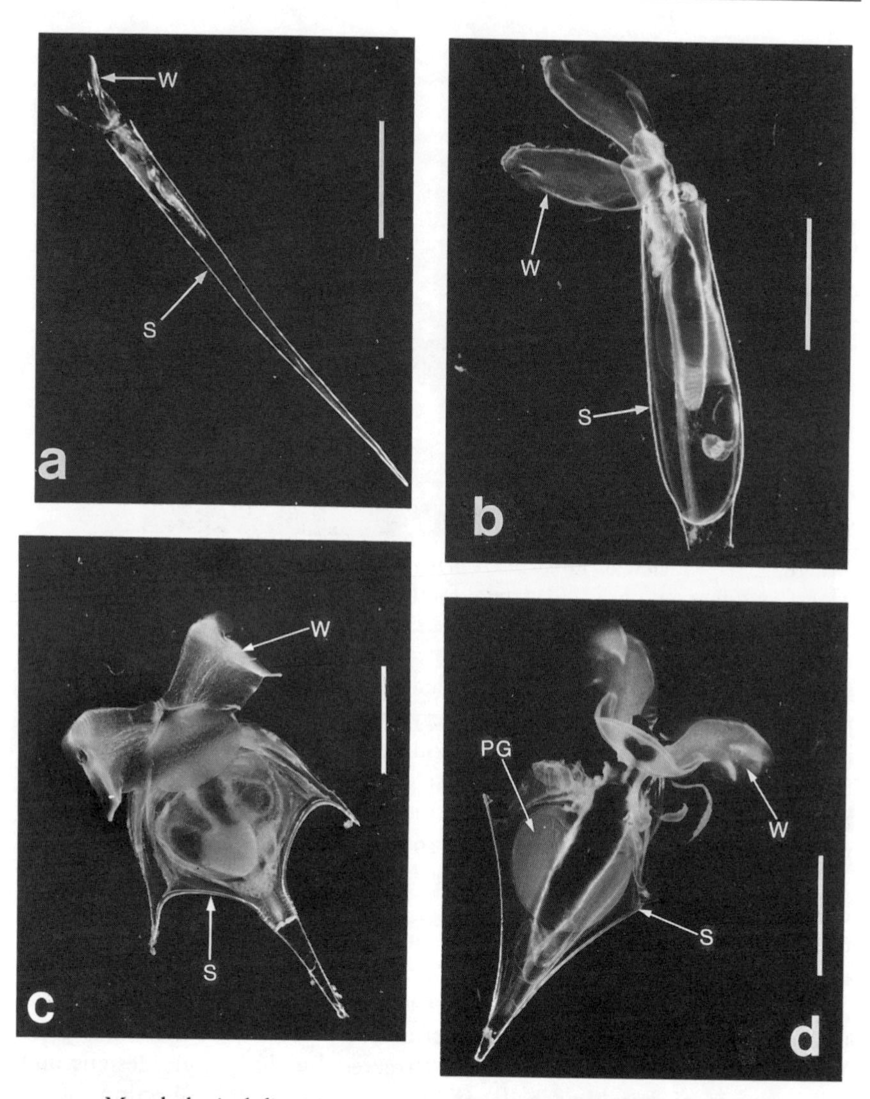

FIG. 17. Morphological diversity in cavoliniids: *a, Creseis acicula*; *b, Cuvierina columnella* (by G. R. Harbison); *c, Diacria major*; *d, Clio pyramidata* (from Gilmer, 1986). PG, pallial gland; S, shell; W, wing. Scale lines represent 5 mm.

muscular wings are formed from the anterior part of the foot; the remainder of the foot is composed of a median, flat, ciliated surface that projects ventrally over the lip of the shell aperture and of two smaller, ciliated lobes on either side of the depressed mouth. The head is not well defined; a small cephalic lobe projects forward from the point where the wings come together, and usually two tentacles (only one in some limacinids) are present on the dorsal surface of the "neck" region. These differ in size, the right tentacle being larger than the left, and histological studies suggest that they may be capable of light perception.

With the uncoiling of the shell in higher euthecosomes, a change in orientation occurs, with the body becoming detorted and secondarily symmetrical (Fig. 17). Thus the mantle cavity of the Limacinidae is in a dorsal position; in all other euthecosomes, it is ventral. This change in orientation is reflected in other organs: the anal opening into the mantle cavity is on the right in *Limacina* and on the left in other species. A single osphradium is present in all species; again, this sense organ is on the left side of the mantle cavity in Limacinidae and on the right side in the Cavoliniidae. Apart from the osphradium and paired tentacles, the only other obvious sense organs are paired statocysts embedded in body tissues and sometimes visible in transparent specimens. A secondary gill is present in the mantle cavity only in the genus *Cavolinia*.

In nature, all *Cavolinia* species have extensive protrusions of mantle covering their exterior shell surfaces and extending into the water (Color Fig. 7; Fig. 18). These mantle lobes were recently described in detail by Gilmer and Harbison (1986), using *Cavolinia uncinata* as their example. Essentially, there are four sets of mantle lobes that extend outside the shell. These include the two lateral mantle appendages first described by Péron and Lesueur (1810), which usually are seen in some state of protrusion in live animals. Mantle appendages are also present in *Diacria* (Gilmer, 1974), but they have not been described in detail. Study of the complete complement of mantle extensions in cavoliniids has been complicated by the great contraction of these lobes; they have only been seen fully extended in undisturbed animals observed by scuba divers. In captive animals and preserved specimens, depending on their handling before and during fixation, the mantle lobes may be reduced to small projecting flaps or, more likely, to tiny nubbins retracted inside the shell. Portions of these contracted lobes often have been described as "corner glands" of the mantle in studies of preserved material (Meisenheimer, 1905b; van der Spoel, 1967). The "balancer" appendage (e.g. Tesch, 1946, 1948; van der Spoel, 1967; Rampal, 1973), which does not appear to act as a stabilizer during swimming but instead directs the flow of water exiting from the mantle cavity (Wells, 1978), may also be part of the external mantle appendages. The mantle lobes appear to be important in both buoyancy regulation and feeding, and they are described in more detail in later sections.

The number of species constituting the Suborder Pseudothecosomata is

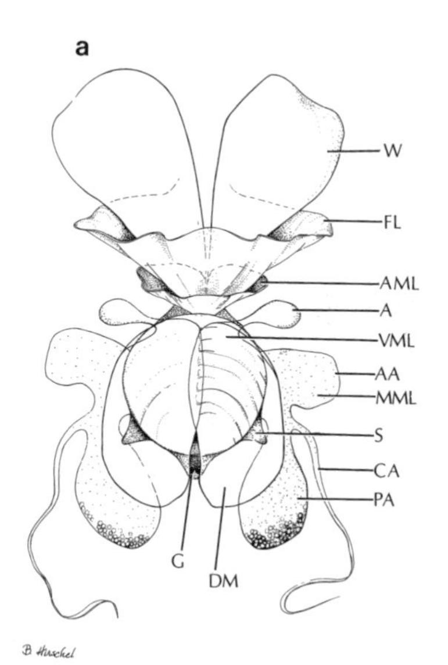

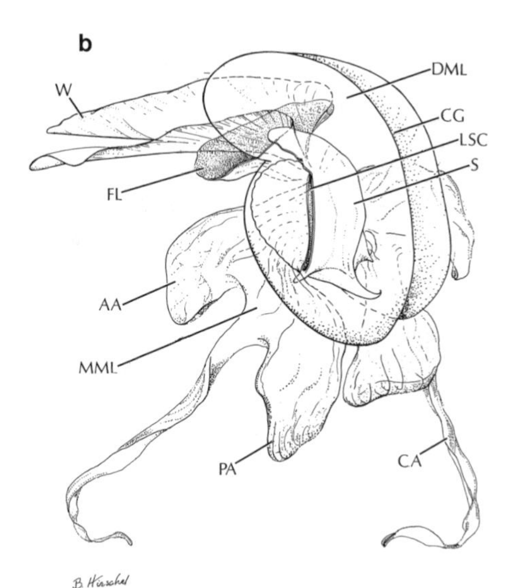

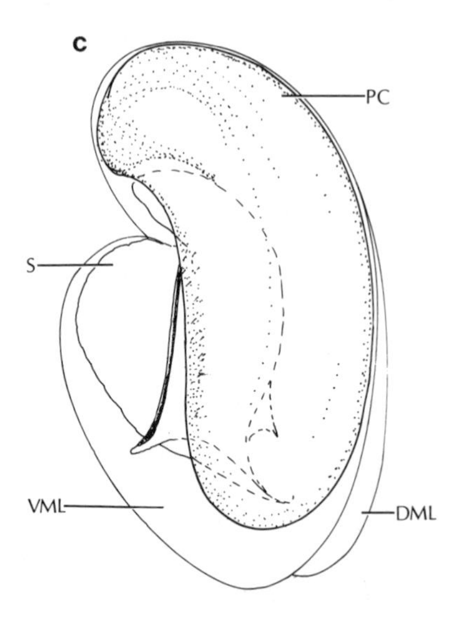

FIG. 18. Mantle appendages of *Cavolinia uncinata,* drawn from field photographs (from Gilmer and Harbison, 1986): *a,* ventral view of the anterior, medial, and ventral lobes of mantle tissue that encase the shell (dorsal mantle lobes omitted for clarity); *b,* dorsolateral view showing the dorsal and medial mantle lobes (anterior and ventral lobes omitted for clarity); *c,* dorsolateral view showing the relative positions of the shell, temporary pseudoconch, and dorsal mantle lobes (medial and anterior mantle lobes omitted for clarity). A, anterior appendage of the ventral mantle lobe; AA, anterior appendage of the medial mantle lobe; AML, anterior mantle lobe; CA, central appendage of the medial mantle lobe; CG, ciliated groove formed by the junction of the dorsal mantle lobes; DM, diaphanous membrane formed by the ventral mantle lobes; DML, dorsal mantle lobes; FL, footlobe supporting mouth; G, groove formed at junction of the ventral mantle lobes; LSC, lateral shell cleft; MML, medial mantle lobes; PA, posterior appendage of the medial mantle lobe; PC, temporary pseudoconch; S, shell; VML, ventral mantle lobes; W, wing.

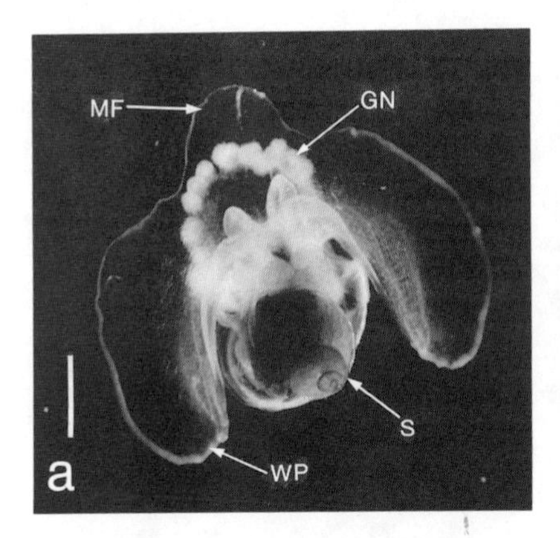

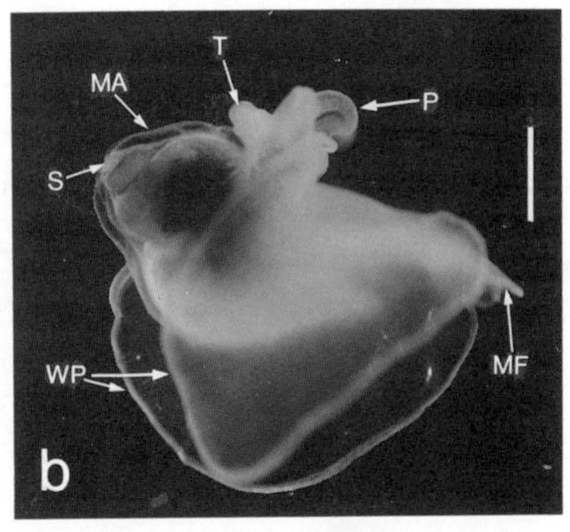

FIG. 19. *Peraclis reticulata*: *a*, posterioventral view with the mantle retracted from the apex of the sinistrally coiled shell, and the footlobes extending ventrally from the posterior margin of the expanded wingplate (from Gilmer and Harbison, 1986); *b*, lateral view with the wingplate extended below the shell. GN, glandular nubs; MA, mantle; MF, median footlobe; P, proboscis; S, shell apex; T, tentacle; WP, wingplate. Scale lines represent 2 mm.

presently uncertain. Taxonomic difficulties have arisen from working with badly preserved material and small numbers of specimens. Many of the descriptions of new pseudothecosome species are vague, unillustrated, and possibly based on artifacts rather than normal characteristics. Further, since life history patterns are essentially unknown in these mollusks, species based solely on differing size may only be descriptions of young and adults of the same species. Thus the list of described species at the end of the chapter includes a high proportion of species of dubious validity; the total number of pseudothecosome species is no less than 13 and possibly as many as 20.

Seven species belong to the most primitive family of pseudothecosomes, the Peraclididae, and they constitute a single genus, *Peraclis* (Color Fig. 9; Fig. 19). They superficially resemble the euthecosome genus *Limacina* in having a small (<8 mm in height), external, sinistrally coiled, calcareous shell and a sinistrally coiled operculum. The shell of *Peraclis*, however, can readily be distinguished from that of *Limacina*, since the columella (or central axis) is drawn out into a spirally twisted rostrum and the shell surface usually is ornamented with spiral lines or a reticulate pattern (absent only in *P. moluccensis*). In nature, the shell of *P. reticulata* and possibly other species of this genus is entirely covered and obscured by thick folds of mantle lining that project from the shell aperture (Gilmer and Harbison, 1986). Living, expanded specimens of *Peraclis* are further distinguished from euthecosomes—as are all the pseudothecosomes—by having the

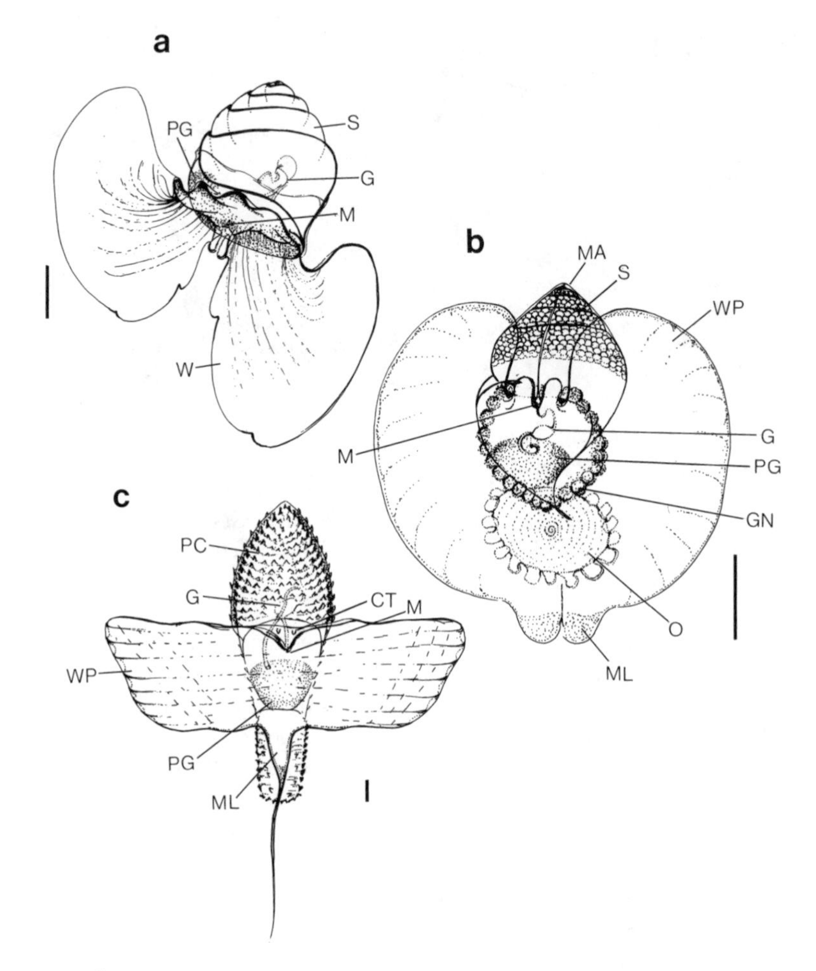

FIG. 20. The comparative morphology of *(a) Limacina, (b) Peraclis,* and *(c) Cymbulia*: all are shown in ventral views with the posterior portion of the shell or pseudoconch directed toward the top of the page. The designations of anterior-posterior and dorsal-ventral axes used here are based on a reinterpretation of the anatomy of *Peraclis* and comparisons with living specimens of *Limacina* and cymbuliids; they agree with the early morphological descriptions of *Cymbulia* made by MacDonald (1885). The orientation axes differ, however, from those originally established by Pelseneer (1888a, 1888b; 1906), which have been commonly employed in the literature (e.g. Meisenheimer, 1905b; Tesch, 1913; McGowan, 1968; van der Spoel, 1976). We have applied standard gastropod nomenclature to describe the orientation of the soft parts with respect to the shell (in *Limacina* and *Peraclis)* or to the pseudoconch (in cymbuliids). Thus the larger blunt end of the pseudoconch in *Corolla* and *Gleba* and the broad pointed end in *Cymbulia* (view *c*) are morphologi-

wings fused into a single wingplate, by the presence of two cephalic tentacles that are symmetrical in size, and by a proboscis formed by the elongation and modification of part of the footlobes. *Peraclis* has been reported to be the only thecosome genus with a true ctenidium in the mantle cavity (Meisenheimer, 1906; Pelseneer, 1906); however, this is retracted and folded mantle tissue.

In 1867, Costa correctly illustrated and labeled an expanded specimen of *Peraclis reticulata* (= *Spirialis recurvirostra*), which clearly showed that the wingplate is located ventrally, with the proboscis arising from its posterior margin. As Gilmer and Harbison (1986) have pointed out, however, most subsequent workers described preserved and contracted animals that may be twisted as much as 160° away from the correct orientation with the shell. Consequently, it was erroneously concluded that the mantle cavity was located dorsally, as in *Limacina*, and not anteroventrally as in all other pseudothecosome species. The mantle cavity of *Peraclis* normally extends broadly across the anterior of the shell aperture, directly over the shell rostrum (Fig. 20). The proboscis is formed from footlobes that extend ventrally from their medial position on the posterior wingplate margin, not on the dorsal margin as often described in the literature. The mouth of *Peraclis* is therefore separated from the greater part of the mantle cavity by the expansive wingplate, and the relative position of the wingplate and mouth to the mantle cavity is thus the same as in the cymbuliids.

The more advanced pseudothecosomes, belonging to the Family Cymbuliidae, are of larger size and have an elongated body. *Cymbulia*, *Corolla*, and *Gleba* do not have an external shell; instead, the skeletal support and protection of the viscera is provided by an internal pseudoconch, or false shell (Color Fig. 10; Figs. 20–22). This firm, but gelatinous, structure is covered by the integument. The transparent pseudoconch contains a cavity that encases the compact visceral mass, but the attachment to the body is

cally equivalent to the apex of the shell of *Peraclis* (or *Limacina*) and, as such, are situated posteriorly. In all pseudothecosomes, the foot and wingplate protrude from the ventral opening of the pseudoconch or shell, with the elongated margin of the wingplate extending anteriorly beyond the border of the pseudoconch or shell rostrum. The proboscis arises from the posterior margin of the wingplate in *Peraclis* and all cymbuliids, and the cephalic tentacles are situated on the posterior surface of the proboscis. With this orientation, the pseudoconch of the cymbuliids is placed in the correct perspective for comparison with gastropods having spirally coiled shells. CT, cephalic tentacle; G, gut; GN, glandular nubs surrounding shell aperture and operculum; M, mouth; MA, mantle covering shell apex; ML, medial lobe of the wingplate; O, operculum (located dorsally to the wings); PC, pseudoconch; PG, pallial gland (located dorsally to the wings); S, shell; W, wing; WP, wingplate. The scale lines represent 2 mm in *a* and *b*, 4 mm in *c*.

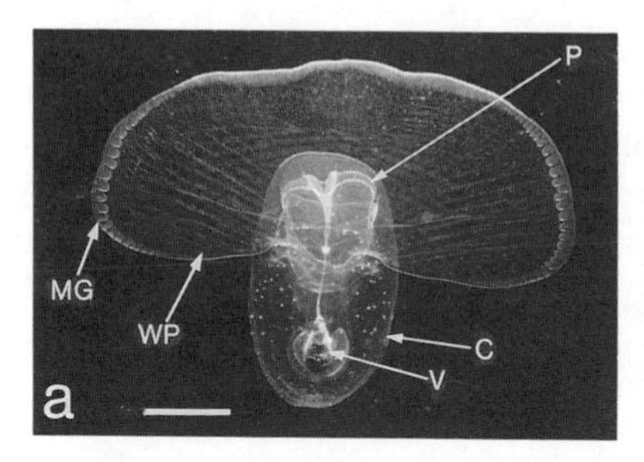

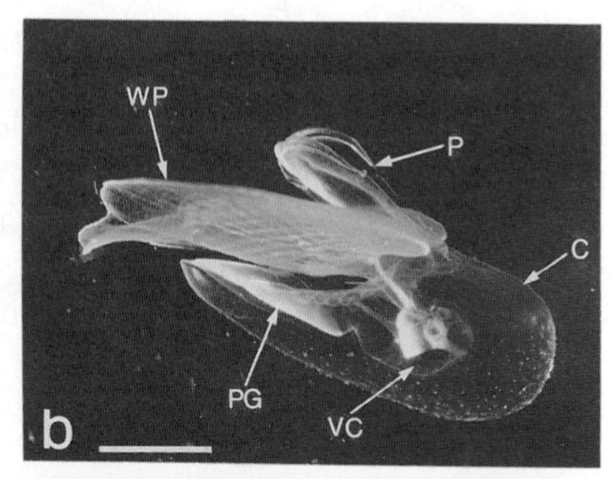

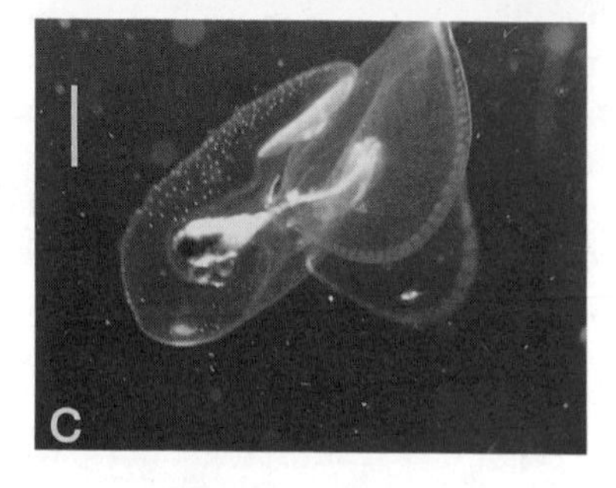

FIG. 21. Three views of *Corolla (calceola?)*, showing the relative positions of the viscera, pseudoconch, wingplate, and proboscis: *a*, ventral; *b*, lateral; *c*, an animal swimming downward toward the lower left. C, pseudoconch; MG, mucous gland; P, proboscis; PG, pallial gland; V, viscera; VC, visceral cavity; WP, wingplate. Scale lines represent 10 mm.

tenuous, and empty pseudoconchs and detached animals are sometimes found in plankton tows. Figure 20 illustrates the similarities of the arrangement and orientation of the pseudothecosome body parts with respect to the pseudoconch in cymbuliids and to the shell in *Peraclis*.

The shape and size of the pseudoconch has been used to distinguish species, with lengths ranging from about 35 mm in the smaller species to a maximum of 80 mm in *Corolla* (Gilmer, pers. obs.). *Cymbulia* (Fig. 20) has an elongate, slipper-shaped, thick-walled pseudoconch that is pointed posteriorly. Rows of spines or tubercles run the length of the conch and terminate in small projections on either side of the anterior blunt end. The ventral aperture leading to the small central cavity is narrow and elongate. The thin-walled oval pseudoconch of *Corolla* (Fig. 21) is covered externally with numerous tubercles and has a broad aperture that may extend to slightly more than half the length of the structure. *Gleba* (Fig. 22) has a thin, oval-shaped conch, but it is much more flattened, has fewer tubercles on the surface, and is often broken away in several pieces from the body. The visceral cavity is large and shallow.

The span of the wingplate of *Corolla* and *Gleba* is approximately 1.5 to 2 times that of the pseudoconch length, whereas *Cymbulia* has a smaller, heart-shaped swimming plate with a span rarely exceeding the pseudoconch

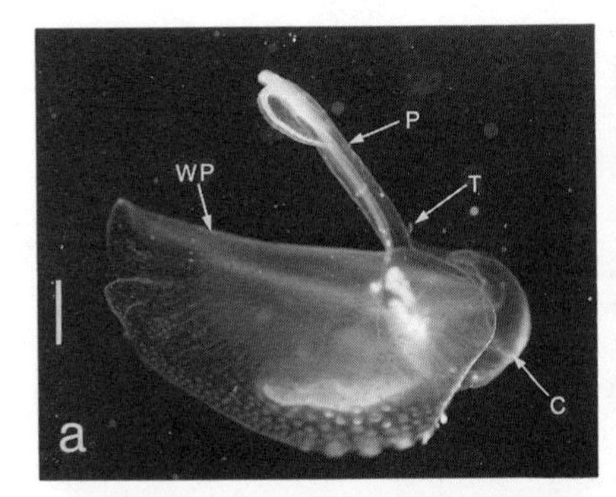

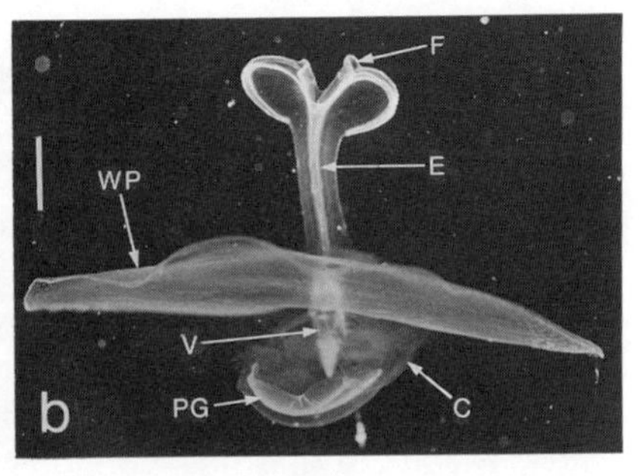

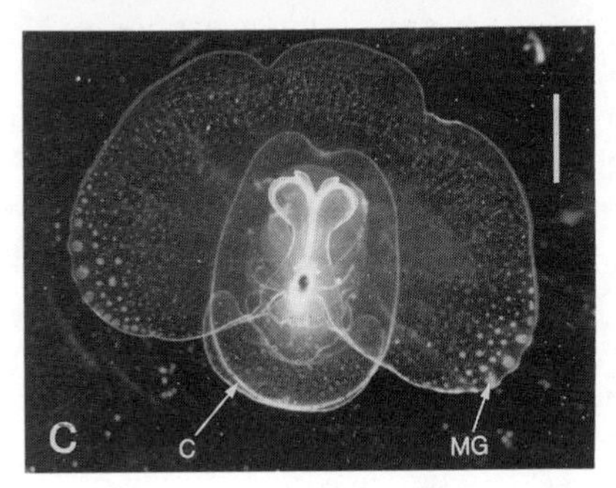

FIG. 22. *Gleba cordata*: *a,* lateral view of an adult, with the proboscis partially extended above the wingplate (from Gilmer, 1972; copyright 1972 by the AAAS); *b,* anterior view, showing the mantle-cavity opening and the pallial gland lying below the wingplate; *c,* ventral (oral) view of a juvenile. C, pseudoconch; E, esophagus; F, footlobes forming the terminal end of the proboscis; MG, mucous glands; P, proboscis; PG, pallial gland; T, tentacle; V, viscera; WP, wingplate. Scale lines represent 10 mm in *a* and *b,* and 5 mm in *c.*

length. The wings of *Corolla* are almost oval, with slight indentations along the anterior and lateral margins. *Gleba*, however, has distinct indentations on the anterior margin, which divide the wingplate into two lobes (Fig. 22c), much as in *Peraclis reticulata* (Fig. 19a). Another notable feature of the wingplate of *Cymbulia* is that the anterior margin is drawn out into a single long tentacle that often projects as much as three body lengths beyond the anterior margin of the pseudoconch (Fig. 20). In *Corolla*, *Gleba*, and probably *Cymbulia*, mucous glands are embedded in the lateral edges of the swimming plate. These glands are extremely large and prominent in *Gleba*; the edges of the wingplate are in fact scalloped by five or six deep indentations that separate the glands (Fig. 22a).

The cymbuliids also are characterized by a pronounced increase in the development of the proboscis. The proboscis of *Cymbulia* is relatively short and fused with the posterior surface of the swimming disk; in *Corolla*, it is slightly longer and only partially fused; and in *Gleba*, the proboscis is free and capable of great extension during feeding (Figs. 22a, b, 29). The mouth lies at the tip of the proboscis and is bordered by ciliated lips formed by modified footlobes. Correlated with the development of the proboscis is a loss of the buccal mass in *Corolla* and *Gleba*, both of which lack the radula, jaws, and salivary glands.

In all cymbuliids, stellate chromatophores are scattered over the body and wing surfaces. These pigment-filled sacs are attached to radially arranged muscle fibers that control rapid changes in the size, shape, and color of the spots (Kölliker and Müller, 1853). The chromatophores of *Gleba* are brown or gold, and in live animals there is a border of color around the edge of the swimming plate and along 10 to 12 narrow ribs radiating toward the center. *Gleba* also has melanophores on the integument covering the pseudoconch; these can be so numerous as to completely color the normally transparent pseudoconch. There are also large dark pigment spots on the extreme processes of the pseudoconch, both anteriorly and posteriorly, in at least one species of *Cymbulia*. In adult specimens of *Corolla*, there are black or dark-brown chromatophores, whereas opalescent spots (iridiocytes?) overlie the tubercles on the surface of the pseudoconch in some smaller individuals (Color Fig. 10). When the animals are disturbed, these pigment spots very quickly change the appearance of more than 50 percent of the pseudoconch from clear to opalescent (Gilmer, pers. obs.). The functional significance of coloration and color change remains unknown.

The genus *Desmopterus* generally is regarded as a separate family of pseudothecosomes (van der Spoel, 1976), but the anatomy of these small animals is so unique and their biology so poorly known that taxonomic position and phylogenetic relationships remain an enigma. In fact, some workers (e.g. Chun, 1889; McGowan, 1968) have preferred to place *Desmopterus* with the gymnosomatous pteropods.

Desmopterus (Fig. 23) lacks many of the anatomical structures present in other pseudothecosomes: there is no shell or pseudoconch, nor is there a mantle cavity, a distinct pallial gland, or a gizzard. But there is a single pair of cephalic tentacles, symmetrical in size, and the wings are fused into a single ventral swimming plate attached to the middle of the cylindrical body. The posterior edges of the wingplate are divided into lobes and, most remarkably, bear two ciliated tentacles that trail behind the swimming animal. The wing tentacles of *D. papilio* are capable of considerable contraction and elongation; they may exceed the wing span in length when fully extended (Chun, 1889). The tentacles are very fragile and are usually broken off in plankton collections. The head is drawn out ventrally into a snout with a terminal mouth, but whether or not this structure is homologous with the proboscis of other pseudothecosomes is uncertain. The body and wings of *Desmopterus* are transparent, but bright-red or reddish-brown pigment spots are scattered over the body and wing tentacles and are grouped on the outer margins of the wingplate (Tesch, 1913; Essenberg, 1919).

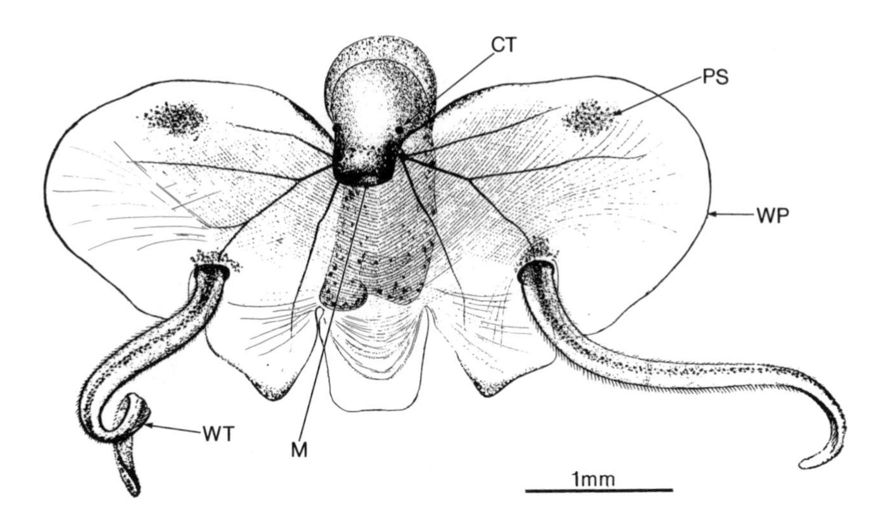

FIG. 23. A ventral view of *Desmopterus papilio* (from Chun, 1889). CT, cephalic tentacle; M, mouth; PS, pigment spots; WP, wingplate; WT, wing tentacle.

Maximum body length is only about 2.5 mm, with a wing span of less than 6 mm. The genus is widely distributed in subtropical and tropical waters of all the major oceans. *D. papilio* is especially common in coastal waters of the Indian Ocean, where more than 3,000 individuals have been collected in a single plankton haul (Sakthivel, 1972a). Despite their great abundance, there are no detailed accounts of observations of living animals.

Sense organs in the Pseudothecosomata include the paired cephalic tentacles with rudimentary eyes, paired statocysts, and a single osphradium (Meisenheimer, 1905b). At least one genus, *Corolla*, has simple light-perceptory organs that appear as numerous, dark, pigmented spots around the periphery of the wing margin (Heath and Spaulding, 1904; Meisenheimer, 1905b). Each organ is composed of a simple lens or vesicle filled with dense fluid and bounded internally by numerous, heavily innervated, pigmented receptor cells.

Swimming and Buoyancy

All euthecosomes swim with the ventral surface slanting upward, using well-developed, paired, muscular wings. A detailed analysis of wing movements has so far been made for only one species, *Limacina retroversa* (Morton, 1954a), but other species appear to swim in a similar manner (Gilmer, 1974; Richter, 1977). Figure 24 shows the successive, dorsal-ventral, scull-

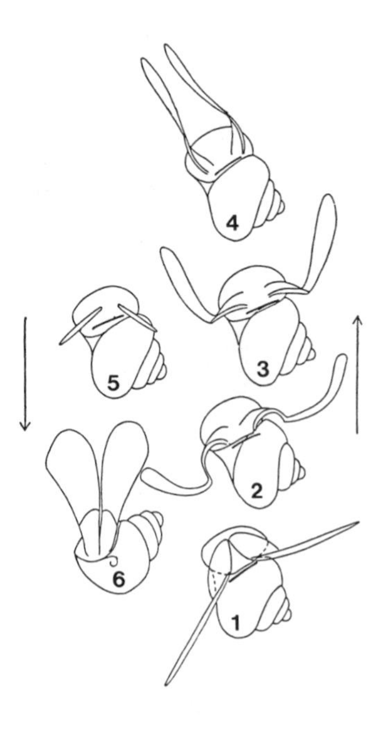

FIG. 24. Swimming in *Limacina retroversa* (redrawn from Morton, 1954a): the drawings illustrate successive positions of the wings in swimming upward (1–4) and in sinking (5 and 6). Position 6 shows the wings held vertically and motionless, in a different view from that of position 5.

ing movements of the wings exhibited by *L. retroversa* in laboratory containers. The downward effector stroke propels the animal upward; the resulting movement is rather jerky and describes a broadly spiral course. When descending, the animal holds its wings parallel above the body and motionless, thus presenting little resistance to the water; consequent sinking is rapid, owing to the weight of the shell. In nature and in specially designed deep tanks, however, both *L. retroversa* and *L. helicina* have been observed to alternate active swimming with periods of slow sinking and neutral buoyancy (Conover and Paranjape, 1977; Gilmer and Harbison, 1986). These species swim upward, then suddenly stop and spread their wings in a flat plane perpendicular to the shell axis and pointed anteriorly (Figs. 16 and 25a). They subsequently sink slowly at a rate of about 0.5 cm/sec, and after sinking a short distance they become neutrally buoyant. Gilmer and Harbison observed through photographic evidence that buoyancy is apparently achieved by the emission of an external mucous web used for feeding. The buoyant period is thus correlated with feeding and lasts at least 6 minutes in most individuals, or as long as they could be observed. Both species will swim away from their feeding webs if disturbed, and maximal swimming rates are 8 and 12 cm/sec for *L. retroversa* and *L. helicina*, respectively (Table 5).

Swimming is facilitated in the Cavoliniidae by the loss of spiral coiling; the center of gravity has been shifted to the structural center of the body, thus increasing maneuverability (Morton, 1964). *In situ* swimming rates of five species of cavoliniids have been recorded by Gilmer (1974), who used a meter stick to measure distance travelled between two points over a determined time interval. These rates, given in Table 5, are in general faster than those of limacinids.

Swimming in pseudothecosomes is accomplished by dorsal-ventral flexion of the wingplate, which beats at a slower rate than do the wings of euthecosomes. Members of the shelled genus *Peraclis* move through the water with the apex of the shell facing ahead. During periods of slow swimming, the wingplate faces upward in the water, and the shell is below. This position is reversed, however, during faster swimming and escape responses; then the animal flips over so that the shell is above and the wingplate below. The maximal swimming speed of *P. reticulata* has been measured at 10 cm/sec (Table 5), which is comparable to rates obtained for euthecosomes.

Morton (1964) analyzed swimming in *Cymbulia*; in this pseudothecosome, the downward stroke of the wingplate surface provides upward movement, while a horizontal component is accomplished simultaneously by the transmission of an undulating wave moving from anterior to posterior over the wing surface. A *Cymbulia* of 20-mm pseudoconch length in the Gulf of

TABLE 5

In situ *swimming rates of thecosomes*

Species	Mean swimming velocity (cm/sec)
Limacina retroversa	8[a]
L. helicina antarctica	12[a]
Creseis acicula	12
Diacria quadridentata	7
Cavolinia longirostris	11
C. tridentata	14
C. uncinata	14
Peraclis reticulata	10[a]
Cymbulia sp.	5–8
Juvenile *Corolla* sp.[b]	13
Adult *Corolla* sp.[c]	40[a]
Gleba cordata	45[a]
Desmopterus papilio	2

SOURCE: Gilmer, 1972, 1974, and unpubl. data; Gilmer and Harbison, 1986.

[a] Maximal swimming rates during escape behavior.
[b] = C. *ovata* of Gilmer, 1974.
[c] = C. *spectabilis* of Gilmer, 1974.

California was observed to move at a relatively slow pace of 5 to 8 cm/sec, although individuals probably are capable of greater speeds when threatened. The cymbuliids normally move through the water like *Peraclis*, that is, with the posterior portion of the pseudoconch facing forward (Fig. 21c). And like *Peraclis*, cymbuliids change orientation during periods of escape or rapid swimming, and then move through the water with the anterior edge of the wingplate leading. Thus all pseudothecosomes normally swim by pushing the shell or pseudoconch through the water; this marks a fundamental difference from the euthecosomes, which pull their shells when swimming. *Corolla* and *Gleba* both have a larger wing surface relative to body size, and the movements of the wingplate in these mollusks have been reported as more leisurely than those of *Cymbulia* (Morton, 1964); these animals are nonetheless capable of very rapid swimming. Adult *Corolla* (dry body weight, 56.0 to 89.5 mg) and *Gleba cordata* (to 106.5 mg) are capable of escape speeds of up to 40 and 45 cm/sec, respectively. Both *Corolla* and *Gleba* accomplish high-speed swimming with the pseudoconch directed upward in the water and the wingplate below.

Because of its very small size, *Desmopterus* has been observed alive only in the laboratory. There the animal often lies motionless, floating near the surface of the water until a disturbance causes it to move in sudden, jerky motions, using its wingplate (Chun, 1889; Essenberg, 1919). This erratic motion is so rapid that it is difficult to visually follow an animal closely, but *D. papilio* is capable of swimming at speeds of at least 2 cm/second (Gilmer, pers. obs.).

Early observations of living euthecosomes indicated that some species appeared to be neutrally buoyant (Rang and Souleyet, 1852; Agassiz, 1866), but in later years it was generally assumed that shelled pteropods had to swim continuously and thereby expend considerable energy in order to offset sinking due to the weight and drag of the shell. This impression has been reinforced by the behavior of euthecosomes in laboratory conditions; captive animals invariably sink rapidly to the bottom of aquaria when swimming ceases.

It is now becoming clear from *in situ* observations that probably all species of euthecosomes and pseudothecosomes are capable of buoyancy regulation, and of maintaining a particular depth position for some time without swimming. Gilmer and Harbison (1986) believe that thecosomes probably spend most of their time passively floating, rather than swimming. Indeed, they never observed thecosomes actively swimming *in situ* during the day unless they disturbed the animals. Detailed anatomical and physiological analyses of buoyancy control remain to be undertaken, but it is clear that different species employ different methods of retarding sinking or achieving neutral buoyancy.

In cavoliniids, shell shape, shell spines, and shell curvature all may act to decrease sinking rates, either by increasing surface area (cf. van der Spoel, 1968) or by causing a motionless animal to turn in such a way as to present the largest surface horizontally to the underlying water. It also seems that all members of this family are capable of extending mantle lining outside the shell (Gilmer, 1974; Gilmer and Harbison, 1986). This ability is least developed in *Creseis*, *Hyalocylis*, *Styliola*, *Cuvierina*, and *Clio*; in these genera, the mantle is extruded only slightly from the shell aperture, and the animals sink slowly when not actively swimming. In addition, the mucoid or gelatinous covering observed surrounding the anterior portions of the shells of some specimens of *Creseis acicula* (Richter, 1973) and *C. virgula* (Gilmer and Harbison, 1986) is perhaps secreted by this extruded mantle lining, and may serve to retard sinking. Species of *Cavolinia* have special slits in the shell through which voluminous folds of mantle are extended in addition to those emerging from the shell aperture (Péron and Lesueur, 1810; Rang and Souleyet, 1852; Meisenheimer, 1905b; Gilmer and Harbison, 1986). Motionless, floating individuals of these species may be entirely encased in external mantle tissue (Fig. 18). Species of *Diacria* also may exhibit external folds of mantle, but they are not as extensive as those of *Cavolinia*. External extensions of mantle lining are seldom seen in captive or preserved animals, but they are always fully extended in animals observed feeding *in situ*. The extended mantle lining must decrease sinking by increasing the surface area of the animal; it may also be that these mantle extensions are involved in buoyancy regulation by ionic exchange within the thin mantle lining (Gilmer, 1974), although this has not yet been investigated.

Pelseneer (1888a) first described another type of flotation structure from specimens of *Cavolinia tridentata*, and it is now known that all species of *Cavolinia* produce a temporary pseudoconch, or false shell (Gilmer, 1974; Bé and Gilmer, 1977; Gilmer and Harbison, 1986). This structure lies in the space created between the extruded dorsal mantle lobes and the shell, and it covers the dorsal exterior shell surface (Fig. 18c). The transparent, gelatinous pseudoconch is neutrally buoyant in sea water, and is similar in other aspects to the permanent pseudoconchs of cymbuliid pseudothecosomes. It is enclosed in folds of extruded mantle, and is presumbably secreted by the mantle. The cavoliniid pseudoconch is not permanently attached to the shell or animal, and is jettisoned when an animal is disturbed, or when the mantle is retracted into the shell. Consequently, these pseudoconchs usually are not observed on captive or preserved specimens, and they may be found as separate structures in net samples. Temporary pseudoconchs have not been found in any species of *Diacria*, but Gilmer and Harbison (1986) do not preclude the possibility that they are present in this genus as well, since pseudoconchs are transparent, difficult to see, and easily detached, and few specimens have been carefully collected by divers. In *Cavolinia*, the combined attributes of a pseudoconch and extended mantle lobes allow animals to achieve suspension during cessation of swimming and during feeding.

Several different buoyancy mechanisms are exhibited by pseudothecosomes. When not disturbed, *Peraclis reticulata* surrounds its shell with extensive mantle lining projecting from the shell aperture (Fig. 19). In cymbuliids, the absence of an external, heavy shell and the development of a neutrally buoyant, permanent pseudoconch generally have been regarded as major evolutionary steps in achieving buoyancy control. In addition, Denton and Shaw (1961) have demonstrated that *Cymbulia*, like some heteropods, has gelatinous tissues and actively replaces heavy sulfate ions with lighter chloride ions. The possibility of ionic regulation in other pseudothecosomes has not been investigated. No information is available on how *Desmopterus* is able to remain floating for long periods of time (Essenberg, 1919), but the morphology of this genus, with its small body and relatively large wingplate and its long, trailing, wing tentacles, presents a large surface area to the underlying water; these features plus the gelatinous nature of its body tissues should significantly retard sinking.

Special flotation devices have not been described for *Limacina*, *Styliola*, *Hyalocylis*, *Clio*, and *Cuvierina*, apart from slight mantle-lining extrusion in the cavoliniids. However, these genera too are able to cease swimming and remain floating with little or no sinking for long periods of time (Agassiz, 1866; Meisenheimer, 1905b; Gilmer and Harbison, 1986). Meisenheimer (1905b) and Richter (1973) suggested that the emission of long mucus strands may account for the flotation capability of these pteropods, and Gilmer and Harbison (1986) attribute buoyancy to the deployment of large mucous webs used in feeding.

In situ sinking rates of living, motionless animals have been measured by Gilmer (1974) and Gilmer and Harbison (1986). *Creseis acicula*, *C. virgula*, *Hyalocylis striata*, and *Clio pyramidata* sink slowly at rates of 0.5 to 1.0 cm/second; the cymbuliids, *Gleba cordata* and adult *Corolla* sp., sink at less than 0.5 cm/second. Sinking was not detectable in *Diacria trispinosa*, *D. quadridentata*, six species of *Cavolinia*, and *Peraclis reticulata*; all of these thecosomes envelop their shells in mantle lobes and appendages, and *Cavolinia* species produce temporary pseudoconchs. The development of buoyancy control by various means and the consequent ability to remain motionless with little or no sinking are obvious attributes in energy conservation, and they appear to be fundamental requisites for the type of feeding employed by euthecosomes.

Thecosomes appear to restrict swimming to two activities: escape from predators and diurnal vertical migrations. Most cavoliniids and all cymbuliids respond to disturbance by divers (and presumably potential predators) by swimming away from the stimulus (Gilmer, 1972; Gilmer and Harbison, 1986). In nature, only *Diacria quadridentata* characteristically responds to disturbance by withdrawal into its shell and passive sinking, a behavior commonly seen in all shelled species in laboratory situations. Most thecosomes disturbed during feeding either quickly pull in the web or use their wings to break free from the surrounding mucus, then swim horizontally or downward from the stimulus. Swimming continues until the animal is 20 cm to several meters away from the disturbance. Maximal *in situ* swimming rates have been measured during escape, and range from 10 to 30 cm/sec in cavoliniids to 40 to 45 cm/sec for *Corolla* sp. and *Gleba cordata*. These cymbuliids are able to sense the presence of a moving diver at a distance of about 1 meter, and most thecosomes can be difficult to capture underwater, since they dodge away from the edge of collection jars. The ability to detect turbulence or other disturbances at some distance and to take evasive swimming action undoubtedly has developed as a predator defense; this behavior also explains the undersampling of this group, particularly of larger and faster-swimming species, by plankton net tows.

Many species of thecosomes show diurnal vertical migration, moving closer to the surface at night and to deeper waters during daylight. This subject was reviewed by Bé and Gilmer (1977) and has been examined most recently by Wormuth (1981). The results of migration studies are difficult to compare because of the use of different types of collecting gear in different localities and seasons. It is clear, however, that some species (e.g. *Creseis acicula*) may show little, if any, change in depth distribution over a 24-hour period, whereas others, such as *Limacina inflata*, may migrate several hundred meters twice each day, which represents a considerable range for such a small animal (Haagensen, 1976). Some larger species (*Cuvierina columnella*, *Clio pyramidata*) typically remain below the euphotic zone during daylight hours and move upward only at night (Wormelle, 1962; Myers, 1968; van der Spoel, 1973a; Haagensen, 1976; Gilmer and Harbison, 1986). In some cases, there may be a reverse diel migration, with populations residing at shallower depths during the day; for example, this has been suggested as occurring at times in *Limacina retroversa* populations in Slope Water off the eastern United States (Wormuth, 1985). To date, the swimming rates of euthecosomes have not been correlated with rates of change in vertical distribution resulting from diurnal migration, but if the swimming rates given in Table 5 can be maintained for extended periods, then euthecosomes may be capable of migrating several hundred meters in 1 to 2 hours.

Food, Feeding, and Trophic Relationships

Early feeding observations of thecosomes were conducted in the laboratory, where animals do not behave naturally and do not survive long. In 1926, Yonge described food capture in several species of cavoliniids and cymbuliids as taking place solely by the collection of particles on ciliary tracts located on the footlobes and wings. He concluded that the large pallial gland of these animals was not involved in feeding, and that pallial feeding and the use of mucus for food collection had been lost during evolution to a planktonic life style. Morton (1954a) later reported that laboratory specimens of a more primitive species, *Limacina retroversa*, collected most of their food from water passing through the mantle cavity. Food particles became entangled in mucus secreted by the large pallial gland, and the resulting food strings were transported out of the mantle cavity and to the mouth along ciliated channels. Both of these described mechanisms implied that feeding could occur while the animals were swimming. In 1974, Gilmer described feeding in eight species of cavoliniids, combining laboratory and the first *in situ* observations. In nature, cavoliniids often had large strands of mucus with attached particles emerging from the exhalant opening of the mantle cavity, and laboratory specimens were observed to ingest similar mucous strands. Thus Gilmer concluded that cavoliniids fed in basically the same manner as Morton had described for *Limacina*, with food particles entering the mantle cavity and being trapped in mucus secreted by the pallial gland. In 1977, Richter suggested that the larger, fast-swimming zooplanktonic prey found in thecosome gut contents could not be captured by being drawn by ciliary currents into the mantle cavities of these pteropods. It also remained unclear how these relatively large and active animals could filter enough food through their small mantle cavities to survive in the low food environment of the open ocean. In the ensuing years, almost 100 more hours of *in situ* observations were accumulated, with divers using dyes for visual aid and better photographic methods. Gilmer and Harbison (1986) have now published a definitive account of feeding in thecosomes, which corrects earlier observations, supports Richter's basic contention, and expands observations to include all cavoliniid genera as well as *Limacina* and the shelled pseudothecosome genus, *Peraclis*. This information, combined with *in situ* accounts of cymbuliid feeding (Gilmer, 1972, 1974), permits a comparison of feeding mechanisms and food selection in the four major thecosome families.

It is now known that all euthecosomes use an external, spherical mucous web many times the size of their bodies to capture and entangle planktonic food (Figs. 25, 26). The web is intermittently withdrawn and conveyed by

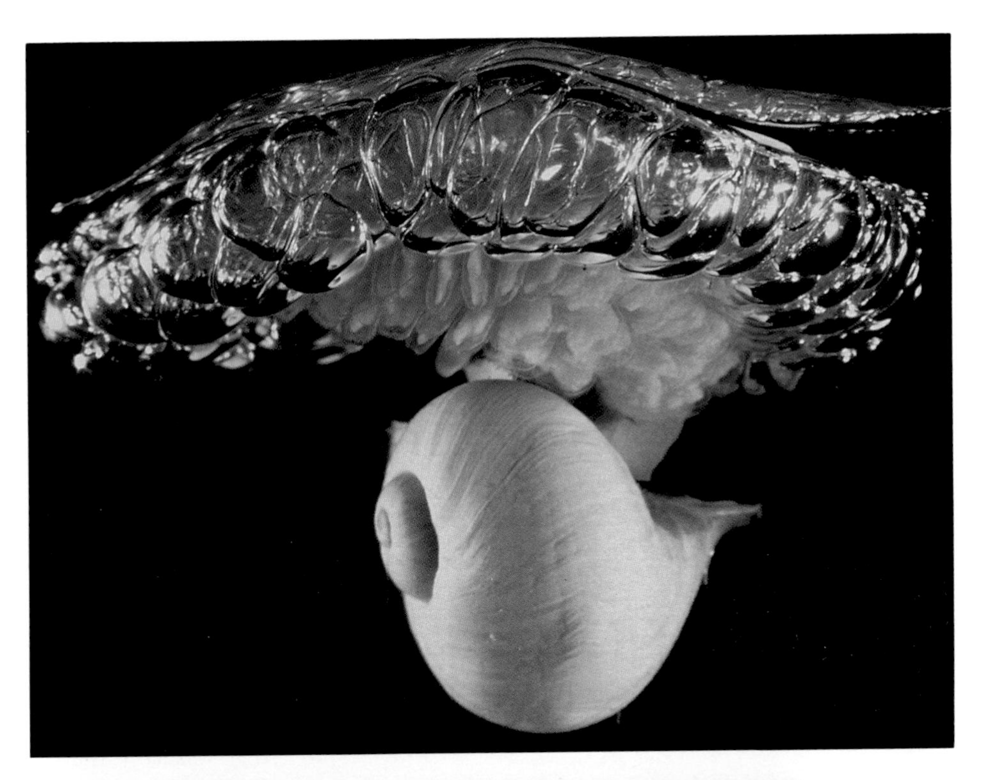

COLOR FIG. 1. *Janthina prolongata,* with egg capsules attached to the underside of its bubble raft. (ca. 2½×)

COLOR FIG. 2. The heteropod *Atlanta peroni,* lateral view. (ca. 7½×)

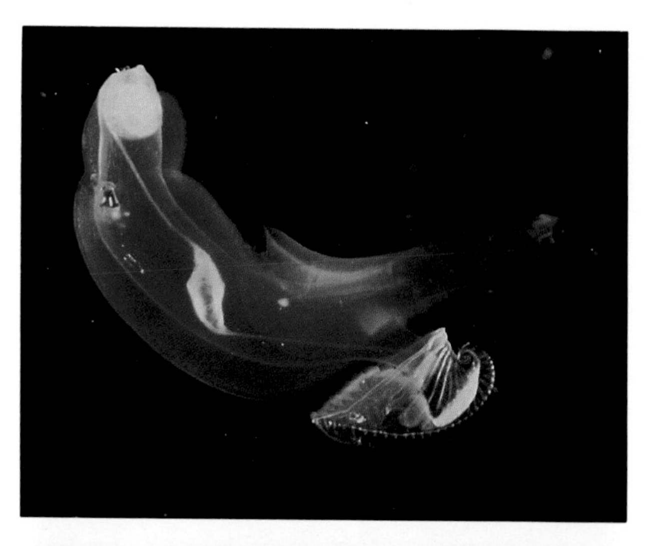

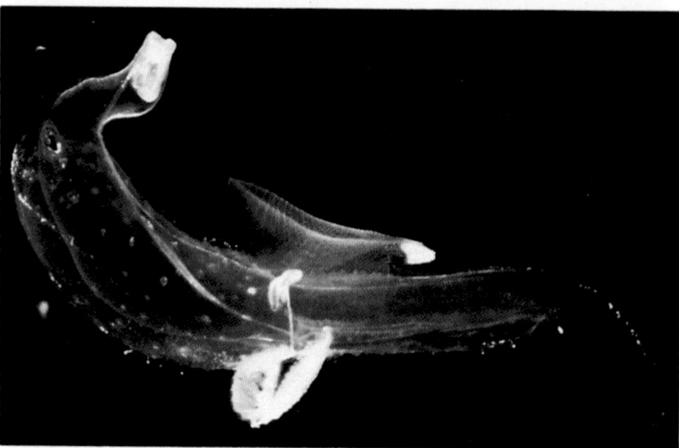

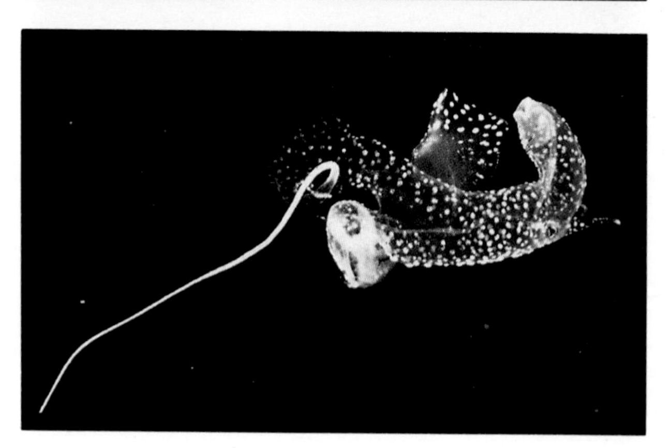

COLOR FIG. 3. The heteropod *Carinaria lamarcki,* lateral view, in its normal swimming posture. (ca. actual size)

COLOR FIG. 4. The heteropod *Pterosoma planum,* lateral view, in its normal swimming posture (photo by R. Reichelt). (ca. actual size)

COLOR FIG. 5. The heteropod *Cardiapoda richardi,* with opalescent pigment spots and long tail. (ca. actual size)

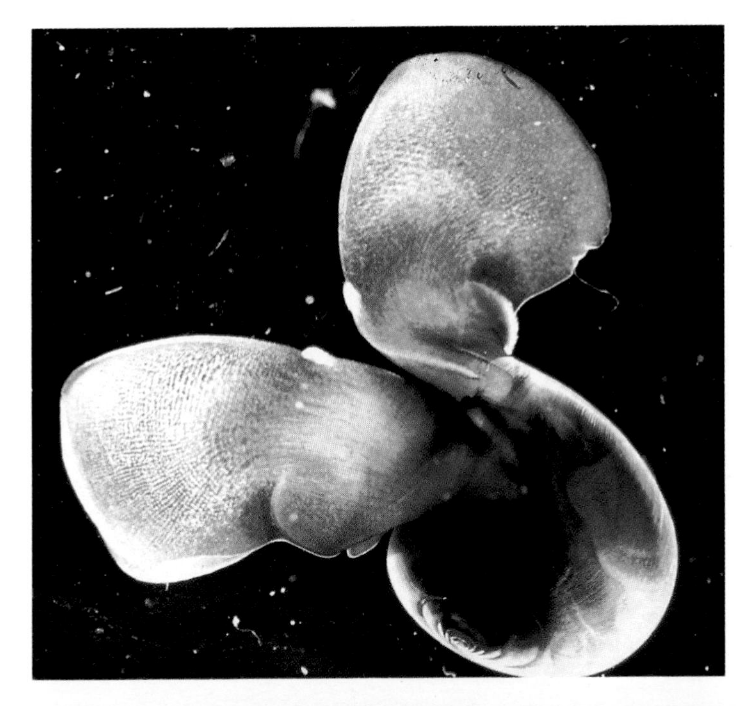

COLOR FIG. 6. The euthecosome *Limacina helicina;* the large, flat pallial gland is visible through the shell on the right side. (ca. 4×)

COLOR FIG. 7. The euthecosome *Cavolina tridentata,* with mantle appendages extended through slits in its shell. (ca. 2×)

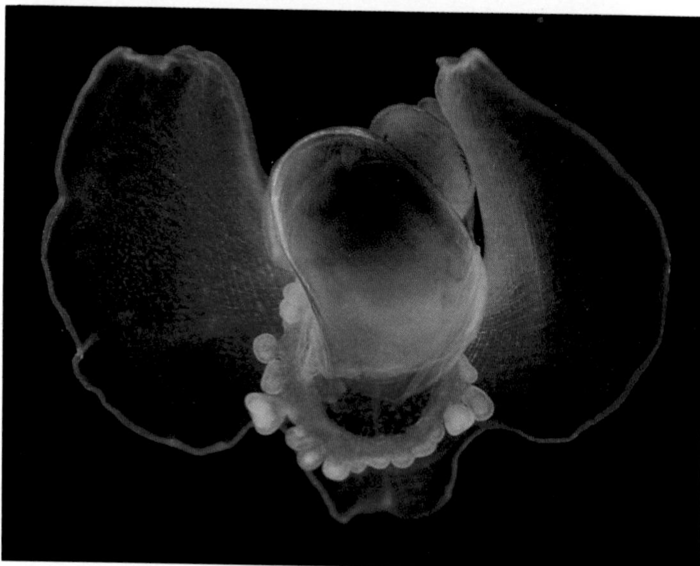

COLOR FIG. 8. The euthecosome *Diacria trispinosa* with the hydroid *Kinetocodium danae* attached to its shell. (ca. 5 ×)

COLOR FIG. 9. The pseudothecosome *Peraclis reticulata*, dorsal view, with expanded wingplate and projecting mantle margin edged with large, white, glandular nubs. (ca. 10 ×)

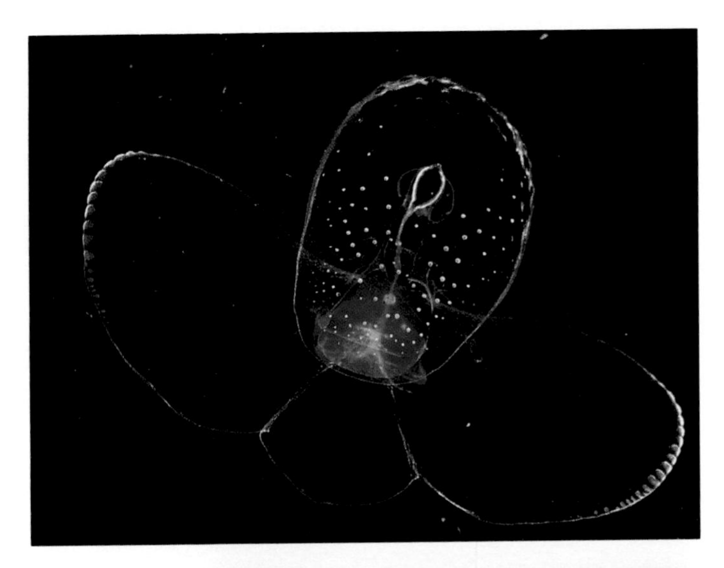

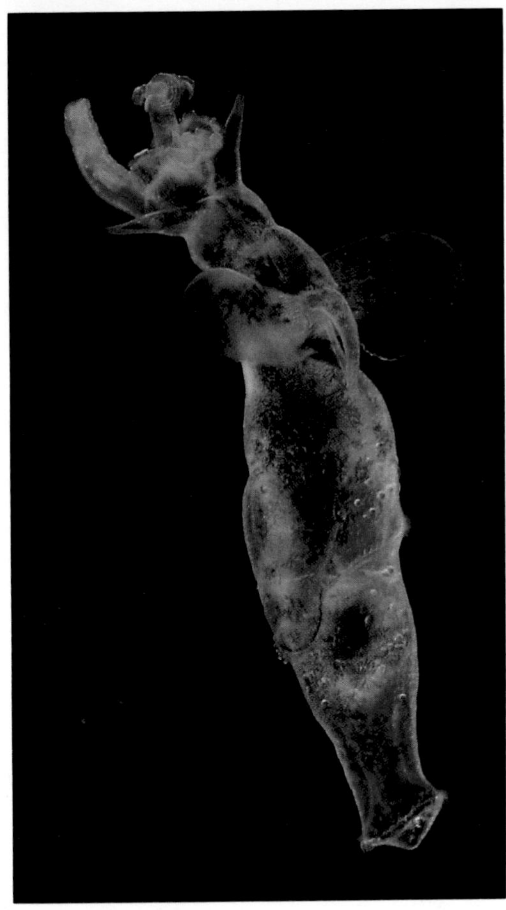

COLOR FIG. 10. The pseudothecosome *Corolla* (*calceola*?); a juvenile with expanded wingplate. (ca. 3×)

COLOR FIG. 11. The gymnosome *Pneumodermopsis* (*canephora*?), ventral view, with its proboscis and suckers partly everted above the elongated anterior tentacles. (ca. 11×)

COLOR FIG. 12. The gymnosome *Clione limacina,* a large specimen from Subarctic water. (ca. 2×)

COLOR FIG. 13. The gymnosome *Hydromyles gaudichaudii,* ventral view (photo by G. R. Harbison). (ca. 7×)

COLOR FIG. 14. The nudibranch *Phylliroë bucephala,* lateral view. (ca. 2½ ×)

COLOR FIG. 15. The nudibranch *Fiona pinnata,* with its egg masses attached to *Velella,* a favored prey. (ca. 9 ×)

COLOR FIG. 16. Two mating individuals of the nudibranch *Glaucus atlanticus*. (ca. 3 ×)

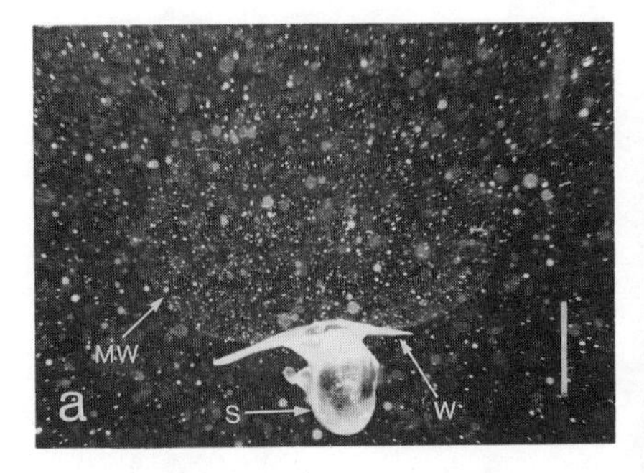

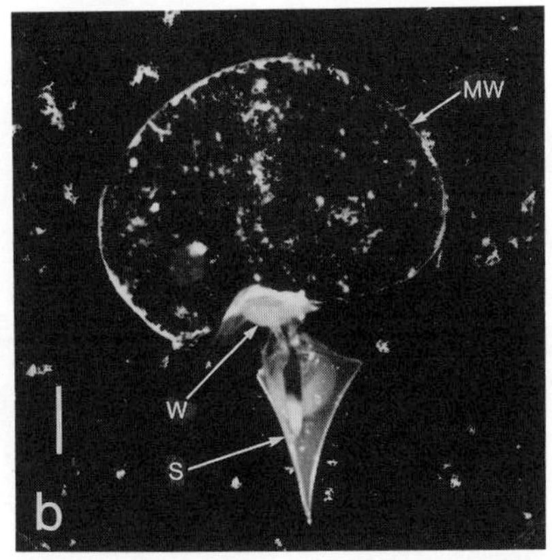

FIG. 25. Feeding in euthecosomes (from Gilmer and Harbison, 1986): *a,* field photograph of *Limacina helicina* in the typical feeding posture, with the mucous web spread above the animal; *b,* nighttime *in situ* photograph of *Clio pyramidata* retracting its feeding mucous web. MW, mucous web; S, shell; W, wing. Scale lines represent 10 mm.

ciliary pathways to the mouth. The feeding mechanism is intimately connected with the ability to regulate buoyancy during feeding, so that wing motion does not interfere with the delicate web. Although euthecosomes generally have been regarded as herbivores feeding upon phytoplankton, small zooplankton also form a significant part of the diet of at least some species. The mantle lining and pallial gland are of paramount importance in mucus production, and ciliary pathways for web deployment and for the collection, sorting, and rejection of food are located on the mantle lining, footlobes, and wings. The mucous strands of the web that accumulate food particles are ultimately pulled into the gut by a small radula consisting of only a few rows of teeth, with three teeth (1-1-1) in each row (Richter, 1979 and Fig. 27). Jaws consist of small cuticular teeth or spines on the lateral walls of the pharynx; they may assist in the ingestion and triturition of food.

The spirally coiled members of the genus *Limacina* are considered the least specialized of the euthecosomes. Gilmer and Harbison (1986) observed two of the largest species, *L. helicina* and *L. retroversa,* while scuba diving along the coast of Greenland. The two species feed in a similar manner, and are probably representative of the whole genus. Feeding animals are motionless in the water with their wings extended (Fig. 25a). Sinking is not detectable in a feeding animal, and this neutral buoyancy is attributed to the deployment and frictional drag of a spherical mucous web that extends above the animal and entangles both passive and swimming food from

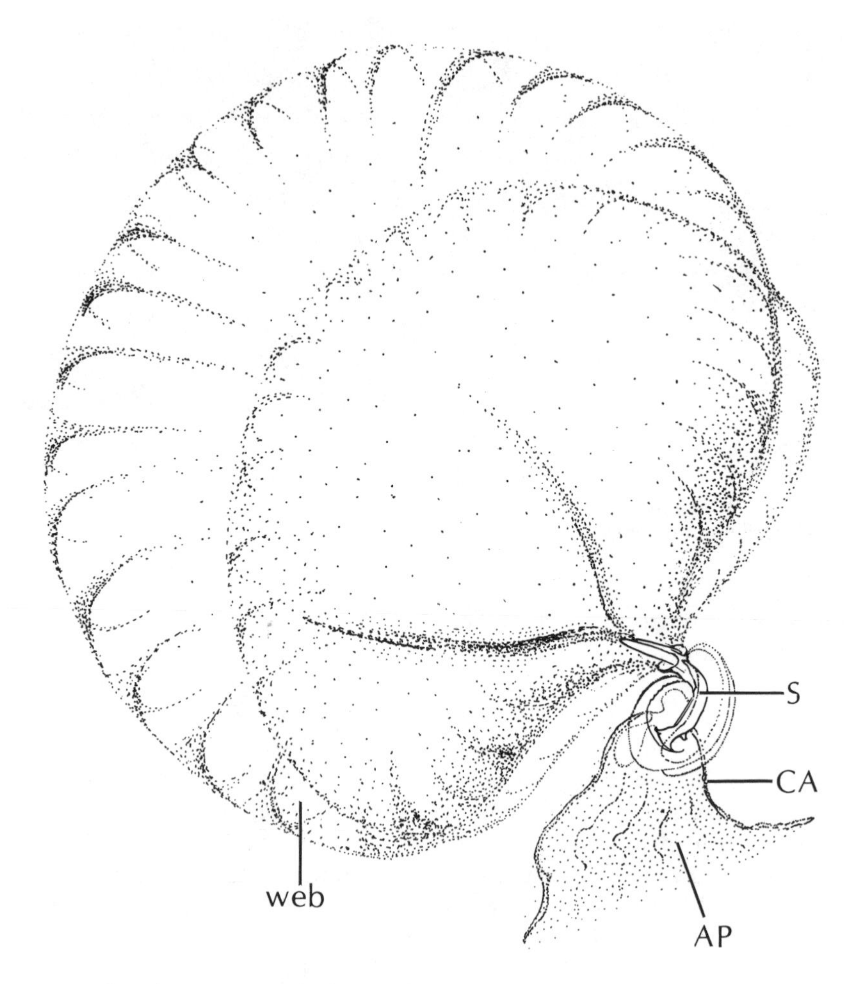

FIG. 26. *Cavolinia uncinata* feeding with a mucous web (from Gilmer and Harbison, 1986): the shell (S) is enveloped by mantle lobes; feces and pseudofeces (AP) accumulate between the central mantle appendages (CA).

the surrounding water. The size of the web varies directly with individual and species size (Table 6). The web can be set within 5 seconds after an animal stops swimming, and this rapid deployment suggests that the mucus is emitted by the very large pallial gland inside the mantle cavity. The retraction of the web is also rapid (>20 sec), and presumably is accomplished by the ciliary currents on the footlobes, as outlined by Morton (1954a). It is not known how long the webs are left in place, although feeding individ-

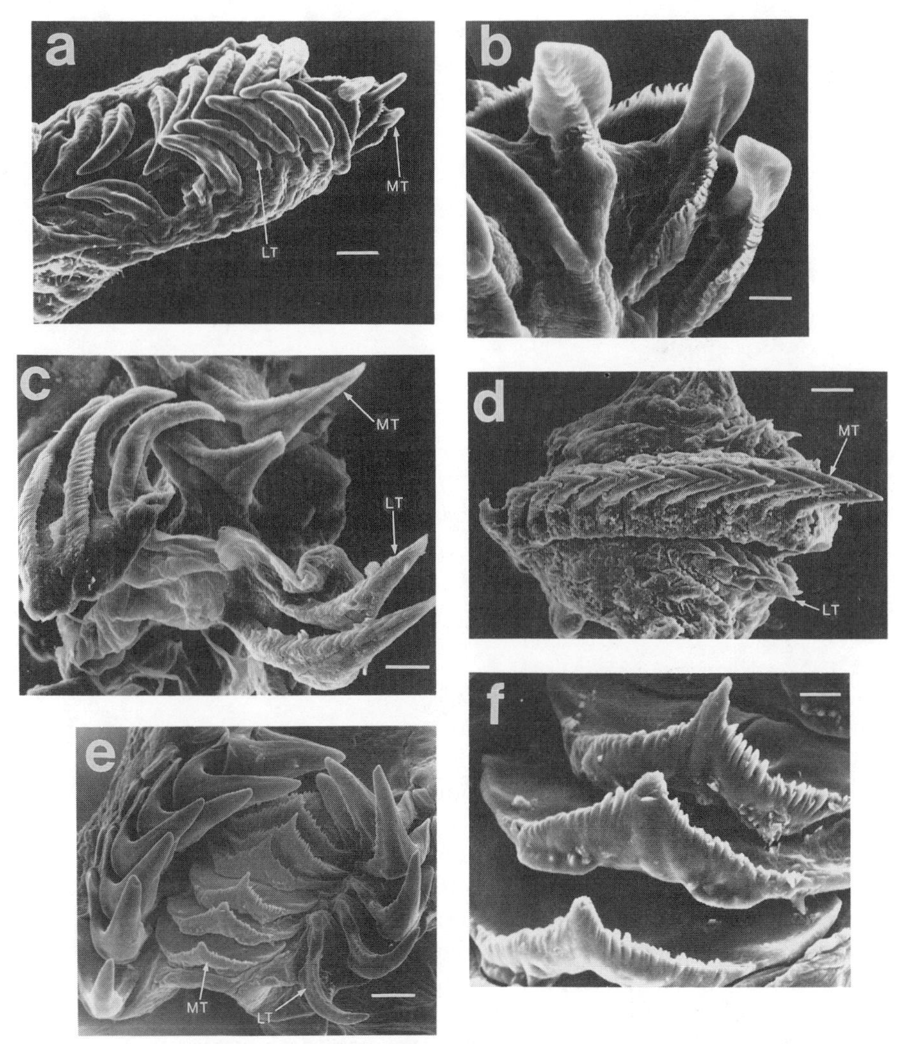

FIG. 27. Scanning electron micrographs of thecosome radulae: *a*, *Limacina helicina*; *b*, enlargement of the median teeth in *a*; *c*, *Cavolinia tridentata* (by J. Linley); *d*, *Cuvierina columnella* (by J. Linley); *e*, *Peraclis bispinosa*; *f*, enlargement of the median teeth in *e*. LT, lateral radular teeth; MT, median radular teeth. The scale lines represent 30 μm in *a*, 8 μm in *b*, 12 μm in *c*, 25 μm in *d*, 40 μm in *e*, and 12 μm in *f*.

TABLE 6

Relationship between shell length and the diameter of mucous feeding webs of euthecosomes

Species	Number observed with webs	Shell length (mm)	Range in web diameter (mm)	Estimated volume of largest web (ml)
Limacina helicina	13	11[a]	40–55	80
L. retroversa	2	6	20	10
Creseis virgula virgula	1	9	30	20
Styliola subula	5	8	30–40	50
Hyalocylis striata	1	9	30	20
Clio pyramidata	8	10	30–40	50
Cuvierina columnella	14	11[b]	30–110	500
Diacria quadridentata	2	3	10–20	10
D. trispinosa	2	10	90–100	300
Cavolinia inflexa	3	5	20–30	20
C. longirostris	11	5	20–40	50
C. tridentata	4	15	100–200	1,000
C. uncinata	9	7	30–90	300

SOURCE: Glimer and Harbison, 1986.

NOTE: Relationships shown are based on measurements taken in the field or from photographs. The volume of the largest web for each species is calculated, assuming that it is a perfect sphere.

[a] Shell diameter.

[b] Excludes shell posterior to caudal septum.

uals have been observed for as long as 6 minutes, nor is it known how often new webs are formed. Presumably the rate of web production is related to the concentration of available and appropriate food and to the size of the feeding animal.

The natural food of *Limacina* has been described only from a few qualitative analyses of gut contents, and thus includes only easily recognizable items; detritus, bacteria, and naked flagellates cannot be identified by this method. Table 7 summarizes food items found in the gut contents of three species of *Limacina*. Phytoplankton and small protozoa predominate in the stomach contents of the epipelagic cold-water species, *L. retroversa* and *L. helicina*. *L. helicoides*, the largest and the only deep-water species, is reported to eat larger protozoa, specifically foraminiferans and radiolarians. It is of interest to note that all of the reported food organisms are provided with hard exoskeletons. These are normally broken open by the crushing action of the gizzard in *Limacina* (Morton, 1954a). This muscular organ, located between the extensible esophagus and the stomach, is lined with five cuticular teeth. Hard skeletal parts of food are crushed by the contraction of the gizzard and the interlocking of the teeth. The natural food of the other four species of *Limacina* has not been described, but presumably the food and feeding mechanisms are similar.

Cavoliniids are generally larger and easier to observe in the field than the limacinids are, and they have thus been studied in more detail with respect to their feeding behavior. In this family, the feeding mechanism is least complex in those genera (*Creseis*, *Hyalocylis*, *Styliola*, *Clio*, and *Cuvierina*) that have conical shells and lack external mantle appendages (Group 2 of Gilmer and Harbison, 1986). When feeding, these pteropods produce spherical mucous webs 30 to 110 mm in diameter (Table 6); the mucus seems to be derived primarily from the large pallial gland. An animal hangs suspended from its web with its wings folded dorsally and with the long axis of the shell inclined up to 45° from the vertical (Fig. 25b). The ventral side and mouth of the animal are directed upward; this feeding posture ensures continued contact of the mouth and ventral footlobes with the web. Feeding animals sink slowly at about 0.5 cm/sec. The web can be drawn in within seconds along ciliary tracts on the footlobes, and it is ingested along with captured food particles at intermittent intervals. Pallial food collection as described by Gilmer (1974) does not appear to take place in the field and is probably a laboratory artifact. In this group, only *Creseis virgula* has been observed to produce strings of pseudofeces, which are moved dorsally between the wings and away from the mouth and web by cilia on the footlobes.

Feeding differs in *Cavolinia* and *Diacria* (Group 1 of Gilmer and Harbison, 1986) in involving the use of mantle lobes that project externally through shell clefts and envelop feeding animals (Figs. 18, 26). An anterior

TABLE 7

A summary of gut-content analyses in the Limacinidae

Species	Food item(s)	Location	Reference
Limacina retroversa	Diatoms Dinoflagellates Coccolithophorids Tintinnids	Subarctic	Boas, 1886a
L. retroversa	Dinoflagellates	Bay of Fundy, Canada	White, 1977
Limacina helicina	Diatoms Dinoflagellates Tintinnids	Subarctic	Boas, 1886a
L. helicina	Diatoms Dinoflagellates Tintinnids	North Pacific	Paranjape, 1968
L. helicina[a]	Diatoms	Antarctic	Hart in Morton, 1954a
Limacina helicoides	Foraminifera Radiolaria	Deep-sea Atlantic	Murray and Hjort, 1912

[a] Cited as *helicoides*.

mantle lobe projects from the shell aperture to form a funnel-like collar around the base of the wings and footlobes. The feeding web apparently is formed by mucus secreted by this mantle lining and by the large pallial gland within the mantle cavity; mucus channeled through the anterior mantle lobe is shaped into a heart-shaped or oblong sphere. The diameter and volume of feeding webs (Table 6) vary with the size of the individual, ranging from 10–20 mm in *Diacria quadridentata* (3 mm shell length) to 100–200 mm in *Cavolinia tridentata* (15 mm shell length). In contrast to Group 2 species, these animals fold their wings ventrally, not dorsally, when feeding, and they appear to be neutrally buoyant, with no perceptible sinking.

Cavolinia can form feeding webs in 2.5 minutes or less. Webs can be ingested in 1 to 3 minutes by being drawn onto the footlobes and toward the mouth in ciliary tracts diagrammed by Yonge (1926) and Gilmer (1974). Some sorting of collected particles occurs before ingestion, since rejected material mixed with mucus moves dorsally between the wings and away from the mouth. A ciliated pathway formed by the junction of the dorsal mantle lobes moves these strings of pseudofeces posteriorly, where they mix with feces in aggregations or strings below the animal (erroneously called "food strings" by Gilmer, 1974). In *Cavolinia tridentata*, compacted pseudofeces may form a continuous string up to 30 cm long. The ciliated medial mantle lobes of cavoliniids (Figs. 18a, b) also function to keep rejected and egested material away from the vicinity of the mouth.

Our knowledge of the natural diets of cavoliniids is limited to items found in the alimentary tracts of only a few species (Table 8). These gut-content analyses have revealed a variety of foods, ranging from phytoplankton and protozoa to larger zooplankton such as molluscan and crustacean larvae. Various types of phytoplankton and protozoa tend to dominate the stomach contents of both adults (Table 8) and young thecosomes (Mironov, 1977) when expressed in terms of percentages of total numbers of food items present, but the larger-size fraction of zooplankton prey may contribute substantially more to nutritional requirements if considered on a volumetric basis (Richter, 1977). Most food items, regardless of size, are protected by carbonate, siliceous, or chitinous coverings. Richter (1977) found that aragonite shells and tests of prey were corroded during passage through the gut of thecosomes. The more resistant calcite tests of coccolithophorids and foraminifera were largely unaffected by chemical change, and were damaged only by the mechanical action of the powerful gizzard. All thecosomes also have a ciliated style sac, and some possibly contain a rotating hyaline rod (Meisenheimer, 1905b; Yonge, 1926; Howells, 1936; Morton, 1954a, 1960). The rod has not been demonstrated to release amylolytic enzymes, as does the crystalline style of certain other molluscan groups, and it is generally regarded as functionally vestigial in thecosomes.

Pseudothecosomes also collect food in external mucous webs, but there are several notable changes in the feeding structures of this group. For one, feeding webs are usually funnel-shaped or flattened sheets and not spherical. Second, all pseudothecosomes have a proboscis that, in the genera *Corolla* and particularly *Gleba*, is long and extensible (Figs. 21, 22). *Peraclis*, *Cymbulia*, and *Desmopterus* have a radula with three to five teeth per row, and small jaws. However, *Gleba* and *Corolla* lack a buccal mass; the radula, jaws, and salivary glands are absent. All genera except *Desmopterus* possess a well-developed pallial gland, gizzard, and style sac.

Feeding behavior has been described in only one shelled pseudothecosome, *Peraclis reticulata* (Gilmer and Harbison, 1986). The shells of feeding individuals are entirely encased in extruded mantle tissue, and the animals lie motionless below a funnel-like sheet of mucus measuring about 5 cm in diameter (Fig. 28a). The mouth and footlobes are in contact with the feeding web, which presumably is ingested by ciliary action and grasping motions of the radula. Fecal pellets measured about 1.0 × 0.5 mm and contained intact cells of a variety of phytoplankton as well as fragments of crustaceans and tintinnids. Stomach-content analyses of *Peraclis reticulata* (Richter, 1977) and *P. apicifulva* (Richter, 1983) have revealed that these species can eat large numbers of young *Limacina inflata* (captured during the prey's descent to deeper waters in the early morning) as well as other juvenile snails and foraminifera. All species of *Peraclis* show great similarity in the anatomy of the buccal mass and footlobes, which suggests that all members of this family feed in a similar manner; the food type ingested may reflect the relative abundance of suitably sized prey in the area.

The external feeding webs of the larger cymbuliids (*Cymbulia*, *Corolla*, and *Gleba*) are more spectacular in size and may extend to 2 m in diameter (Fig. 28b) (Gilmer, 1972, 1974, pers. obs.). In these species, the formation of a web begins with the cessation of swimming and the release of copious amounts of mucus by glands along the periphery of the wingplate, and probably also by the voluminous pallial gland. The mucus is carried in strands along ciliary tracts that extend over the swimming plate and the length of the proboscis and terminate in the lateral grooves surrounding the mouth. When the sheetlike web is completed and spread in the water, the animal lies motionless and inverted below, with the elongate proboscis maintaining contact with the web (Fig. 29). The animal and its web sink slowly (<0.5 cm/sec), or not at all, and food is collected passively by becoming entangled in the mucus. Ensnared food particles are pulled toward the lateral grooves of the proboscis, where the web and food are consolidated into a string by ciliary action. Some food and mucus are rejected as pseudofeces.

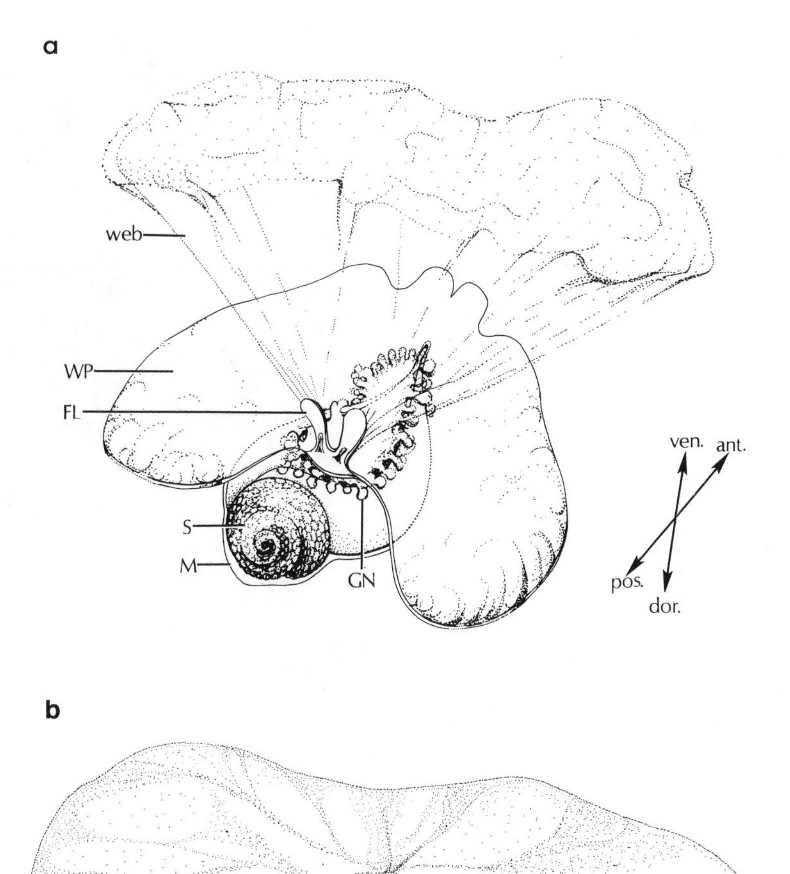

b

MW
WP
PF

FIG. 28. Feeding in pseudothecosomes: *a*, posterioventral view of *Peraclis reticulata*, drawn from field photographs, showing the feeding web entering the mouth between the footlobes (from Gilmer and Harbison, 1986); *b*, drawing of *Gleba cordata* made from field observations, showing the relative sizes of the animal and its mucous feeding web. ant., anterior; dor., dorsal; FL, footlobe; GN, glandular nub; M, mantle; MW, mucous web; PF, pseudofeces; pos., posterior; S, shell; ven., ventral; WP, wingplate. The scale line in *b* represents 200 mm.

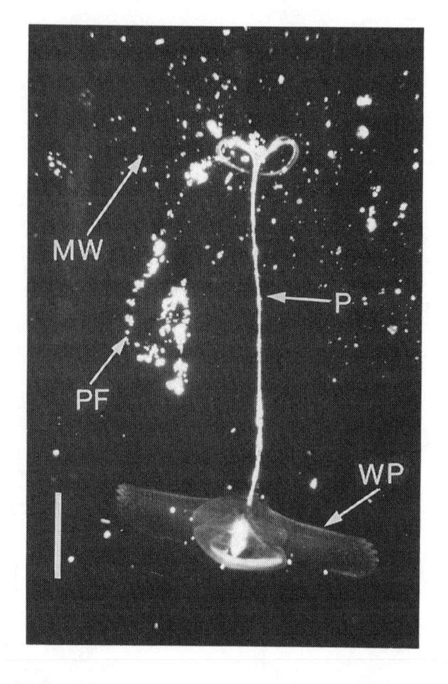

FIG. 29. *Gleba cordata* with its proboscis extended in the feeding position (from Hamner, 1974). MW, mucous web; P, proboscis; PF, pseudofeces; WP, wingplate. The scale line represents 5 mm.

The remainder is pulled into the mouth and drawn through the esophagus to the muscular gizzard. The gizzard apparently retains the same function as in euthecosomes, that of crushing hard tests or exoskeletons of food (Howells, 1936).

With the development of a long proboscis, these cymbuliid pseudothecosomes are capable of freeing the body and wings from the mucous webs and strings that surround euthecosomes and, to some extent, *Peraclis*. Thus there is less interference with the ciliary currents that transport mucus, and the animals can quickly free themselves of the web if necessary. Further, these animals have effectively expanded the area of food collection by producing larger webs, and feeding has become more efficient, especially in tropical oceanic waters where phytoplankton and zooplankton densities are relatively low.

Microscopic analysis of cymbuliid feeding webs revealed mucous strands varying from 1 to 6 μm in width and criss-crossing to form pores of approximately 250 to 4,000 μm^2 (Gilmer, 1974). Pore size, however, is probably not significant in food capture, since both smaller and larger prey can be trapped by adhesion to the mucus. On a numerical basis, 50 percent of the captured food of *Gleba* consisted of detritus and nanoplankton, including bacteria and small phytoplankton; however, smaller numbers of protozoa and crustacean larvae of 300 to 800 μm in diameter also were collected, and constituted the major food source on a volumetric basis (Fig. 30). In

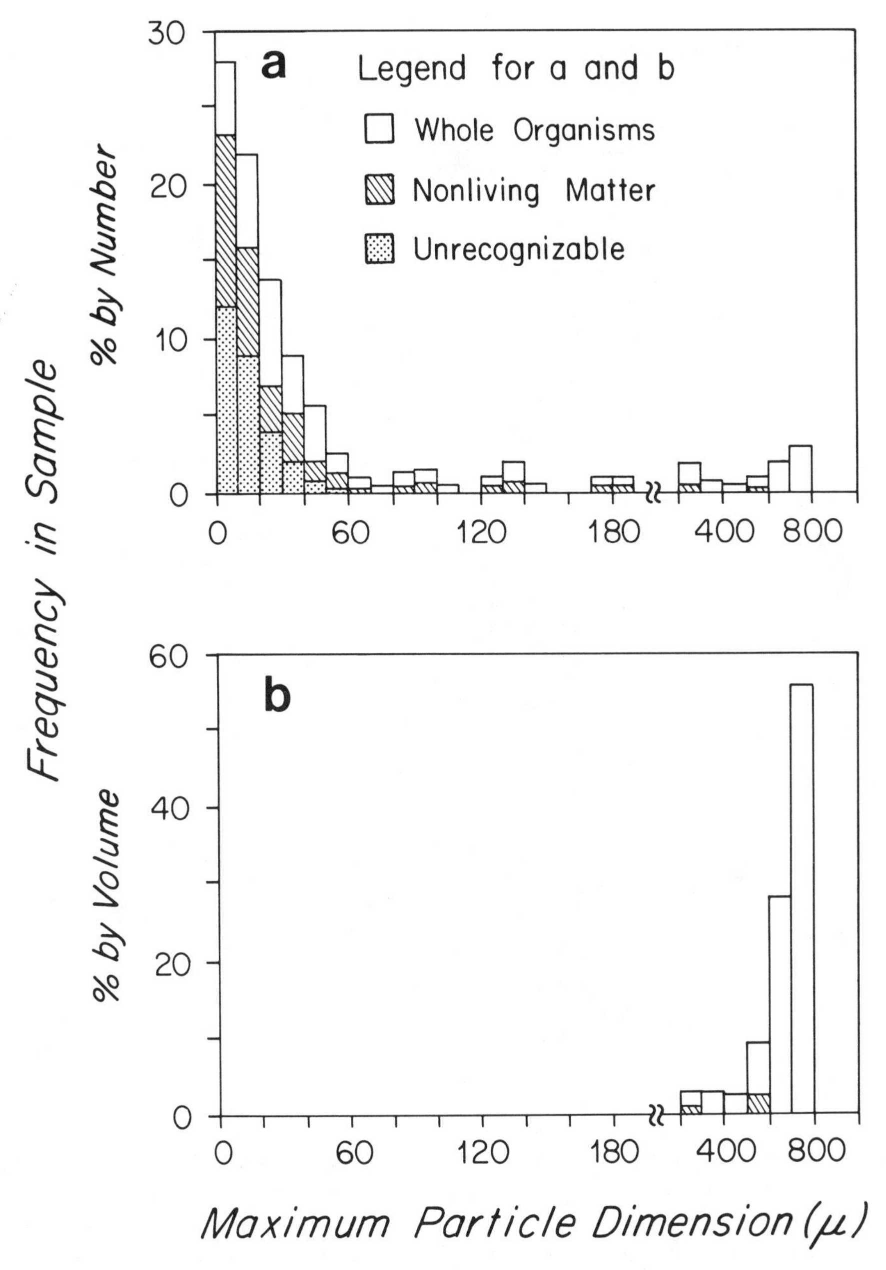

FIG. 30. Particle-size distribution of food captured in the feeding webs of *Gleba cordata,* based on a total of 450 measured particles: *a,* size distribution on a numerical basis (from Gilmer, 1974); *b,* size distribution on a volumetric basis, calculated by assigning ideal volumes to diameter dimensions (original).

TABLE 9

A summary of gut-content analyses in the Pseudothecosomata

Species	Food item(s)	Location	Reference
Peraclis apicifulva	Foraminifera Young *Limacina inflata*	Tropical seas	Richter, 1983
P. reticulata	*Globigerina* (300 μm)[a] *Atlanta* sp. (150 μm)[a] *A. helicinoides* (350 μm)[a] *Limacina inflata* (200–400 μm)[a]	?	Richter, 1977
Cymbulia peroni	Diatoms Flagellates Heteropods Copepods Chaetognaths	Adriatic Sea	Tesch, 1913; van der Spoel, 1976
C. valdiviae	Foraminifera	?	Richter, 1977
Corolla calceola	Diatoms Foraminifera Copepods Heteropods	?	Peck in Meisenheimer, 1905b
C. spectabilis	Diatoms Copepods	Monterey Bay, California	Heath & Spaulding, 1904
Corolla sp.	*Globigerina* (400 μm)[a] Other foraminifera (100 μm)[a] *Limacina inflata* (200 μm)[a] *Creseis acicula* (150 μm)[a] Gastropod veligers (150–400 μm)[a] Copepod remains (150 μm)[a]	?	Richter, 1977
Gleba cordata	Foraminifera Tintinnids Gastropod larvae Crustacean larvae Small copepods	Florida Current	Gilmer, 1974 and unpubl. data
Gleba sp. (= *Tiedemannia*)	Dinoflagellates Coccolithophorids Foraminifera Radiolaria Tintinnids Crab larvae	?	Boas, 1886a

[a] Sizes refer to maximum dimensions.

fact, all food particles smaller than 180 μm accounted for less than 1 percent of the total volume of food captured by *Gleba*. Gut analyses of other cymbuliids (Table 9) agree in reporting that a mixture of food types is ingested; these range in size from phytoplankton and protozoa to heteropods, copepods, chaetognaths, and crab larvae. On the other hand, examination of fecal pellets produced by *Corolla spectabilis* revealed mostly small-sized particles (Silver and Bruland, 1981). Pellets contained primarily diatoms of 5- to 100-μm sizes, which were also the numerically dominant phytoplankton in near-surface waters. Smaller numbers of dinoflagellates (5–20 μm), coccolithophores (<10 μm), and fecal pellets produced by small crustaceans were also present. The prevalence of small-sized particles, many of which were intact in the feces, may be less indicative of what constitutes an important food source than of what materials pass through the gut undigested. As with other thecosomes, less numerous but larger prey can be nutritionally more important than small particles that may be trapped incidentally in the mucous webs (Richter, 1977; Gilmer and Harbison, 1986).

It is highly likely that the type of feeding mechanism used by euthecosomes and pseudothecosomes would never have been discovered except through direct scuba observations. Even hand-collected specimens do not behave normally in the laboratory, nor do they survive for more than a few days, and the large size of cymbuliid feeding webs precludes the use of average-sized laboratory containers. *In situ* observations of feeding behavior have now been made of all thecosome genera except *Desmopterus*. It is possible that *Desmopterus* does not feed like other thecosomes, since it is the only genus that lacks a gizzard, and in which the pallial gland is absent or rudimentary. These features suggest that mucus production is not important in food capture, and that food does not consist of items with hard external tests or exoskeletons. Mironov (1977), however, found both phytoplankton and foraminiferans in the gut contents of one specimen of *Desmopterus* (*papilio*?); five other specimens had empty guts.

In addition to determining the mode of feeding and food type in *Desmopterus*, several other points concerning the natural foods of thecosomes remain to be investigated. Gut-content analyses have been conducted on only about 30 percent of the total number of thecosome species, and in general these analyses have been qualitative descriptions based on a few specimens. To assess the nutritional contribution of various ingested food types to the diet, the gut contents should be described on both numerical and volumetric bases, and should be compared with fecal material to establish which ingested food types tend to pass through the gut undigested. An evaluation of the impact of thecosomes on the trophic ecology of areas where they are reasonably abundant will ultimately depend on knowing something of the metabolic requirements and feeding rates. The difficulty of maintaining thecosomes in laboratory conditions has hampered the collection of relevant data, but the following studies have attempted to examine certain aspects of metabolism and population dynamics in these planktonic gastropods.

The chemical composition of euthecosomes has been analyzed by Curl (1962), Beers (1966), and Omori (1969); their results are summarized in

TABLE 10

The major chemical constituents of Euthecosomata

Species	Collection area	Dry weight/ wet weight (%)	C (%)	N (%)	C/N (ratio)	P (%)	Ash (%)	Carbohydrate (%)	Reference
Limacina inflata	North Pacific	36.4	17.0	1.5	11.1		46.6		Omori, 1969
L. helicina		25.0	29.0	6.0	4.9		28.5		
Clio pyramidata		31.1	20.3	2.9	7.0		42.8		
Cavolinia longirostris		—	22.0	3.5	6.3		39.3		
AVERAGE		30.8	22.1	3.5	7.3		39.3		
Pteropods (combined spp.)	Sargasso Sea	25.6	22.7	3.5	7.0	0.30		1.12–0.55	Beers, 1966
Limacina retroversa	North Atlantic		28.3 (56.0)[a]	4.1	6.9	0.58	64.2		Curl, 1962

NOTE: Values are expressed in terms of percentages of dry weights, which are inclusive of shell weights.
[a] As percent of organic weight (ash-free dry weight).

Table 10. All values are expressed in terms of percentages of dry weights, which include shell weights; thus the ash content and dry-weight to wet-weight values are high. The values for carbon, nitrogen, and phosphorus are apparently very low in comparison to those for other zooplankton, but this is because these also reflect comparison with total dry weights inclusive of shell weights. The one carbon value that is corrected to percentage of ash-free dry weight (56.0 percent) is closer to carbon-content values for other zooplankton. Nevertheless, the carbon-to-nitrogen ratios are within the normal range for marine zooplankton, and all values are remarkably consistent between animals collected in different areas and analyzed by different researchers. The only carbohydrate values are comparatively high, and showed a gradual decline from 1.12 percent in February to 0.55 percent in June. This may suggest some capacity for food storage, although no storage cells or fat layers are known to occur in these animals (van der Spoel, 1967).

Ikeda (1974) and Ikeda and Fay (1981) measured oxygen uptake and ammonia excretion in several thecosomes and obtained O:N ratios of between 11 and 14 for *Clio cuspidata*, *Diacria trispinosa*, *Cavolinia globulosa*, and *C. uncinata*, and between 12 and 13 for *Limacina helicina antarctica*. Gilmer (1974) measured respiration in *Cavolinia longirostris* that was hand-collected to minimize collection trauma. Oxygen uptake per unit body weight decreased with increasing size; larger animals (0.5 mg dry body weight excluding shell) consumed about 3 μl O_2/mg/hour, whereas smaller individuals (0.1 mg) used almost 16 μl O_2/mg/hour. There was no significant difference in the respiration rates of animals maintained at 20° or at 26° C. These rates may be similar to those obtained by Ikeda (1974) and by Nival et al. (1972) for *Cavolinia inflexa* and a species of *Creseis*, but the values are difficult to compare, since these researchers included shell weights in the dry weights of animals. This decision may have produced unusual values and results, since, in any one species, shells of similar dimensions can vary appreciably in thickness and weight. The relationship between ash-free dry body weight and shell dimension or weight is nonlinear (Conover and Lalli, 1972). Gilmer's regression equation for respiration in *C. longirostris* was expressed as: $\log \hat{Y} = 11.1 - 8.6 \log X$, where $\hat{Y}$ is the uptake of oxygen in μl O_2/mg dry body weight/hour and X is body dry weight excluding shell weight, in mg.

Smith and Teal (1973) investigated the effects of temperature and pressure on the respiration of three epipelagic species, *Diacria trispinosa*, *Cuvierina columnella*, and *Clio pyramidata*, and one deep-water species, *Limacina helicoides*. They found a fundamental difference in oxygen uptake between the two groups. The respiration of epipelagic species was directly influenced by temperature alone over their normal depth ranges; at greater depths, pressure caused increased respiration rates and ultimately death in

D. trispinosa and *C. columnella* that were subjected to more than 50 atmospheres (500 m). The bathypelagic species, *L. helicoides*, was only affected by temperatures above 10° C, and showed a constant oxygen-uptake rate at pressures between 50 and 150 atmospheres (500 to 1,500 m). These results are similar to the respiration patterns reported for planktonic crustaceans, which suggests that epipelagic species have lower metabolic demands at their lower range of temperature and depth, whereas bathypelagic species remain unaffected by both temperature and pressure over their normal depth range (Pearcy and Small, 1968; MacDonald et al., 1972). Therefore, epipelagic thecosomes may minimize energy losses to respiration by moving into colder, deeper water during the daytime, as suggested for other vertical migrators (McLaren, 1963).

The energy required by a species to support its respiration represents a major fraction of the energy obtained from food, but to accurately calculate an energy budget for any thecosome species, certain other data are required that presently are either entirely unknown or can only be roughly estimated. The filtering rates of thecosomes have not been determined. Ideally, these could be measured in laboratory experiments following procedures employed with other suspension-feeding zooplankton; practically, this technique is presently unrealistic, since these animals do not exhibit normal behavior in culture. Alternatively, filtration rates could be estimated from further *in situ* observations on the sizes of webs and the volumes of water filtered and on the frequency of web deployment. Ingestion rates would then depend on a knowledge of ingested and assimilated food types and on an estimate of the amount of nutritionally important foods available in natural conditions.

At present, a provisional energy budget is possible for only one species, *Cavolinia longirostris*, for which the most data are available on metabolism and feeding (Gilmer, 1974). In presenting these data and estimates of energy requirements (Table 11), several assumptions have been made: (1) that the metabolic rate is constant; (2) that the filtration rate equals one web produced per hour; (3) that the maximal volumetric size of the web equals 50 ml; and (4) that the animal is 100-percent efficient in removing particles from the water and in ingesting collected particles. Further, the carbon concentrations were determined from water collected in the vicinity of the animals used for the respiration measurements, but these concentrations represent only the smaller-size fraction (<150 μm) of the particles capable of being collected by this thecosome; the abundance of larger potential prey was not estimated. The results suggest that *C. longirostris* requires a minimal daily food intake equivalent to 7 to 16 percent of its dry body weight; these values lie within the range of daily ration requirements for filter-feeding copepods (Conover, 1978). Depending on the natural food avail-

TABLE II

Calculation of minimal energy requirements and assimilation efficiency of carbon for Cavolinia longirostris

Mean dry body weight, excluding shell	0.45 mg
O_2 consumption/individual of 0.45 mg dry weight	0.06 ml O_2/day
1 ml $O_2 \equiv 5$ g-cal	
Therefore, 0.06 ml $O_2 \equiv 0.3$ g-cal	
Food required for maintenance on a pure fat diet:	
9.45 cal/mg fat	
Therefore, $\frac{0.3 \text{ g-cal}}{9.45 \text{ g-cal/mg fat}}$	= 0.032 mg fat/indiv./day
or	
$\frac{0.032 \text{ mg fat}}{0.45 \text{ mg dry wt}} \times 100$	= 7.1% of dry body weight/day
Food required for maintenance on a pure carbohydrate diet:	
4.1 cal/mg carbohydrate	
Therefore, $\frac{0.3 \text{ g-cal}}{4.1 \text{ g-cal/mg } CH_2O}$	= 0.073 mg CH_2O/indiv./day
or	
$\frac{0.073 \text{ mg } CH_2O}{0.45 \text{ mg dry wt}} \times 100$	= 16.2% of dry body weight/day
Assumed filtration rate of 1 50-ml web/hour	= 1,200 ml/day
In situ particulate carbon (<150 μm)	82–125 μg C/l
Carbon available/individual:	
Minimal: 1.2 l filtered/day × 82 μg C/l	= 98 μg C/indiv./day
Maximal: 1.2 l filtered/day × 125 μg C/l	= 150 μg C/indiv./day
Assimilation efficiency of available carbon:	
Carbohydrate contains 40% carbon	
Therefore, at minimal C concentration:	
$\frac{98 \text{ μg C/indiv./day}}{0.4}$	= 245 μg CH_2O/indiv./day
and	
$\frac{73 \text{ μg } CH_2O\text{/indiv./day}}{245 \text{ μg } CH_2O\text{/indiv./day}} \times 100$	= 29% assimilation efficiency
and at maximal C concentration:	
$\frac{150 \text{ μg C/indiv./day}}{0.4}$	= 375 μg CH_2O/indiv./day
and	
$\frac{73 \text{ μg } CH_2O\text{/indiv./day}}{375 \text{ μg } CH_2O\text{/indiv./day}} \times 100$	= 19% assimilation efficiency

SOURCE: Gilmer, 1974; Gilmer and Harbison, 1986.

ability, the minimal daily ration of *Cavolinia* represents a daily intake of 245 to 375 μg of carbohydrate per individual. The assimilation efficiency of carbon is estimated to be between 19 and 29 percent, which is intermediate with the ranges determined for other marine herbivorous zooplankton (Gaudy, 1974; Conover, 1978). The assimilation results agree with qualitative observations that much of the fecal matter appears to be undigested and includes many intact diatoms and coccolithophores (Gilmer, pers. obs.). Although these calculations are based on a number of assumptions and must be considered as preliminary, they represent the first attempt to quantify feeding in euthecosomes based on the new knowledge of collection of food in mucous webs.

Similar approaches have been taken in the few studies on the metabolism and energetics of pseudothecosomes. Biggs (1977) determined the respiration and excretion rates of hand-captured *Corolla* (*spectabilis*?) and *Gleba cordata* (Table 12). In both species, oxygen consumption decreased with increased body size, and excretion rates were very low for all size categories. Gilmer (1974) also measured respiration rates of *Gleba cordata* and made estimates of minimal food requirements for this species. The mean dry body weight of his specimens was 80 mg, and oxygen consumption amounted to 1.23 ml O_2/day. Thus, *Gleba* requires 0.6 to 1.5 mg of food/individual/day, depending on whether the food consists of fat or carbohydrate, respectively (see Table 11 for a sample calculation); or, each individual requires at least 0.7 to 1.8 percent of its dry body weight/day for maintenance. These values are low compared with those of *Cavolinia longirostris* (Table 11), but they may reflect the larger size and slower swimming movements of the pseudothecosome.

A consideration of the trophic ecology of any group must also include a discussion of predators, and for euthecosomes these include a wide array of larger carnivorous zooplankton as well as fishes, marine mammals, and birds. Known zooplanktonic predators of shelled pteropods include chaetognaths (Lebour, 1932), heteropods (see Chapter 3), ctenophores and medusae (Fraser, 1970; Lalli and Gilmer, pers. obs.), siphonophores (Purcell, 1981), and species of *Peraclis*, a pseudothecosome (Richter, 1977, 1983). One pelagic cephalopod, *Argonauta boettgeri*, is reported to prey upon *Cavolinia tridentata* (Okutani, 1960). Some of the more intriguing prey-predator relationships are those that involve gymnosomatous pteropods as

TABLE 12

Respiration and ammonium-excretion rates of Corolla *sp. and* Gleba cordata

Species	Size (mg protein)	Number	Average respiration rate (ml O_2/mg protein/hour)	Average excretion rate (μg NH_4^+/mg protein/hour)
Corolla sp.	0.1–1.0	5	18 ± 2.9	—
	1.1–10	8	16 ± 3.0	0.2 ± 0.1
	10.1–100	3	11 ± 2.3	0.3 ± 0.2
Gleba cordata	1.1–10	1	16.5	0.5
	10.1–100	1	8.7	0.2

SOURCE: Biggs, 1977.

highly specialized predators of both euthecosomes and pseudothecosomes; these are discussed in detail in Chapter 5. At least one pseudothecosome, *Corolla*, is also attacked by *Oxycephalus*, a hyperiid amphipod known to prey on many other gelatinous zooplankton (Gilmer, pers. obs.; Harbison et al., 1977).

Limacina helicina was originally described by Martens (1675) and Phipps (1774) during voyages to Arctic waters, where the species was first recognized as a common inhabitant of whaling grounds and as a food item of baleen whales. The same species has been reported to be eaten occasionally by ringed seals (Dunbar, 1942; McLaren, 1958). Marine birds also prey on surface-dwelling thecosomes (Meisenheimer, 1905a; van der Spoel, 1967). Pteropods may even be captured and eaten by filter-feeding benthos; Arnaud (1973) found two *Limacina retroversa* in the buccal cavity of a stalked barnacle, *Lepas anatifera*.

Because of the commercial importance of many marine fishes, there are more records of euthecosomes being eaten by fish than by other types of predators. In the North Atlantic, fish predators of *Limacina* include herring and mackerel (Lebour, 1932; White, 1977) and the larvae of cod and redfish (Bainbridge and McKay, 1968). Hardy (1924) estimated that *L. retroversa* contributes about 2.2 percent of the yearly food supply of North Sea herring. Salmon in the North Pacific Ocean consume *L. helicina* (Ito, 1964; LeBrasseur, 1966; Takeuchi, 1972). Russell (1960) found four species of *Cavolinia* in the guts of yellowfin tuna and lancetfish, and the latter also will eat *Clio pyramidata* (Kubota and Uyeno, 1970). Juvenile dolphin fish consume tropical species of *Cavolinia* as well as other prey (Richter, 1983), and small myctophid fish may feed on pteropods as well as on shelled heteropods (Hartmann and Weikert, 1969). It is most likely that other fish species also eat thecosomes when available. None of the above associations appears to be specific, and the total biomass of pteropods eaten is usually small compared to other items in the fish diets. Nevertheless, there are two interesting consequences that may arise when fish ingest large quantities of thecosomes.

The first of these is a condition known as "black gut." It occurs in commercial pelagic fishes, such as herring, mackerel, and Pacific chum salmon, and may occur in groundfish such as sole and cod, when these fish encounter and feed on dense concentrations of *Limacina retroversa* or *L. helicina*. *Limacina* contains a dark pigment that leaches out into the gut and muscles of the fish, and an unpleasant odor is produced in association with this staining (Fraser, 1962; Sipos and Ackman, 1964; Lippa, 1965). Ackman et al. (1972) have determined that these pteropods acquire and accumulate dimethyl-β-propiothetin from certain phytoplankton species. This is passed on to fish feeding on the pteropods, with the result that the compound breaks down to dimethyl sulfide. It is this degradation product that produces the fetid odor associated with black gut. Neither the pigment change nor the accumulation of dimethyl-β-propiothetin or dimethyl sulfide is harmful to the fish, but obviously fishermen have difficulty in marketing discolored fish with a noxious odor. In some areas, the problem may occur frequently enough to warrant research directed either toward predicting a possible seasonal occurrence or toward ways in which fishermen themselves can determine whether commercial fish are associated with swarms of *Limacina*.

Limacina also has been implicated as a vector of dinoflagellate toxins causing the death of fish. In 1977, White reported on the cause of a herring kill in eastern Canadian waters. The stomach contents of affected fish contained *Limacina retroversa* almost exclusively. Evidence suggested that this pteropod had been feeding on the dinoflagellate *Gonyaulax excavata*, which was the dominant phytoplankton species in the area at the time. *Gonyaulax* is well known as a producer of a toxin causing paralytic shellfish poisoning. The toxin does not harm benthic mollusks that feed on the dinoflagellate, but the ingestion of affected shellfish can cause human fatalities. In the case of the Canadian herring kill, it appears quite likely that *Limacina* acted as a pelagic vector of the toxin, resulting in the death of fish that consumed large quantities of the pteropod. White (1981), however, has shown that the transmission of dinoflagellate toxins to fish can be effected by other herbivorous zooplankton, such as copepods, barnacle nauplii, cladocera, and tintinnids. The vector species, therefore, depends on the relative abundance of the various herbivorous zooplankton species and/or prey selection by the fish.

Finally, thecosomes affect the trophic ecology of an area through the production and abandonment of feeding webs (Gilmer, 1974) and the release of pseudofeces and fecal material. Drifting mucous webs and mucous strings of pseudofeces and uncompacted fecal material contain aggregations of small organic particles that may serve as feeding surfaces for small grazers such as copepods. They also provide a substrate and nutrients for the growth of bacteria. The fate of the large fecal pellets produced by *Corolla spectabilis* has been examined by Bruland and Silver (1981). The pellets are tightly coiled and dense, and measure about 0.2 mm in diameter by 6 mm in length. They sink rapidly, at rates of 440 to 1,800 m/day, and thus help speed the transport of biogenic materials to deep water.

Reproduction and Development

Considerable attention has been directed toward morphological and histological studies of the reproductive system of thecosomes. All of the species examined to date are protandrous hermaphrodites, maturing and functioning first as males, then as females. Fewer studies have been made of reproductive behavior and of embryonic and larval development, but the majority of thecosomes release free-floating egg masses from which veliger larvae hatch. There are several notable exceptions, however, of species in which a free-swimming larval stage is partially or completely suppressed.

Family Limacinidae

Reproduction in the genus *Limacina* has been studied by Meisenheimer (1905b), Hsiao (1939a, 1939b), Morton (1954b), and Lalli and Wells (1973). The following description is from the review by Lalli and Wells (1978). Five of the seven species of *Limacina* (*L. bulimoides*, *L. helicina*, *L. lesueuri*, *L. retroversa*, and *L. trochiformis*) share a common type of reproductive system, which is illustrated in Figure 31. In the mature male stage, sperm produced by the ovotestis are stored in the distended hermaphrodite

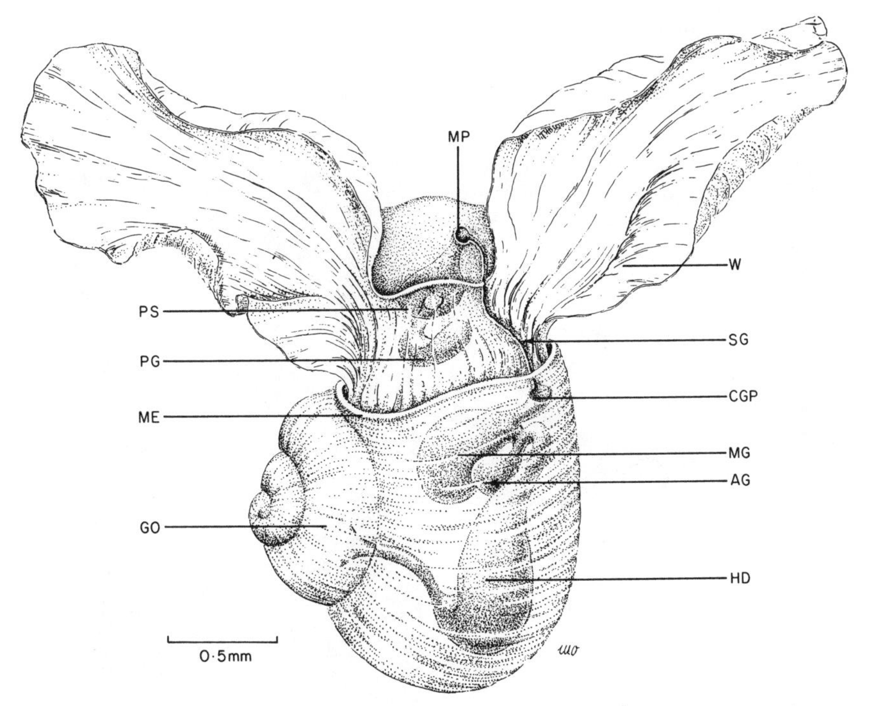

FIG. 31. A composite illustration of the typical male and female reproductive anatomy of the genus *Limacina* (from Lalli and Wells, 1978). AG, albumen gland; CGP, common genital pore; GO, area of gonad; HD, hermaphrodite duct; ME, mantle edge; MG, mucous gland; MP, male pore; PG, prostate gland; PS, penial sheath; SG, sperm groove; W, wing.

duct. This duct leads forward to two female accessory glands, which are incompletely developed in the male phase, and then opens into the mantle cavity by the common genital pore through which sperm, and later eggs, are discharged. A ciliated sperm groove runs forward on the external surface of the body to the male pore on the right side of the head. This pore marks the opening for the penis, which is normally retracted into a sheath. There are a bifurcated accessory copulatory organ and a prostate gland, both attached to the base of the penis.

Copulation in *Limacina helicina* involves a reciprocal exchange of spermatophores (Figs. 32a, b) (Gilmer, pers. obs.). Just prior to mating, two individuals face each other with the apices of their shells pointing in the same direction, but with their shell apertures opposite, so that the wings are paired right to right and left to left. During this time, the animals swim in a spiral pattern for as long as 1 minute while maintaining their same relative positions. Swimming ceases once the animals couple, and mated pairs appeared to be neutrally buoyant during underwater observations. The accessory copulatory organ of each individual is extended around the left wing of the partner and attached to the mate's operculum or other exposed body area. This apparently strengthens the connection between mates, pre-

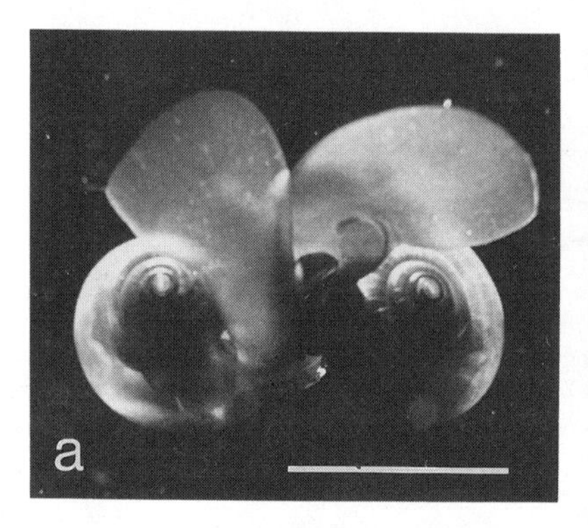

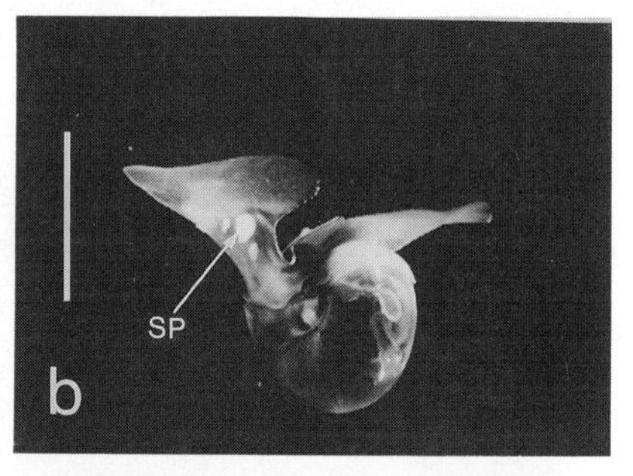

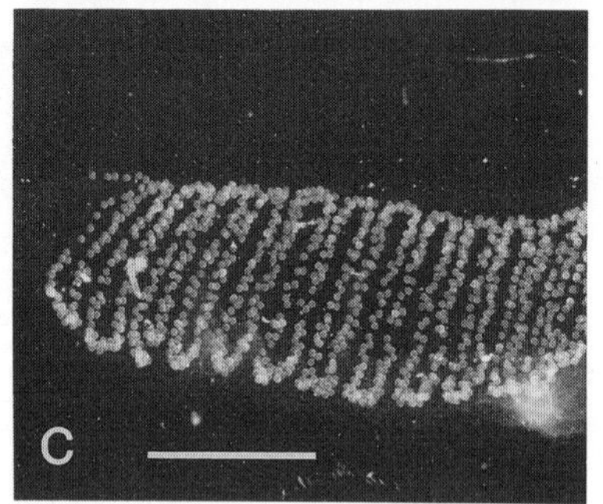

FIG. 32. Reproduction in *Limacina helicina: a,* mating individuals; *b,* an individual with a spermatophore (SP) attached to its wing; *c,* an egg mass. Scale lines represent 5 mm in *a* and *b,* 2 mm in *c.*

venting inadvertent separation. It is unclear how the penes are positioned during mating, but reciprocal fertilization occurs with the placement of a spermatophore on the underside of the right wing of each partner (Fig. 32b). Very few preserved specimens of *L. helicina* are found with an attached spermatophore, indicating that they may be easily dislodged or may only remain attached for very brief periods. Spermatophore exchange also has been documented in *Limacina inflata* (Lalli and Wells, 1978), but has not yet been described in other species of this genus. It is believed that fertilization is reciprocal between males or between individuals that are in the transition between male and female. It is not known whether self-fertilization is possible. Sperm received during copulation apparently are stored until the individual becomes a mature female.

In mature females of six species of *Limacina* (excluding *L. inflata*), the penis and prostate gland are partially or wholly resorbed, the gonad contains oocytes and mature eggs, and the female accessory glands are fully developed. The eggs are fertilized within the hermaphrodite duct, then pass through the albumen gland, where each is provided with an egg capsule. The encapsulated eggs are embedded in mucus from the female mucous gland and, with the exception of *L. helicoides*, are released as an egg mass through the common genital pore.

Egg masses have been described for *Limacina retroversa* by Lebour (1932), and for *L. helicina* (Fig. 32c) by Massy (1920) and Paranjape (1968). In both species, the spawn consists of transparent, free-floating egg ribbons from which free-swimming veligers hatch. Egg-mass sizes range from 2 to 4 mm length in *L. retroversa* to 10 to 12 mm for large *L. helicina*. Egg masses of *L. bulimoides*, *L. lesueuri*, and *L. trochiformis* remain undescribed, but all three species are known to have a free veliger stage and the spawn is presumed to be similar. Lalli and Wells (1978) conducted egg counts of *L. helicina* and *L. retroversa*, and the results are presented in Table 13. *L. retroversa*, from the most northerly extent of its range in southern Davis Strait, produces a mean of 260 eggs per female. Very large females of *L. helicina*, collected from eastern Canadian Arctic and Subarctic waters, lay an average of 6,000 eggs. In comparison, smaller females of the same species, but from warmer water off Vancouver Island in the North Pacific, produce an average of only 600 eggs (Paranjape, 1968). All of these counts were conducted over the course of a few to 20 days, and it is not known how often an individual will continue to produce eggs over longer time spans.

Lebour (1932) described veliger development in *Limacina retroversa* from the English Channel (Fig. 33). Unshelled veligers hatch as early as the second day after spawning, and they rapidly acquire a caplike shell and

TABLE 13

Numbers of egg masses and eggs deposited by females of Limacina helicina *and* L. retroversa

Datum	*L. helicina*	*L. retroversa*
Number of specimens	14	14
Size of females	6–10 mm (shell diameter)	1.8–2.0 mm (shell length)
Number of egg masses/female		
Range	1–18	1–7
Mean	7.7	2.9
Dimensions of egg masses		
Range	1.0 × 1.0–16.0 × 6.0 mm	1.0 × 0.5–9.0 × 0.5 mm
Mean	6.1 × 3.1 mm	4.3 × 0.5 mm
Number of eggs/female		
Range	524–10,051	83–650
Mean	5,936	260

SOURCE: Lalli and Wells, 1978.

bilobed velum. Older larvae, collected by plankton nets, have paired statocysts and eyes and an elongate foot with an operculum. As development proceeds, the shell acquires its characteristic spiral form, and two lappets appear at the sides of the foot; these will develop into the swimming wings. The velum gradually disappears as the wings begin to take over the locomotory role, and metamorphosis is complete when shell diameter measures about 0.4 mm. Larval development is similar in *L. helicina* (Mileikovsky, 1962; Paranjape, 1968), except that veligers (Fig. 34) hatch 2 (Paranjape, 1968) to 6 days (Lalli and Wells, 1978) after the eggs are spawned, when they already have a shell. The time required from hatching to metamorphosis remains unknown in both species, although Paranjape (1968) kept veligers alive for 30 days, and wing development had not yet begun at that time.

Limacina helicoides, the largest and only bathypelagic species of the genus, is also protandrous, and has a reproductive anatomy similar to the general *Limacina* pattern described above. This species differs from all others of the genus, however, in being ovoviviparous, a phenomenon initially reported by Bonnevie (1913) and later studied by Tesch (1946) and van der Spoel (1964). There is neither a free-floating egg mass nor a free-swimming veliger stage. Eggs are fertilized in the hermaphrodite duct, provided with an egg capsule in the albumen gland, and are retained in the large mucous gland, where they undergo development. Juveniles eventually break free of the mucous gland and are released into the female's mantle cavity and then into the sea. Although females may contain a large number of eggs (van der Spoel, 1964), the maximum number of young present in the mucous gland or mantle cavity at any one time is less than ten per female (Lalli and Wells,

1978). The newly released young are miniature adults with a brown shell of two whorls measuring approximately 5.0 mm in diameter, well-developed wings, and no trace of a velum (Lalli and Wells, 1978).

Reproduction and development in *Limacina inflata*, as described by Lalli and Wells (1973, 1978), differ in fundamental ways from those of all other species of the genus. Mature males (ca. 1.0 mm in shell diameter) lack a penis, but have a very large prostate gland that is involved in the production of spermatophores. The mechanism of spermatophore release and transfer remains unknown, but transfer occurs only between mature males or im-

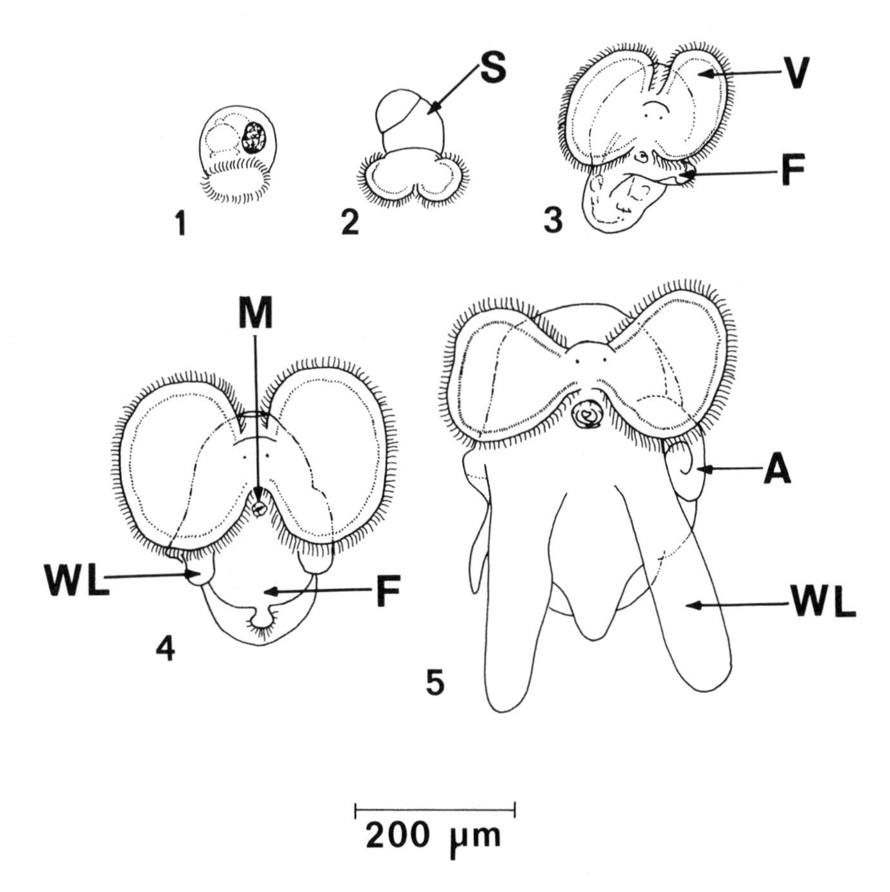

FIG. 33. Successive developmental stages of *Limacina retroversa*, from early larvae (1, 2) to a metamorphosing individual (5) (redrawn from Lebour, 1932). A, shell apex; F, foot; M, mouth; S, shell; V, velum; WL, wing lappet.

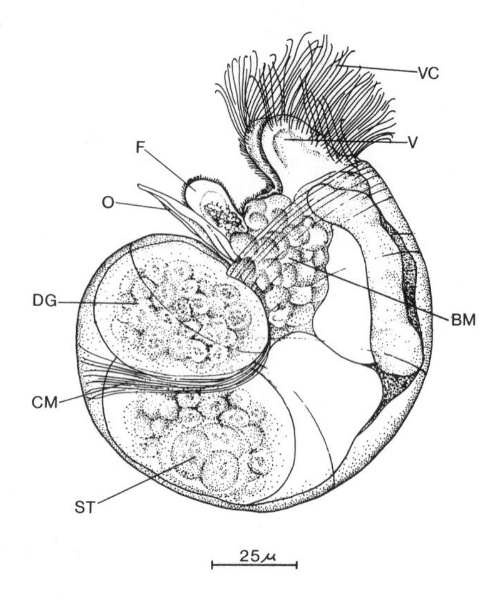

FIG. 34. A seven-day-old veliger larva of *Limacina helicina* (from Paranjape, 1968). BM, buccal mass; CM, columellar muscle; DG, digestive gland; F, footlobe; O, operculum; ST, stomach; V, velum; VC, velar cilia.

mature females, and the process seems to be reciprocal, with each partner acting as both a donor and a recipient of a single spermatophore. A spermatophore is attached by a long projection to the common genital pore of the mate, with the bulk of the spermatophore filling the mantle cavity (Fig. 35a). Sperm and prostatic fluid move from the spermatophore to the hermaphrodite duct, where they are stored until the individual becomes a mature female. Mature females of *L. inflata* (1.1–1.3 mm shell diameter) differ from those of other sibling species in lacking albumen and mucous glands and in exhibiting brood protection of the young. The eggs are fertilized within the hermaphrodite duct, then are released into the mantle cavity of the female, where they attach to the mantle lining. Embryonic development proceeds in the mantle cavity, with early shelled veligers becoming detached from the lining and swimming freely within this spacious chamber (Fig. 35b). Brooding females, which constitute approximately 5 percent of a total population, contain an average of 45 ± 20 young in various stages of development. Brood protection continues until the young are released as late veligers measuring 0.067 mm in shell diameter. It is not known how long growth and development continue in the free-swimming stage (Fig. 35c), but metamorphosis is complete at a shell size of about 0.4 mm diameter.

The spirally coiled limacinids grow by a gradual deposition of new shell at the apertural margin. Growth appears to be linear, and the animals increase in dimension throughout life as new shell whorls are added. Thus the size of the shell (either diameter or height) correlates with gonadal development and sexual stage (Hsiao, 1939a, 1939b; Morton, 1954b; Wells, 1976a; Lalli and Wells, 1978). Few long-term studies, however, have been made of natural populations of *Limacina*, and there is little reliable information on times of spawning, growth rates, and length of the life cycle in different localities.

L. retroversa, which inhabits boreal and temperate waters, shows evidence of seasonality in reproductive activity. This species spawns in the Gulf of Maine from April or May through September (Hsiao, 1939b). Young

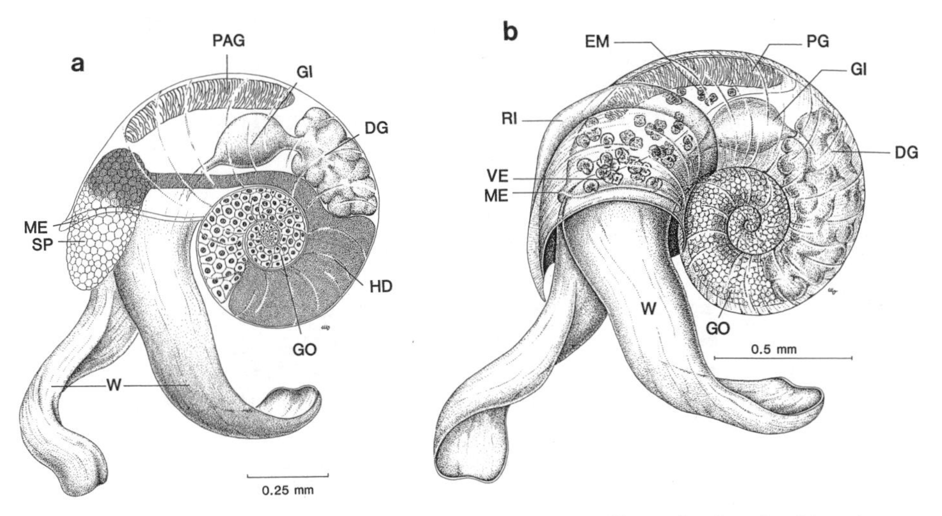

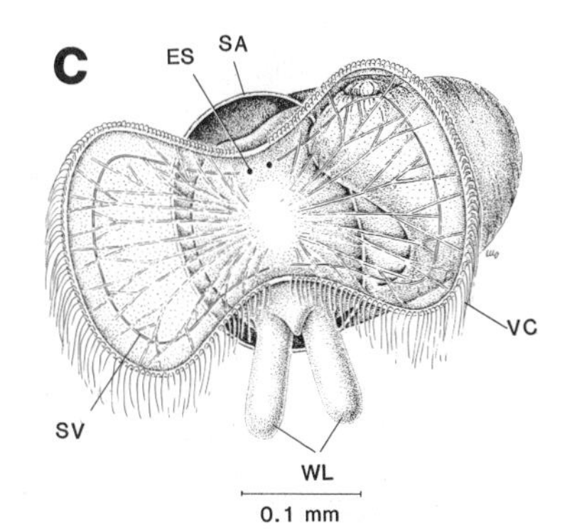

FIG. 35. Reproduction in *Limacina inflata*: *a*, the reproductive system of mature males or immature females, with a spermatophore partially filling the mantle cavity (from Lalli and Wells, 1978); *b*, a female brooding embryos in the mantle cavity (from Lalli and Wells, 1973); *c*, a free-swimming veliger just prior to completion of metamorphosis (from Lalli and Wells, 1973). DG, digestive gland; EM, early embryonic stages; ES, eyespot; GI, gizzard; GO, developing oocytes in gonad; HD, spermatophore fluid in hermaphrodite duct; ME, mantle edge; PAG, pallial gland; RI, rib of shell; SA, inflated aperture of shell; SP, attached spermatophore; SV, subvelar ridge; VC, preoral cilia of velum; VE, shelled veligers; W, wings; WL, wing lappets.

produced in the spring and summer undergo faster growth and show sexual development at smaller sizes than those spawned in the fall (Hsiao, 1939b; Redfield, 1939), probably as a result of warmer temperatures and increased food. Specimens measuring 0.6 mm in shell diameter in early December grow to a size of about 1.5 mm by early June, a growth rate of 0.15 mm/month; young spawned in the spring grow from 0.3 mm in mid-April to 1.5 mm by early July, or 0.4 mm/month. The maximum size attained by females is about 2.7 mm shell diameter, and the life cycle is thought to be less than 1 year. Northeast of this area, along the southern coast of Nova Scotia, peak periods of spawning of *L. retroversa* coincide with both spring and autumn phytoplankton blooms (Conover and Lalli, 1972). The same species, however, is reported to release eggs during most of the year in the English Channel, with a peak in the summer (Lebour, 1932).

The cold-water species, *L. helicina*, spawns primarily during May through July in the central Arctic Ocean (Kobayashi, 1974). In populations of this area, females seem to be sexually mature at a size of 0.8 mm shell diameter, and they attain a maximal size of 3.7 mm. The life cycle is approximated at 1.5 to 2 years. There is a curious difference in the maximal size of *L. helicina* collected from eastern Canadian Arctic and Subarctic waters; in these regions, the average size of females is 8 to 9 mm shell diameter, with a maximum of 11 mm (Lalli and Wells, 1978). This suggests a very different growth rate and length of life cycle in these populations, which possibly is related to differing environmental factors. There is less information on populations of *L. helicina* in the Antarctic Ocean, but Massy (1920) found the largest specimens (to 6 mm diameter) in the summer (December to February), and spawning was observed in January.

The warm-water species, *Limacina inflata*, *L. trochiformis*, and *L. bulimoides*, are known to reproduce year-round off Barbados in the Caribbean Sea (Wells, 1976b). These species do not have discrete cohorts, and therefore changes in size-frequency curves with time cannot be used to calculate growth. Instead, Wells (1976a) employed a paper probability method to estimate growth rates and length of life cycles (Table 14). Depending on the species, metamorphosis was estimated to occur 1.5 to 3 months after hatching, and the attainment of sexual maturity within 7 to 8 months. Life cycles thus could be completed in less than 1 year. This is supported by observations on *L. inflata* that become trapped in cold-core Gulf Stream rings; Wormuth (1985) reported that such populations may increase within 5 months by an order of magnitude relative to surrounding Sargasso Sea densities. It is not clear, however, whether this is a normal response to more favorable conditions or whether the temporary increase in numbers is related to the abrupt release of brooded young, to changes in predation pressure, or to a combination of factors. The growth rates of *L. bulimoides* and

TABLE 14

Growth rates and times required to attain maturational stages in tropical species of Limacina

Species	Averaged growth rate (mm/mo)	Time required to reach maturational stage (in months)		
		Veligers to metamorphosis	Veligers to sexual maturity	Veligers to maximum size
L. bulimoides	0.15	2	7	9
L. inflata	0.12	3	8	8
L. trochiformis	0.10	1.5	7	9

SOURCE: Wells, 1976a.

L. trochiformis given in Table 14 are based on increases in shell length and thus are not strictly comparable to the rates obtained by Hsiao (1939b) and Redfield (1939) for *L. retroversa*, which were based on shell diameter. The lengths of life cycles in various species of limacinids, however, are consistent with data obtained for other zooplankton, with tropical species generally having shorter life cycles than species living in cold waters.

Family Cavoliniidae

Reproductive anatomy and histology in this family have been studied by Meisenheimer (1905b), Bonnevie (1916), van der Spoel (1967, 1973a), and references cited therein. Apart from minor species differences, the reproductive tract is basically the same as described above for limacinids. The animals are protandric hermaphrodites, first maturing as males and then passing through a hermaphroditic phase to the mature female stage. In females, spermatocytes may still be present in the gonad, but the male organs are reduced or absent and accessory female glands are fully developed. The ovotestis, hermaphroditic duct, female accessory glands, and seminal receptacle are located in the trunk with a single or, rarely, double genital pore opening to the mantle cavity on the right side; the branched copulatory organ and prostate gland are situated in the head, with the male pore opening near the right tentacle. These separate regions of the reproductive tract are connected by a ciliated sperm groove, which is external and open in all species except *Cavolinia longirostris*, in which it has become a closed tube. A notable difference from limacinids is that the penis of some cavoliniids is armed with chitinous stylets.

Copulation in cavoliniids has been observed in the field between individuals of *Creseis acicula*, *C. virgula*, *Diacria trispinosa*, and *Cavolinia uncinata* (Gilmer, pers. obs.) and is described here for the first time. Copulation in *Creseis* appears to be simultaneous and reciprocal between males or hermaphrodites, with both partners having enlarged copulatory organs. The shell lengths of mated individuals are given in Table 15. The generally small sizes of both partners, compared with the maximum size of the species, also confirms that copulation can take place early in the life cycle before the animals become mature females. Mating takes at least 2 minutes, but longer *in situ* observations have not been made, and it is not known how pairing is initiated or how long it lasts. In both species of *Creseis*, mated pairs do not swim, and their wings are locked together in an alternating pattern that places the genital openings on the dorsal region of the neck in close contact. The penes connect externally, indicating that sperm are transferred into the region that is temporarily expanded beyond the genital opening and that presumably contains a portion of the seminal receptacle. Mated pairs of *Cavolinia uncinata* also cease swimming during mating, but in this species individuals pair back to back, with the dorsal neck regions and genital openings in close proximity (Fig. 36).

In contrast to the close physical coupling between mates of *Creseis* and *Cavolinia*, males of *Diacria* have complex and massive copulatory organs, which separate paired animals by about 1 cm during mating (Fig. 37a). Copulation in this species involves the reciprocal exchange of a spermatophore, which is attached to the ventral shell surface near the aperture (Figs. 37b, c). A long stalk extends from the main body of the spermatophore into the mantle cavity, and sperm presumably move to the genital opening along this route. The attachment of the spermatophores appears to be temporary, since they are easily detached and in fact have never been reported from preserved material. Although this is the only cavoliniid presently known to produce spermatophores, it is likely that they also occur in other species.

In addition to cross-fertilization, it has been suggested that a different mode of reproduction occurs in *Clio pyramidata* and *C. polita*. Van der

TABLE 15
Size relationships of mating cavoliniids

Species	Number (pairs)	Sizes of mating individuals (shell length in mm)	Maximum size of species (shell length in mm)
Creseis acicula	2	11.5 and 14.0	33
		12.5 and 13.5	
C. virgula conica	5	4.6 and 4.8	10+
		4.6 and 5.8	
		5.4 and 5.6	
		5.4 and 5.6	
		8.3 and 9.2	

SOURCE: Gilmer, unpubl. data.

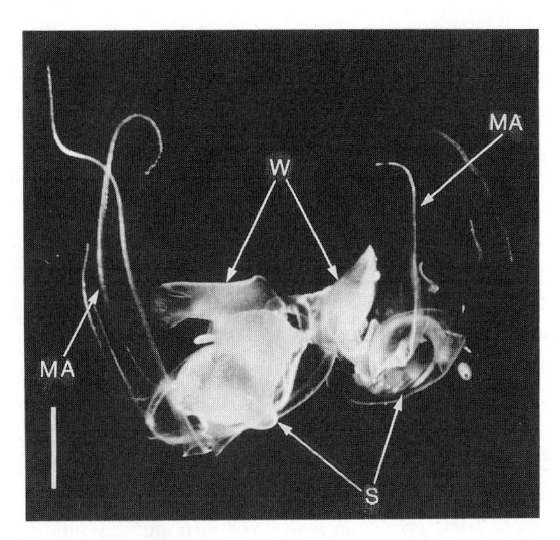

FIG. 36. The mating posture of *Cavolinia uncinata*. MA, mantle appendages; S, shells; W, wings. The scale line represents 5 mm.

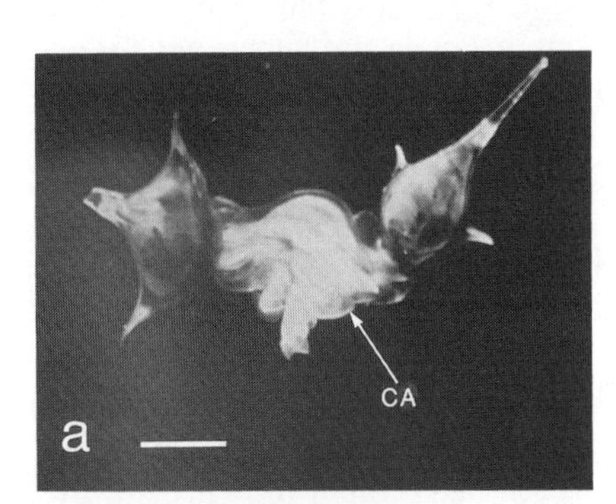

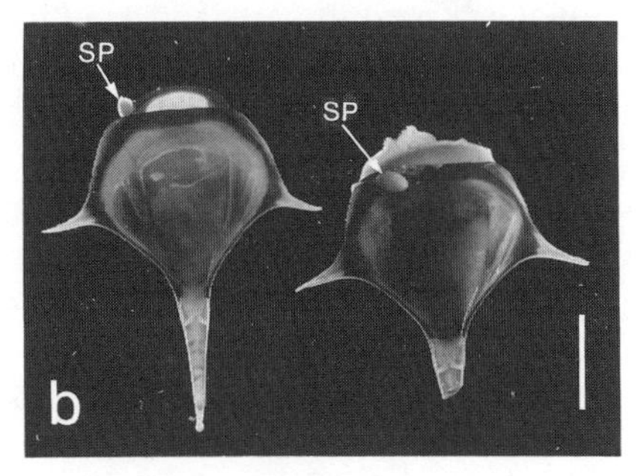

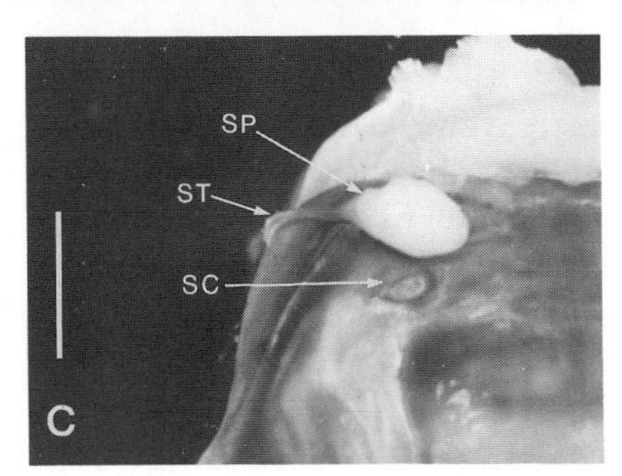

FIG. 37. Copulation and spermatophore transfer in *Diacria trispinosa*: *a*, mating individuals with entwined copulatory structures (CA); *b*, individuals after mating, each with an attached spermatophore (SP); note the shell breakage in one specimen, a common occurrence in this and other species; *c*, enlargement of the spermatophore (SP) and its stalk (ST), and a scar (SC) from the earlier attachment of a spermatophore. Scale lines represent 4 mm in *a* and *b*, 2 mm in *c*.

Spoel (1962, 1967, 1973b, 1979) described "aberrant" individuals of these species that were found in normal, full-grown shells but that differed markedly in body morphology from normal specimens. He suggested that these individuals arose from strobilation, or asexual transverse splitting of a single normal female, and represented resting stages that were capable of developing into normal adults during favorable environmental conditions. More recently, Pafort-van Iersel and van der Spoel (1986) reinvestigated this process in studies of living normal and abnormal specimens obtained in plankton tows. Schizogamy had occurred in most (25 of 27) of the abnormal specimens that were collected, and was induced in one normal individual exposed to the hormone thyroxine. The division process apparently begins with a loss of pigmentation and a change in the shape of the posterior region of an adult. Subsequently, the anterior body region (complete with mouthparts, gut, and wings but lacking reproductive organs) detaches from this posterior area and swims out of the shell; this shell-less, asexual animal is referred to as a "primary specimen" in van der Spoel's original articles. The remaining posterior region is referred to as a "metamorphosed specimen" or "aberrant" individual; it develops two neomorphic wings before schizogamy and also eventually swims out of the original shell; this individual lacks mouthparts and a gut but contains male and female gonads and accessory reproductive structures and is presumed to be capable of self-fertilization. The authors believe that schizogamy occurs under unfavorable environmental conditions, such as those formed at the boundaries of different water masses. They assume that the metamorphosed individuals will, on encountering conducive conditions, reproduce more rapidly by self-fertilization than by the usual cross-fertilization that requires location of a mate.

Several aspects of this hypothesis have not been resolved, and remain puzzling. For one, the asexual, shell-less individuals resulting from the anterior half of the original pteropod have not been found in plankton tows or observed *in situ*. Shelled "metamorphosed" or "aberrant" individuals have been found in both live and preserved collections from plankton tows (Fig. 38), but these too have not been seen in the field by divers, nor have shell-less aberrants been observed. These facts suggest either that these individuals are very rare in most localities or that they have a very brief period of existence. Metamorphosed individuals, however, were reported to remain alive for at least 2 weeks in laboratory conditions (Pafort-van Iersel and van der Spoel, 1986). It is well known that morphological artifacts can result in individuals that have been damaged by contact with, or compaction in, plankton nets and in specimens preserved rapidly by chemicals (Fig. 39). In fact, Gilmer (1986) has produced animals that closely resemble aberrant

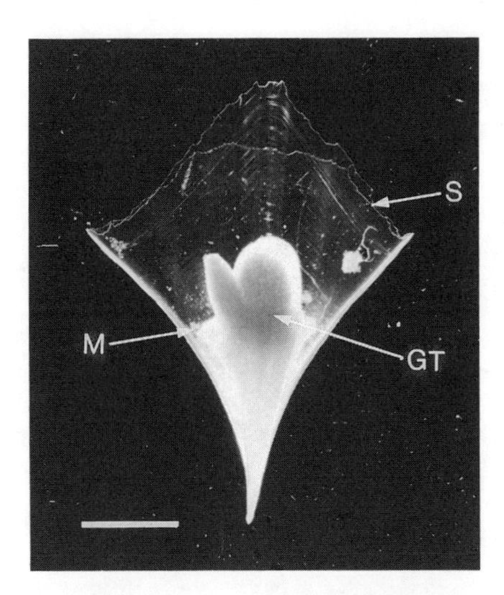

FIG. 38. A specimen of *Clio pyramidata* after net collection. The top portion of the individual ("primary specimen") has been severed and has left the shell; the remaining portion ("metamorphosed" or "aberrant" individual) consists of gonad and mantle tissue and may migrate out of the shell. Pafort-van Iersel and van der Spoel (1986) suggest that this is a natural process of asexual reproduction that occurs in up to 6 percent of a population, especially near the borders of the species' range. GT, gonad tissue; M, mantle; S, shell. The scale line represents 3 mm.

individuals by adding formaldehyde to living animals of a certain size class; their histological structure, however, has not been compared with that of aberrants to ascertain whether the resemblance is more than superficial. Finally, self-fertilization and reproductive capacity in aberrants has not been demonstrated, except by inference from histological studies, and it has not yet been demonstrated that self-fertilization is possible in normal specimens of these or other species of pteropods.

The majority of cavoliniids inhabit warm seas where seasonal changes in the environment are slight. Under such circumstances, reproduction is continuous, and the eggs and veliger larvae are present in the water column throughout the year (van der Spoel, 1973a; Wells, 1976b). Most cavoliniid species lay free-floating, mucoid egg masses or egg strings (Fig. 40a) that are released rapidly from the genital opening while the spawning animal is quiescent in the water. The number of eggs spawned probably varies with the species, age, and size of spawning individuals and geographic location, but data are too few for an evaluation. Observations of animals from the Gulf Stream show that *Cavolinia longirostris* generally spawns two to three egg masses over a 12-hour period in response to a slight rise in temperature after collection (Gilmer, pers. obs.). The egg masses contain between 130 and 200 eggs each, and the eggs measure approximately 100 μm in diameter at the time of release. Egg masses of *C. uncinata*, *C. tridentata*, and *Creseis acicula* from the same region contain similar numbers and sizes of eggs.

The subsequent development of embryos and larvae has been described for only a few cavoliniid species (e.g. Fol, 1875; van der Spoel, 1967). Hatching generally occurs within several days after spawning, and early ve-

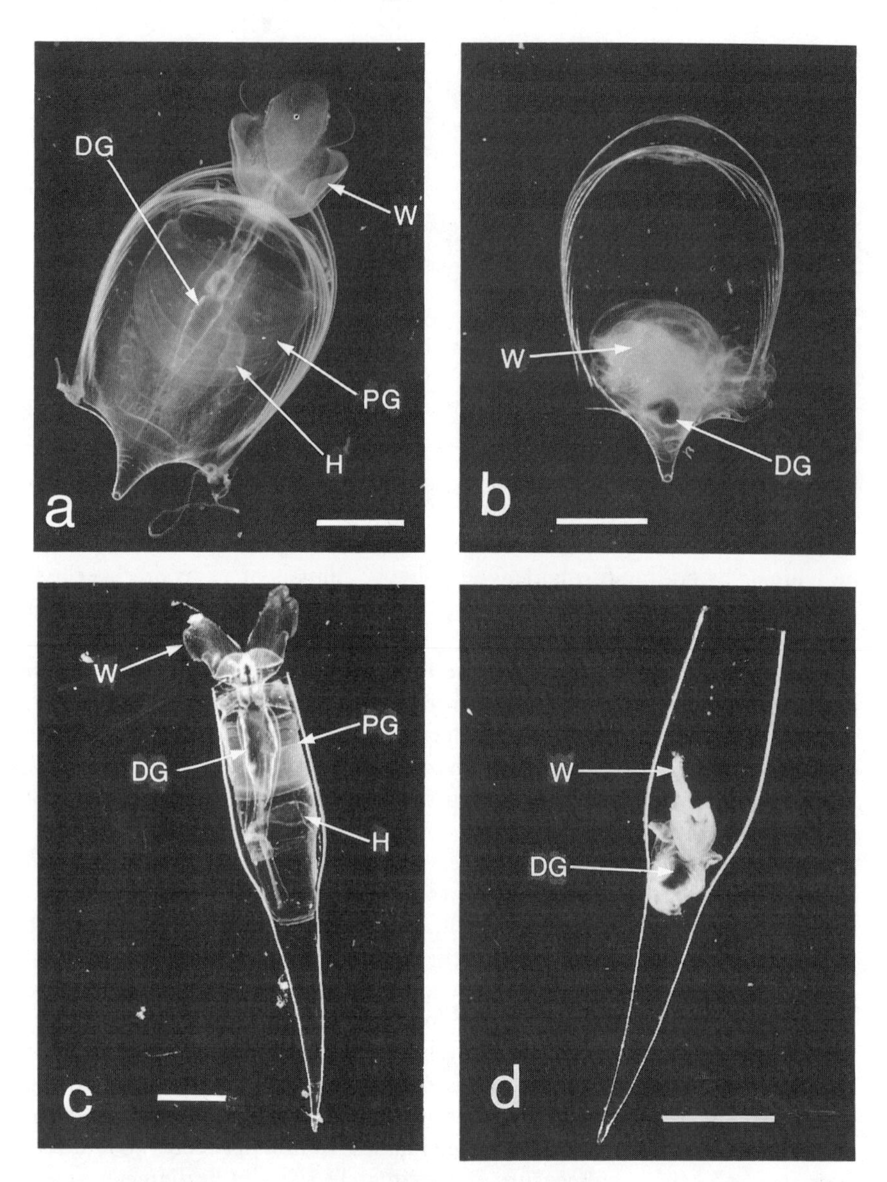

FIG. 39. Morphological changes resulting from chemical preservation: *a,* a living specimen of *Cavolinia gibbosa*; *b*, the same specimen as in *a* after preservation in weak formaldehyde; *c*, a young living specimen of *Cuvierina columnella* prior to the separation of the caudal shell (from Gilmer, 1986); *d,* the same specimen as in *c* after preservation in weak formaldehyde (from Gilmer, 1986). DG, digestive gland; H, heart; PG, pallial gland; W, wing. Scale lines represent 3 mm.

ligers are provided with an uncoiled shell, bilobed velum, small foot, and prominent paired statocysts. Subsequent development is similar to that of the limacinids: the shell increases in length and the velum regresses as the wings develop from the sides of the foot.

Little attention has been given to time-series studies of development, but Gilmer (pers. obs.) has followed the young of *Cavolinia uncinata* from the time of spawning through metamorphosis. Embryos (Fig. 40a) measured 400 μm in length and had a well-developed velum 68 hours from the time the eggs were spawned. At 26° C, most larvae hatched and were free-swimming by 92 hours (Fig. 40b). One day later, the velum started to regress as the wings began to be used for swimming. Most larvae had lost

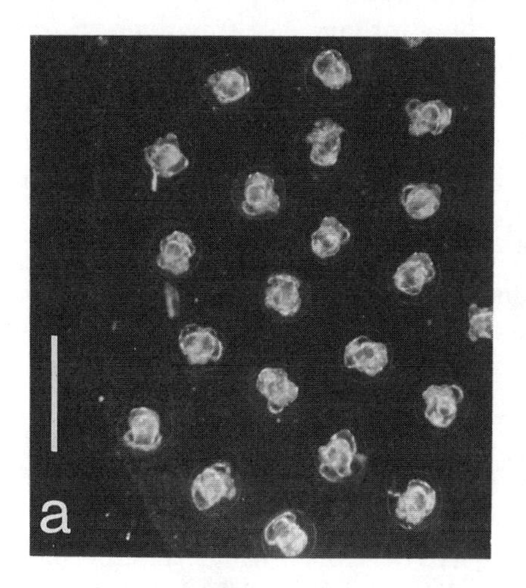

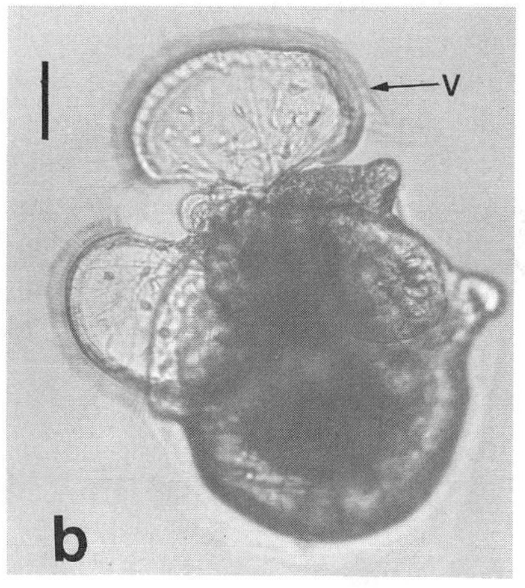

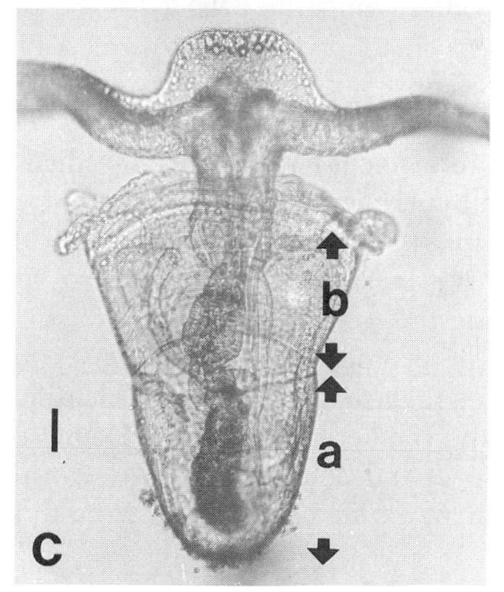

FIG. 40. Embryonic and larval development in *Cavolinia uncinata*: *a,* embryos within an egg mass at 50 hours after spawning; *b,* a free-swimming veliger at 3.5 days after spawning; *c,* a juvenile 10.5 days after deposition of eggs (area a is veliger shell; area b is shell deposited after metamorphosis). V, velum. Scale lines represent 500 μm in *a,* 50 μm in *b* and *c.*

their velum by day 6, and the shells of newly metamorphosed juveniles measured 600 μm in length on day 10 (Fig. 40c). Between days 6 and 15, the larval shells increased in length by 35±3 μm to 290 μm and diameter increased by about 10 μm to 240 μm. The young of *C. longirostris* followed a similar course of development and showed protoconch length increases of 40±10 μm, and of apertural expansion in width of 10±6 μm; at the end of 13 days, several large individuals had attained shell dimensions of 380 μm (length) and 360 μm (aperture width) (n = 16). In this species, the majority of shell growth occurred between the sixth and ninth days after spawn release.

The time required for metamorphosis from the veliger to the juvenile stage appears to vary. Fol (1875) reported that metamorphosis in the laboratory occurred within 5 days after hatching for larvae of *Cavolinia tridentata*, and within 6 days for those of *Hyalocylix striata*. Wells (1976a), using an indirect method, estimated that the larvae of *Creseis virgula* in Caribbean populations required 0.6 to 1.5 months to attain metamorphosis. Jägersten (1972), however, observed embryos in the egg masses of an unidentified species of *Cavolinia*, and reported that both a velum and wings were visible by the second day after the first cleavage. Hatching occurred 5 days after spawn release, and the velum disappeared within 5 minutes after the young emerged from the eggs. This short veliger stage approaches direct development.

Only three deep-water species of cavoliniids have been reported as being ovoviviparous or exhibiting brood protection of the young. Tesch (1946, 1948) found encapsulated larvae adhering to the mantle lining of females of *Clio chaptali*. Apparently there is no free-swimming veliger stage in this species, since some of the young already had developed wings. Tesch (1948) also reported finding numerous oval bodies, which he believed to be embryos, in the mantle cavity of a single specimen of *Clio campylura*, a rare and poorly known thecosome. More recently, van der Spoel (1970) found embryos developing within an accessory sexual gland of *Clio recurva*. The mechanisms of embryo retention and nourishment have not been studied in detail in any of these mesopelagic or bathypelagic species.

The shells of young thecosomes consist of two distinctive parts: the rounded or ovoid embryonic shell, which is laid down before hatching, and the pyramidal shell, which forms during the free-swimming veliger stage. Together, these shell areas constitute the protoconch. In many cases, the protoconch morphology is distinctive enough to permit species identifications (see van der Spoel, 1967; Troost and van der Spoel, 1972; Haagensen, 1976; Bé and Gilmer, 1977; van der Spoel and Boltovskoy, 1981), but the characteristic features of the protoconch may differ considerably from those of the adult shell (compare Figs. 17b and 39c). After metamorphosis, an adult shell (or teloconch) is secreted on top of the protoconch, and in many species the protoconch will be shed naturally or broken off from the permanent adult shell. The transition is accomplished by the detachment of the columellar muscle from the base of the protoconch, retraction of the body into the adult shell, and reattachment of the columellar muscle to the base of the teloconch (Fol, 1875; van der Spoel, 1967; Bé et al., 1972). A new shell layer (the closing membrane or caudal septum) is secreted between the protoconch and teloconch, either before or after the protoconch is lost.

Considerable attention has been given to the development of juvenile cavoliniids, primarily because very small individuals often are found in adult-sized shells in preserved collections (Bonnevie, 1913; Tesch, 1946; van der Spoel, 1967). This has posed the question of how the shell is formed by an animal that appears to occupy only a small volume of the internal shell space, and has led to various interpretations and hypotheses concerning growth. Van der Spoel has published numerous papers on these points (e.g. 1967, 1974 and references cited therein) and has concluded that cavoliniids pass through morphologically unusual stages in the course of development. These stages are referred to as "minute" and "skinny" forms and have been described for all cavoliniid genera. The "minute" and "skinny" developmental stages are distinguished primarily by the presence of a small *juvenile* animal that does not appear capable of occupying the entire volume of its large *adult-sized* shell. However, live juveniles normally do fill the shell, which, although it may be of adult dimensions, is very thin. Gilmer (1986) has shown that "minute" and "skinny" stages can be produced by contraction during preservation. Figures 39c and 39d show the differences that result when a thecosome is subjected to a rapidly applied preservative without the prior, slow, addition of chemical relaxants. Since rapid preservation with formalin is commonly applied to routine oceanographic collections, and since juveniles contract more strongly that adults, it is not surprising to find van der Spoel's "developmental stages" only in preserved material. Further, although juveniles are found in shells with length and width dimensions similar to those of adults, the juvenile shells are thinner and the ratio of body weight to shell weight is the same in both juveniles and adults (Fig. 41). There does not seem to be any reason to suppose that shell deposition in pteropods differs from that of other mollusks; the mantle appears to be the primary organ responsible for the production of the shell.

The shell microstructure of cavoliniids is fundamentally different from that of the limacinids (Fig. 42), and their growth patterns are also strikingly unlike those of the spirally coiled thecosomes. The adult shells of cavoliniids have an outer prismatic layer and a thicker inner shell layer composed of

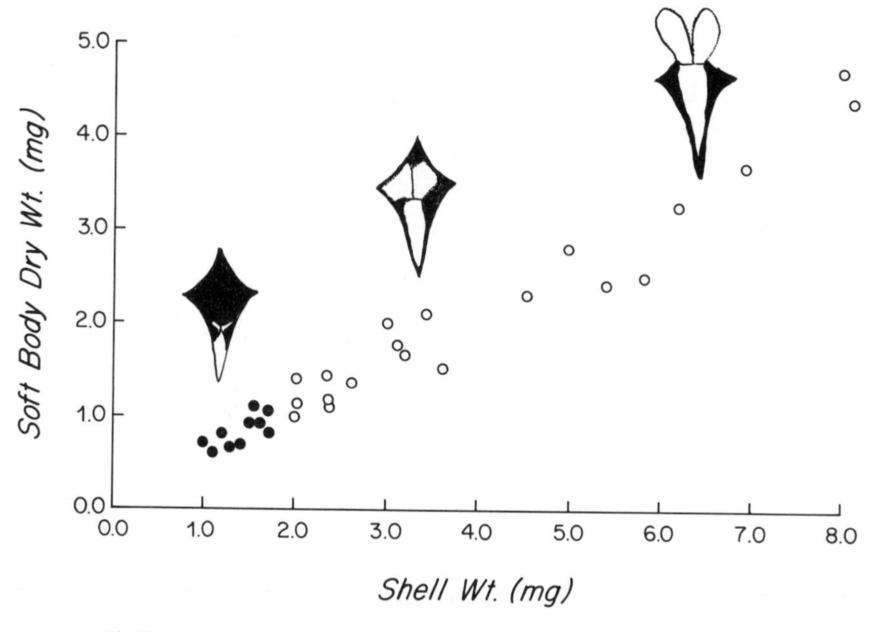

FIG. 41. Shell weight versus tissue body weight (dry) for adult *Clio pyramidata* of similar size dimensions (length, width) collected in the western North Atlantic (modified from Gilmer, 1986). Shell-wall thickness increases with age and accounts for the differences in shell weight. The drawings indicate the relative contraction of soft parts with increasing age when subjected to preservation. ●, preserved animal showing severe contraction, i.e. animal occupied less than 1/3 of shell volume; ○, preserved animal showing little contraction.

DNA-like, helically arranged, aragonite rods (Fig. 42b); this is referred to as helical (Bé et al., 1972) or spiral (Rampal, 1977) structure. The growth patterns of these shells are such that it is not possible to use simple length/width measurements to establish age or developmental state. Growth in members of the Family Cavoliniidae occurs in two phases as established by Bé et al. (1972) for *Cuvierina columnella*: first, the shell attains its maximum length and final shape (Phase 1, Fig. 43); second, the entire shell thickens inwardly and apparently continues to do so throughout the remainder of an animal's life (Phase 2, Fig. 43). New shell is added to the inner surface in the form of aragonite rods coiled in a helix perpendicular to the shell surface (Fig. 42b). Only slight increases in shell length occur during growth phase 2, which is due to the deposition of a secondary prismatic layer at the shell aperture (Fig. 43); this addition accounts for only a small fraction of new shell material compared with that added in the thickening helical layer.

This growth pattern means that juvenile cavoliniids have shells that are of adult dimensions but with very thin, transparent walls. Shells of older animals become opaque as they thicken, but there is little change in the external shell measurements. (All sizes of empty, sedimented shells eventually turn chalky and opaque because of chemical changes.) For example, a typical *Cavolinia tridentata* reaches its maximum shell measurements when the shell dry weight is about 16 to 17 mg and the body dry weight is about 3 mg. In older individuals, which have the same size of shell, the shell dry weight may be as much as 85 mg. Thus the shell weight may increase fivefold after the final shell height and width have been attained (Gilmer, 1974). Therefore, shell size is not a reliable indicator of age in cavoliniids, as it is in limacinids. Weight, rather than shell dimensions, should be used to estimate the age of individual cavoliniids.

Several approaches to growth rate and age determination in cavoliniids have been employed. Some workers have tried to detect growth patterns and age by the examination of fine features on the external shell surface. This approach assumes that the growth rate is not uniform when the shell is formed in the early life of a cavoliniid, but that it varies with temperature, food supply, or other factors. Thus periods of shell formation may record daily environmental changes, such as temperature differences experienced during diel vertical migration, or seasonal environmental events, such as differences in productivity. Van der Spoel (1969), for example, measured surface features in transparent shell segments of *Clio pyramidata* by passing them through a microscope field. Since there are no laboratory measurements of shell growth much beyond the metamorphosis of any species, his interpretation of what constitutes a day's growth increment is very subjective and based on the assumption that temperature differences encountered daily could be read from the shell surface. He estimated that *Clio pyramidata lanceolata* may reach 15 to 18 mm in length in 52 days, while a smaller form, *C. pyramidata convexa*, could attain a length of 6 to 7 mm in 26 days. It should be stressed, however, that the type of growth exhibited by cavoliniids makes age determination by counting increments on the shell surface difficult to interpret and, at best, is only a crude estimate of growth during the animal's early life.

Van der Spoel's growth estimates in *Clio* do show some agreement with recent isotopic O^{18}/O^{16} work that also attempts to detect small temperature changes recorded in the chemistry of the shell. Jasper and Deuser (pers. comm.) approximated the calcification rates of thecosome shells collected from deep sediment traps in the Sargasso Sea that were sampled every 60 days over a complete annual cycle. Nonmigratory species, such as *Creseis acicula* and *C. virgula conica*, best reflected the seasonal trends in surface

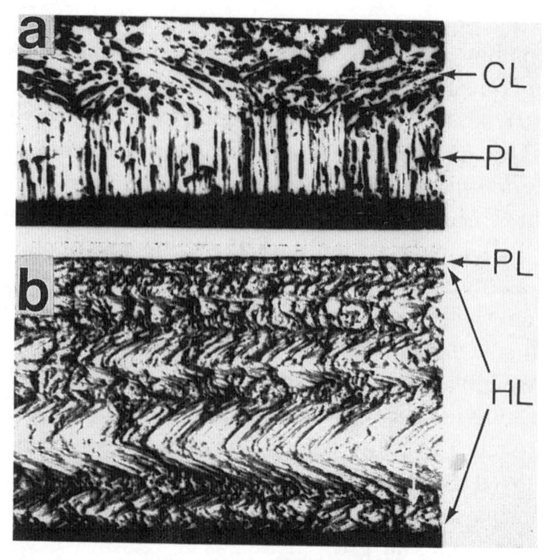

FIG. 42. Microstructure of thecosome shells (from Bé and Gilmer, 1977): *a,* the crossed-lamellar/prismatic microstructure characteristic of the spirally coiled euthecosome Family Limacinidae and the pseudothecosome Family Peraclididae (1450×); *b,* the helical microstructure of cavoliniids, overlain by a prismatic layer of only a few microns in thickness (550×). CL, crossed-lamellar layer; HL, helical layer; PL, prismatic layer.

temperature present during each sampling period; in these species, the calcification of the juvenile caudal shell probably takes place within 60 days, which is within the range calculated by Wells (1976a). Four migratory species (*Styliola subula, Clio pyramidata, Cuvierina columnella,* and *Diacria trispinosa*) did not show good correlations with seasonal changes, which presumably reflects their daily migration into deeper cold water. None of the shells analyzed in the study were of adult length/width dimensions. Grossman et al. (1986) used both O^{18}/O^{16} and C^{13}/C^{12} ratios to compare seasonality effects and depth distributions in several species of thecosomes. They suggested that four adult *Cuvierina columnella* of intermediate shell weight were 3 to 5 months of age, since they contained isotope ratios that reflected seasonal isotope levels characteristic of the period prior to their collection. They also found that the caudal portion of the shell in both *C. columnella* and *Diacria trispinosa* was slightly enriched with C^{13} relative to the remainder of the shell. This may indicate either that the animals migrate out of the euphotic zone to deeper water as they age or that increases in metabolism and growth rate occur during the formation of the adult shell; both factors could reduce the amount of "heavy" carbon incorporated into the shell.

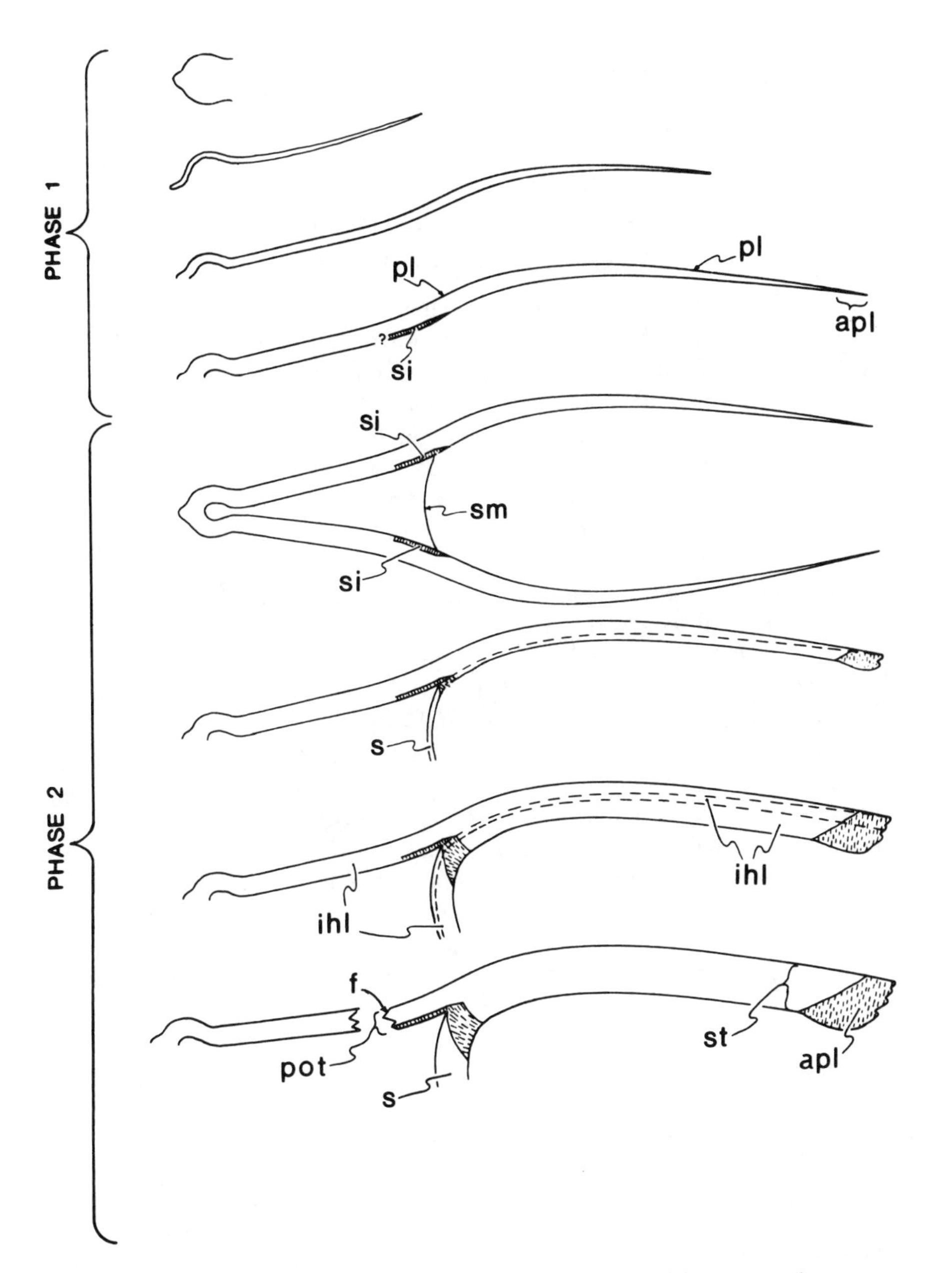

FIG. 43. Growth of the shell of *Cuvierina columnella* (modified from Bé et al., 1972). apl, apertural prismatic layer; f, natural fracture of caudal portion of shell; ihl, inner helical layer; pl, prismatic layer; pot, phase 1 thickness; s, caudal septum; si, septal insertion; sm, septal membrane; st, shell thickness.

Bé et al. (1972) devised a mathematical expression that they employed to approximate the age and growth rate of *Cuvierina* from comparisons of the relative thickness of the phase 1 shell (Fig. 43) with the total shell thickness. The formula, which assumes a uniform growth rate, is:

$$APO + APO \times ST/POT = MA,$$

where APO is the age of the phase 1 shell in days (estimated), ST is the total shell-wall thickness measured just posterior to the apertural layer, POT is the thickness of the phase 1 prismatic layer near the septal growth area (9–10 μm measured), and MA is the minimum age of the shell in days. Their calculations indicated that individuals having a shell thickness of 25 μm would be between 90 and 180 days old, and older specimens with shells 67 μm thick would be 200 to 400 days old. Bé et al. considered that the higher estimates were more realistic; if this is so, then the shell thickness of this species increases by 0.16 μm each day.

Few long-term oceanographic studies have been made of cavoliniid populations in one geographic area in order to determine spawning times, growth rates in nature, and lengths of life cycles. In fact, the value of this approach is questionable when applied to drifting populations that do not usually remain long in any one area and to populations that do not have seasonal breeding periods and therefore produce mixed cohorts. The approach is further complicated by the fact that the use of only one mesh size in collection nets will under- or overestimate the relative proportions of certain size categories. From the information presently available, it seems that reproduction in warm-water cavoliniid species occurs throughout the year with differing peaks of maximal activity in different species (van der Spoel, 1973a; Wells, 1976b). No information on breeding periods is available for cool-water cavoliniids, which might be expected to demonstrate seasonality.

Data on growth rates have also been derived from size-frequency diagrams based upon preserved collections taken off Bermuda over a period of 10 months (van der Spoel, 1973a) or off Barbados over 2 years (Wells, 1976b). Both studies suggest that the life cycles of five species (*Creseis virgula*, *Clio pyramidata*, *Cuvierina columnella*, *Diacria trispinosa*, and *Cavolinia gibbosa*) are approximately 1 year in length. These authors, however, differ in their interpretations of the fate of reproductively active females. Wells (1976a) believed that female *Creseis virgula* die after spawning, whereas van der Spoel (1973a) suggested that spawned females of *Clio pyramidata* redevelop as males and begin a second sexual cycle.

Definitive studies of sexual cycles, reproductive capacity, growth rates, and other aspects of development require the laboratory culture of these animals, combined with long-term studies of natural populations. No cavoliniid species has yet been maintained in the laboratory for more than a few days. It would also be of interest to reevaluate the early studies of chromosomes in pteropods. Variability in chromosome numbers between species led Zarnik (1911) to suggest that some cavoliniids and pseudothecosomes may be polyploids that have arisen from a doubling or trebling of chromosomes.

Suborder Pseudothecosomata

The reproductive tract anatomy of pseudothecosomes is described by Meisenheimer (1905b) as being exactly like that of the euthecosomes. The animals are protandric hermaphrodites, but very little is known about reproductive behavior.

Copulation in *Peraclis bispinosa* was briefly described from a pair of preserved, and presumably contracted, specimens (Lalli and Wells, 1978). The animals were paired at an angle of about 180°, with the shell apertures pressed together. The accessory copulatory organ of each individual was everted and attached to the operculum of the partner. Both specimens were mature males, supporting the belief that thecosomes mate before attaining full development of the female genital system. Copulation also has been observed once in live *Corolla*, though no details of the coupling are known.

Pseudothecosomes that have been gently collected will normally spawn in the laboratory within several hours in response to slightly warmer conditions. Gilmer (pers. obs.) induced several specimens of *Peraclis reticulata* to spawn single egg strings that were approximately 25 mm in length. The eggs measured 90 to 100 μm in maximum dimension at the time of release and were spaced about 600 μm apart, so that each string contained 35 to 40 eggs. After 34 hours, the eggs had gone through several cleavages and measured about 150 μm; by 48 hours, free-swimming veligers began to emerge. Development time is presumably somewhat longer in the natural environment, where temperatures are colder.

Species of *Corolla* and *Gleba* also have been observed to spawn, both in the laboratory and in the field (Gilmer, pers. obs.). The eggs are spawned either in mucous strings of up to 0.5 m length or in bundles linked together with mucus (Fig. 44). These egg masses remain attached to the genital opening of the parent for some undetermined time. In *Corolla* (*calceola*?), the numbers of eggs released per spawn varied directly with the size of the parent from about 900 eggs in small individuals (65 mm wing span) to about 4,500 in large animals (110 mm span).

Embryonic and larval development are known in detail only for *Cymbulia peroni* and *Gleba cordata*. Fol (1875) described the embryonic development of *Cymbulia* from fertilization to the early veliger stage at 6.5 days

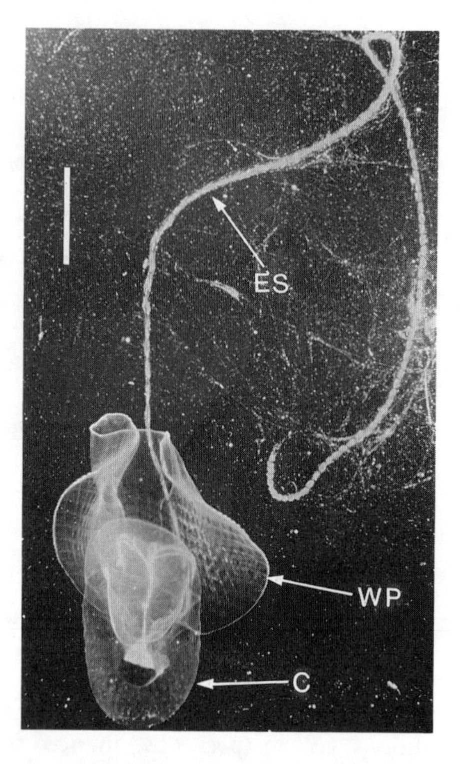

FIG. 44. Spawning of *Corolla* (*calceola*?). C, pseudoconch; ES, egg string; WP, wingplate. The scale line represents 10 mm.

after deposition of the egg mass. By this time, the veligers had a bilobed velum, a small sinistrally coiled shell, an operculum on the foot, and large statocysts. Older, free-swimming veligers of the same species have been described by Krohn (1860) and Thiriot-Quiévreux (1970). Prior to metamorphosis, the larvae are characterized by a multilobed velum and a wide-mouthed shell of 1.5 whorls measuring 0.72 mm in maximum diameter. As in euthecosome veligers, the foot develops two rudiments that will develop into the swimming plate. Metamorphosis begins with the disintegration of the velum and the concomitant growth of the wing rudiments. This is followed by the casting of the shell and operculum. The pseudoconch begins to develop between the integument and pallial gland, but its final form is attained after metamorphosis. The development of *Gleba cordata* (described under the name of *Tiedemannia neapolitana* by Krohn, 1860) follows essentially the same general course (Fig. 45). The larvae of *Gleba* can be distinguished from those of *Cymbulia peroni*, however, by several distinctive features: reddish-brown chromatophores are present on the surface of the developing wingplate; the veliger shell is not so tightly coiled, so that the inner surface of the last half whorl is free and not attached to the initial whorl; and the cilia on the velar lobes are proportionately longer.

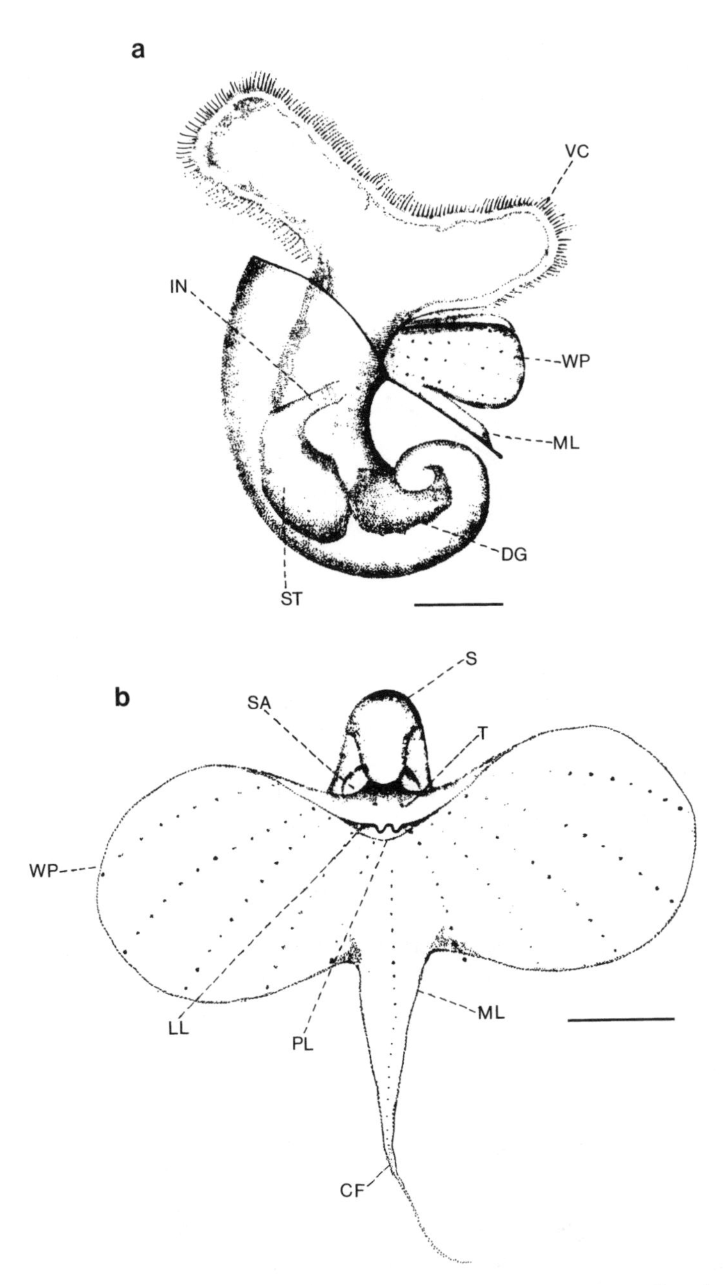

FIG. 45. Development in *Gleba* sp. (from Krohn, 1860): *a*, a young veliger larva; *b*, a metamorphosing individual in ventral view. CF, contractile filament of medial lobe; DG, digestive gland; IN, intestine; LL, lateral footlobes (precursors of the lateral lips of the proboscis); ML, medial lobe of wingplate (with attached operculum in *a*); PL, posterior footlobe; S, larval shell; SA, shell aperture; ST, stomach; T, tentacle; VC, velar cilia; WP, wingplate. Scale lines represent 500 μm in *a*, 5 mm in *b*.

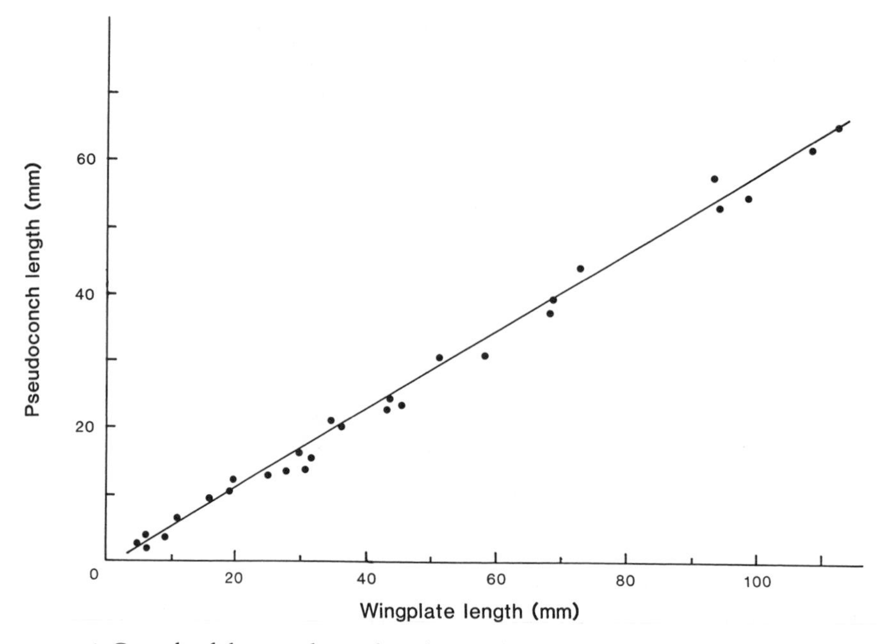

FIG. 46. Growth of the pseudoconch and wingplate of *Corolla* (*calceola*?) from size measurements of specimens collected in the western North Atlantic (Gilmer, unpubl. data).

Preliminary growth data are available only for *Corolla* (*calceola*?) collected in Slope Water of the northwest Atlantic. In this species, growth seems to involve a gradual increase in the length of the pseudoconch relative to the wing span (Fig. 46). More work of this nature is needed on cymbuliids in order to assess the validity of species based primarily on size differences.

This brief discussion of reproduction and development in pseudothecosomes points to the obvious need for more research and observations. It will not be possible to make meaningful comparisons among the different species of pseudothecosomes and with the euthecosomes until more data are available on these topics.

Epifauna and Parasites

In a fluid environment with no permanent solid substrates, the shells of living pteropods offer a site of attachment for certain hydrozoan polyps (Color Fig. 8; Fig. 47). Table 16 provides a list of euthecosome species and their associated hydroid epifauna.

The associations between hydroids and pteropods appear to be very specific. None of the hydrozoans is reported to occur on any other substrate, and each is associated with a specific pteropod species, with the exception of *Clytia* ("*Laomedea*") *striata*. This species frequently is present on *Diacria* shells but is very rarely found on *Cuvierina* (Kramp, 1922; Tesch, 1946). Hirohito (1977) considers that *C. striata* is a valid species confined to pteropod shells. Cornelius (1982), however, regards *C. striata* as conspecific with *C. linearis* (described by Thornely, 1899), polyps of which attach to barnacle shells, other hydroids, and intertidal rocks.

Taxonomic problems also exist with the other hydroid species, owing partly to incomplete descriptions and partly to the fact that only one species has been linked with a medusoid dispersal stage. Picard (1956) succeeded in rearing medusae liberated by the polyps of *Campaniclava cleodorae*, and identified these as *Pandea conica*; the latter name has priority and should be applied to both the hydroid and medusoid stages. *Perigonimus sulphureus* is perhaps the least known of these hydroids, and although it has been placed in the genus *Perigonella* by some workers, its taxonomic position remains uncertain (Rees, 1956), as does that of the other hydroid species.

The nature of the hydroid-pteropod associations is not well known. The association presumably begins with the attachment of a free-swimming planula larva to the pteropod shell. After the development of the initial polyp, additional feeding and reproductive polyps are budded off from a stolon that spreads over the shell surface.

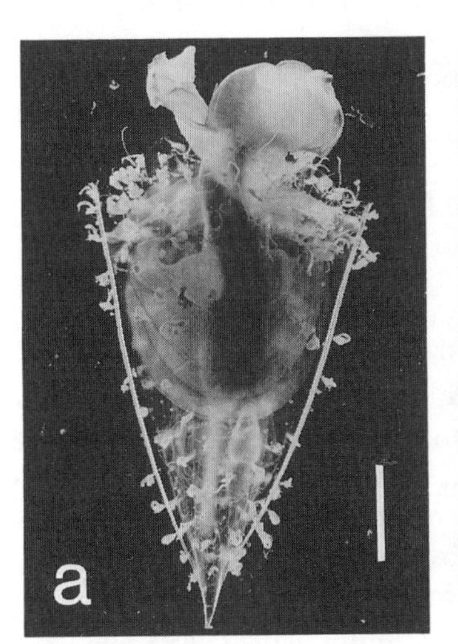

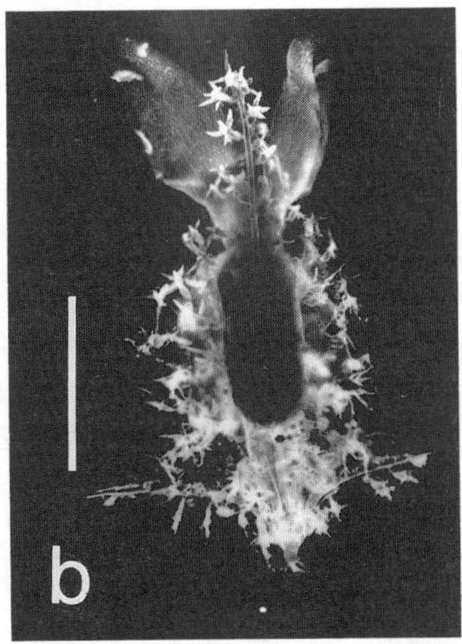

FIG. 47. Hydroid epifauna of euthecosomes: *a, Clio recurva* with the hydroid *Campaniclava clionis* growing on its shell surface; *b, Clio cuspidata* with the hydroid *Pandea conica* attached to its shell. Scale lines represent 5 mm.

TABLE 16

Hydrozoan epifauna associated with euthecosome species

Euthecosome species	Hydroid species name	Medusa species name
Clio cuspidata	*Campaniclava cleodorae* (Gegenbaur, 1854)	*Pandea conica* (Quoy & Gaimard, 1827)
C. recurva	*Campaniclava clionis* Vanhöffen, 1910	
Diacria trispinosa	*Kinetocodium danae* Kramp, 1922; *Clytia* ("*Laomedea*") *striata* (Clarke, 1907)	
Cuvierina columnella	*Clytia striata*	
Cavolinia tridentata	*Perigonimus sulfureus* Chun, 1889	

There has been some conjecture about the food sources of attached hydroids. Feeding polyps of *Clytia striata* have up to 20 long, thin tentacles provided with nematocysts, and specimens attached to *Diacria trispinosa* (illustrated by Richter, 1983) have been observed to capture small zooplankton in the laboratory (Lalli, pers. obs.). Similarly, the feeding polyps of *Pandea conica* are nutritionally independent of their pteropod host and use their eight tentacles to capture planktonic food (Picard, 1956). The possession of nematocyst-bearing, elongate tentacles by *Campaniclava clionis* suggests that this species too obtains food from the water column. Steche (1906 in Picard, 1956) suggested that polyps of *Perigonimus sulphureus* captured and fed upon eggs released by *Cavolinia tridentata*, but it is doubtful whether the eggs of the host would be sufficiently abundant to constitute the only nutritional source for the hydroid. It is possible, however, that all of these hydroids may obtain at least some food by removing particles trapped in the mucous feeding webs of their pteropod hosts.

On the other hand, Kramp (1922) believed that the feeding polyps of *Kinetocodium danae* are not capable of capturing free-swimming zooplankton prey because of the "degenerate" nature of their two to six tentacles. He suggested that the feeding polyps, which are confined to the anterior portions of the shell of *Diacria trispinosa*, feed parasitically on the surfaces of the footlobe and wings of the pteropod host, using the mouth to remove epidermal tissues. Kramp's conjecture, however, was based on histological studies of admittedly badly preserved and contracted material. Specimens of *K. danae* examined by Lalli (pers. obs.) include large polyps that extend well beyond the area of the wings and foot of *Diacria*. Large hydranths have up to seven, short tentacles that clearly contain nematocysts and cannot be considered as degenerate. The possession of nematocysts suggests that *Kinetocodium* is not parasitic but is capable of obtaining food either from the capture of free-swimming zooplankton or by symbiotically removing food from the feeding web of the pteropod. It is also possible, however, that the nematocysts may function solely in a protective capacity.

It is interesting that hydrozoan polyps have been found on only five, or possibly six, of the 34 euthecosome species: van der Spoel (1967) stated that hydroids can be found on *Diacria major*, although Tesch (1946) stated that they are present only on *Diacria trispinosa* forma *minor*. The affected species are among some of the larger euthecosomes (maximum shell length: 10 to 30 mm), but there is no obvious reason why other species that are equally abundant and in this size range (e.g. *Clio pyramidata*) are not also infested. Tesch (1946) suggested that the mantle appendages of *Cavolinia* serve to keep the shell surface free of attached organisms, and in fact only one species of this genus (*C. tridentata*) is reported to harbor a hydroid. Gilmer (1974), however, has shown that all species of *Diacria* also have lateral mantle lobes, yet *D. trispinosa* frequently harbors hydrozoans, whereas *D. quadridentata* does not. At present, there does not seem to be any obvious correlation between those thecosomes that never bear hydroids and any anatomical structure (e.g. extended mantle lining or temporary pseudoconch) that conceivably could prevent the attachment and growth of epifauna.

Some further problems arise from these hydroid-euthecosome associations. Anatomical studies suggest that the hydroids produce medusae, yet apparently only one (*Campaniclava cleodorae*) has been linked with its free-swimming medusoid stage (*Pandea conica*). It is possible that some of the other hydroid species have vestigial medusae that are not released from the gonophores. In either case, it would be of interest to know how larvae produced by the medusae locate and attach to a specific host at metamorphosis.

Most hydroids and hydromedusae are confined to neritic areas where there are solid substrates for the attachment of the sessile stage. The attachment of hydroids to pteropod shells results in the ability of these species to expand their distributions into oceanic areas. Hydroids may benefit by having access to food caught in pteropod feeding webs, but the question remains whether there is any advantage to the thecosome host. The nematocysts of the hydroid may confer some protection from predation to both the coelenterate and the pteropod, but it is unlikely that this would balance the obvious disadvantage to the thecosome host: heavy infestations of hydroids on a shell must surely increase frictional drag and otherwise interfere with swimming and buoyancy.

A number of euthecosomes, including *Styliola subula* and species of *Creseis*, *Clio*, and *Cavolinia*, have been reported as acting as intermediate hosts for parasitic copepods. These are believed to be pennellid (formerly Lernae-

TABLE 17

Infestation of euthecosomatous pteropods by postembryonic stages of the parasitic copepod Cardiodectes medusaeus *(Wilson)*

Host species	Number examined	Number infested	Incidence (%)	Maximum number of larvae per host
Cavolinia tridentata	63	24	38	7
Clio recurva	62	49	79	66
Clio cuspidata	28	7	25	6
Clio pyramidata	746	284	38	20
Cuvierina columnella	42	23	55	8
Diacria trispinosa	15	6	40	8

SOURCE: Perkins, 1983.

oceridae) copepods, for which the final hosts are fish (Ho, 1966; Stock, 1973). The most complete study of thecosomes that act as intermediate hosts for a parasitic copepod is that of Perkins (1983). She recovered post-embryonic stages of *Cardiodectes medusaeus* (Family Pennellidae) from the mantle cavities of six species of euthecosomes (Table 17). These planktonic mollusks and the heteropod *Carinaria cristata* are the only recorded intermediate hosts of this copepod; although an additional 14 species of holoplanktonic mollusks were examined, none harbored *Cardiodectes*. Free-swimming copepodites enter the mantle cavity of the intermediate host and attach to the mollusk's mantle lining. Development proceeds through the chalimus stages and copulation takes place within the intermediate host. Fertilized female copepods leave the molluscan host to seek out a definitive host, which may be one of a variety of myctophid (lanternfish) species.

Rediae of trematodes have been found in single specimens of *Cuvierina columnella* (Bonnevie, 1916) and *Cavolinia longirostris* (Vande Vusse, 1980), leading the latter author to suggest that pteropods may be important first intermediate hosts for trematodes that complete their life cycles in pelagic fish. Finally, a single parasitic amphipod, *Brachyscelus rapacoides*, has been collected from a *Cavolinia longirostris*, but the usual hosts of this parasite are medusae (Harbison et al., 1977).

Few pseudothecosomes have been carefully examined for epifauna or parasites. Hochberg and Seapy (1985 and pers. comm.), however, report that parasitic ciliates have been found in *Corolla*, digeneans in *Cymbulia*, and cestodes in *Gleba*. In addition, hyperiid amphipods (*Lycaea* spp.) have been observed on both *Corolla spectabilis* and *Gleba cordata* (Harbison et al., 1977). No damage was visible on the pseudothecosomes, and the nature of the association with these parasitic amphipods is unclear.

Pteropods as Ecological and Palaeoecological Indicators

More data have been amassed on the geographic distribution of euthecosomatous pteropods than on any other aspect of their ecology. Reviews of the world distributions of all of the species have been compiled by van der Spoel (1967) and Bé and Gilmer (1977). Distributional studies correlated with hydrographic data have led to the recognition that certain species are restricted to a narrow range of environmental conditions that can be defined in terms of combinations of temperature, salinity, depth, or biotic factors. Those species that are particularly sensitive to change in environmental conditions can be used as biological indicators of the particular water-mass types they inhabit. Changes in the distribution of an indicator species will reflect directional fluctuations in the flow of a particular water mass. This concept often has been applied to other zooplankton species (particularly foraminifera, copepods, and chaetognaths) that are restricted to narrowly defined environments and are sufficiently abundant to be sampled routinely.

Shelled pteropods have not been so frequently used as biological indicators as some other zooplanktonic groups, primarily because pteropods are often less abundant in plankton tows than such groups as foraminifera and copepods. Nevertheless, the literature on this topic is extensive, and only a few selected examples are given here to illustrate the use of thecosomes as indicator species of present water masses and of past changes in oceanic conditions. A more thorough review of this subject has been undertaken by Furnestin (1979).

Zooplankton distributional studies have received particular attention in the area off Cape Hatteras, North Carolina, because of the proximity of several water masses of markedly different hydrographic conditions. These include the tropical Gulf Stream, the subtropical Sargasso Sea, and the cold continental Slope Water that is a southward extension of the Labrador Current. Myers (1968) and Chen and Hillman (1970) analyzed pteropods collected from these areas and selected the following as indicator species: *Limacina retroversa*, which was the only pteropod present in cold (6° to 16° C), low-salinity Slope Water; *L. inflata*, which comprised 60 percent of the total pteropods in Sargasso Sea water of about 18° to 22° C; and *L. trochiformis* and *Creseis virgula conica*, which were the dominant species of warm (23° to 28° C) Gulf Stream water. Seasonal changes in the distribution of these indicator species were in agreement with seasonal fluctuations in the flow and vertical structure of the water off North Carolina. Gulf Stream movements could be traced by the presence of *L. trochiformis* and *C. virgula*. In the area bounded by 35°30′ N to 37°10′ N and 74° W to 74°30′ W, there is vertical stratification of the water in summer. A thin layer of Gulf Stream water with attendant tropical species overlies a thick layer of cold Slope Water containing *L. retroversa*. This stratification breaks down in winter, and *L. retroversa* is then found throughout the upper 250 m of the water column. A similar seasonal shift occurs off Bermuda (Chen and Bé, 1964). In summer, tropical Gulf Stream water, dominated by *L. trochiformis* and *C. virgula*, overlies Sargasso Sea water. In winter, subtropical species (indicated by the dominance of *L. inflata*) become distributed throughout the upper 500 m as vertical stratification breaks down.

Some caution must be applied, however, in using oceanic thecosomes as indicator species in nearshore regions, since different patterns of vertical migration may affect distributions. Vecchione and Grant (1983) have suggested that thecosomes that normally coexist in warm oceanic areas off the east coast of North America may separate along continental slope boundaries, with weak diurnal migrators (e.g. *Limacina trochiformis*) moving onto the shelf in occasional intrusions and strong migrators (*L. inflata*) remaining offshore in water depths greater than 100 m. This would imply that the detection of Sargasso Sea and Gulf Stream water by the presence or relative abundance of these two species of *Limacina* may be artificial.

The area of the Antarctic Convergence in the southern Atlantic has also been of interest because of the separation of distinct water masses and a relatively sharp distinction in Subantarctic and Antarctic thecosome species (Chen, 1966, 1968a). *Limacina retroversa* is mostly concentrated in the upper 200 m of Subantarctic surface water north of the Convergence, where the temperature range is approximately 3° to 6° C (Fig. 48). Fewer individuals of this species extend to 5° to 8° of latitude south of the Convergence zone. *Clio antarctica* is a less abundant species that is restricted to Subantarctic intermediate water (between 200 and 800 m depth) north of the Convergence; this water originates from the mixing of sinking Antarctic surface water and the warmer Subantarctic surface water. Two pteropod species are classified as indicators of Antarctic waters: highest densities of *Limacina helicina* are found south of the Antarctic Convergence in Antarctic surface water (-1° to 1° C) extending to depths of 100 to 250 m; *Clio sulcata* is concentrated at a depth of 200 m in this area. Though these separations of Antarctic and Subantarctic species appear to be sharp, it is important to realize that these pteropods can be transported considerable distances away from their main centers of concentration.

For example, Boltovskoy (1971, 1975) has found both *Limacina helicina* and *L. retroversa* at a latitude of about 36° S off the coast of Argentina; their co-occurrence with subtropical species of pteropods is indicative of a northward flow of Subantarctic water along the coast and mixture of this water with southward-moving subtropical water. Boltovskoy (1975) in-

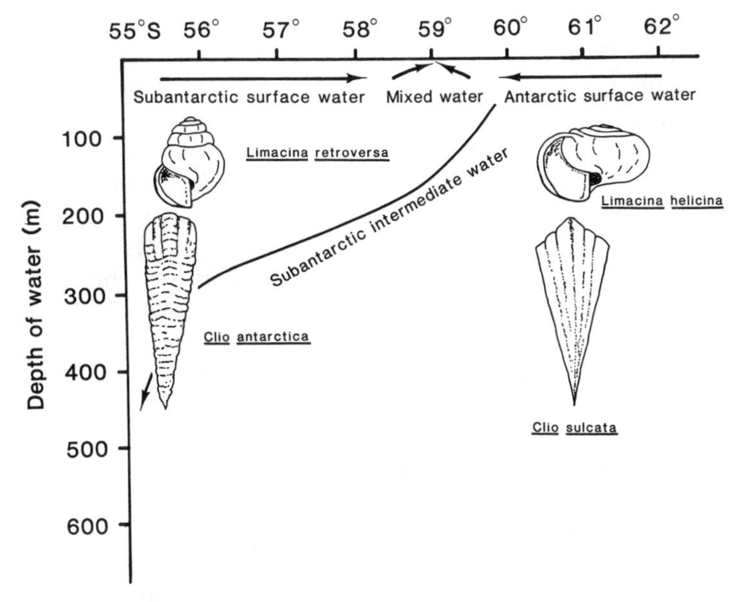

FIG. 48. Vertical distribution of thecosome species in western Subantarctic and Antarctic waters (redrawn from Chen, 1968a).

cluded a distributional analysis of foraminifera and chaetognaths in this area, resulting in the determination of faunistic groupings of the three groups of zooplankton. This approach determines the percentages of co-occurrences of different species (not all of which will be indicator species), and permits a composite of species to be used as a group index to existing environmental conditions, rather than basing water-mass identification on isolated species. This is undoubtedly a more realistic appraisal, since the definition of an environment should be based on biotic conditions, such as relative abundance of prey and predators and the presence or absence of competitors, as well as on abiotic factors, such as temperature and salinity. This approach may also result in a better understanding of trophic relationships and factors determining biological productivity in particular water types. Further, it reduces error due to the patchy distribution of a single species, or to failure to capture certain species when only one type of collecting gear is used. It also allows for measures of the relative abundance of species, which may prove significant in areas of water mixing.

The characterization of mixed plankton communities as indicators of water masses was undertaken in the Pacific Ocean by Fager and McGowan (1963) and McGowan (1971, 1974). Associations of pteropod species with other planktonic species have resulted in the distinction of eight or nine major faunal provinces in the pelagic zone of the Pacific; these are generally in good agreement with physically defined water-mass boundaries and known circulation patterns. Similarly, analyses of distributional patterns in the Gulf of Thailand and the South China Sea have revealed thecosome species associations that are linked to particular abiotic environmental characteristics and with other zooplankton species (Rottman, 1976, 1978).

Vertical water movements can also be traced by changes in the depth distributions of zooplankton. Stepien (1980), for example, used the known depth ranges of certain species of pteropods, chaetognaths, and euphausiids to analyze the vertical movements of water in the Straits of Florida. The presence in deep water (>600 m) of epipelagic pteropod species, which either do not migrate vertically or are feeble migrators, was taken as an indication of periodic downwelling of shallow water.

The use of biological indicators in identifying and tracing water masses requires taking plankton samples over a wide geographic area and at different times of the year. This is both time-consuming and expensive in terms of man-hours and ship-time. Ideally, nets of different mesh sizes should be used to sample the size range of different species representatively, and collection problems are compounded by those species that undergo extensive diel vertical migrations. The sorting and identification of species from large numbers of tows is slow and tedious, and results in a considerable time lag before the analyses are complete. The use of biological indicators in the identification of water masses is now becoming obsolete, since new techniques, such as satellite imagery, can provide precise identification of surface-water masses much more quickly and over vast expanses of ocean. A contemporary importance of indicator species of pteropods lies in their use for the detection of changes in hydrographic conditions over geologic time through the examination of shifts in species composition in sediments produced by the accumulation of pteropod shells.

Pelseneer (1888a) was the first to record pteropod shells in sediments collected by the *Challenger* Expedition, and the term "pteropod ooze" was coined at that time. This is now defined as any pelagic sediment containing at least 30 percent calcium carbonate of organic origin, of which pteropod and other pelagic mollusk shells are the important constituents (Sverdrup et al., 1942). Pteropod ooze is found primarily in tropical and subtropical zones, and for physico-chemical reasons is usually restricted to depths not exceeding 2,800 m in the Atlantic Ocean or 500 m in the Pacific (Fig. 49). Pteropod species older than Pleistocene (2 to 3 million years ago) have not been recorded from deep-sea sediments (Chen, 1968b), though fossil species are known from older uplifted sediments on land (e.g. Curry, 1965). Therefore, assemblages of shells in ocean-floor sediments contain many of the same species as those living in the water column today. Since the environ-

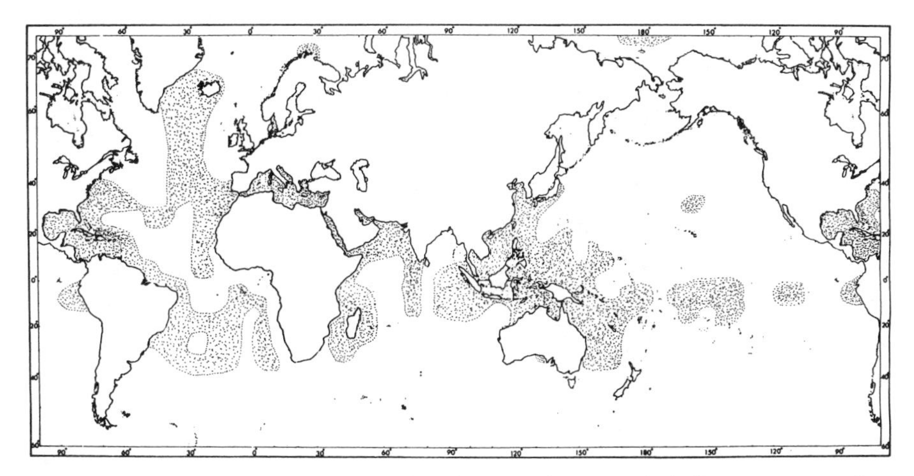

FIG. 49. The known and probable distribution of pteropod shells in marine sediments (based on data from Herman, 1971; Berner, 1977; and Berger, 1978).

mental requirements are known for many thecosomes, changes in pteropod sediments may provide clues about past hydrological and climatic changes that have occurred over very long time periods.

In analyzing sedimentary shell records, it is important to know whether the deposited shells accurately mirror the planktonic populations from which they originated. This can be assessed by comparing shell assemblages collected from very recently deposited, undisturbed sediments in shallow water with living pteropods collected in plankton tows in the same area. Wells (1975), for example, found 20 species of euthecosomes occurring in the same relative abundance in the plankton and in sediments collected from 150 m depth off Barbados (Table 18). In older and deeper sediments, some shifts in relative abundance and species composition could be expected through the faster dissolution rates of smaller shells or the winnowing effects of currents. The general species composition in Recent sediments, however, should reflect the environmental conditions and the populations that prevailed in the overlying water column at the time of deposition.

In 1964, Chen compared pteropod species obtained from sediment samples from the Bermuda Pedestal with those collected in plankton tows in the same area. Of the 22 species and subspecies identified in the sediments, only one—*Limacina retroversa*—was not collected in the overlying water. This is a common species of the North Atlantic, and its appearance in Tertiary deposits off Bermuda probably indicates past intrusions of cold water into this area. Similarly, Rottman (1980), finding striking disparities between the present planktonic distributions of certain pteropod species and their distributions in surface sediments, was able to detect relatively

TABLE 18

A comparison of the relative population densities of dominant (>1 percent of total pteropods) euthecosomes in the plankton and in sediments off Barbados, West Indies

Species	Plankton (percent)	Sediment (percent)
Limacina inflata	61.5	66.8
Creseis virgula conica	23.0	13.3
L. trochiformis	6.7	9.0
C. virgula virgula	3.0	3.0
L. bulimoides	2.2	2.2
C. acicula	1.3	2.5
Styliola subula	0.4	1.8

SOURCE: Wells, 1975.

recent fluctuations in nearshore oceanographic conditions and water movements in the South China Sea.

Intensive palaeoecological studies have been carried out in the Mediterranean Sea, where pteropod shells are well preserved in an environment of relatively shallow water depths and warm bottom-water temperatures. Several authors (e.g. Froget, 1967; Chen, 1968b; Blanc-Vernet et al., 1969; Herman, 1971, 1981) have detected climatic changes from shifts in the species of thecosomes contained at different depths in Mediterranean sediments. The shallowest core depths represent a Postglacial period and contain warm-water species that are similar to those currently living in this sea; *Limacina inflata*, *Creseis acicula*, and *Styliola subula* are dominant species in this faunal group. At deeper core depths, Glacial periods are indicated by the presence of shells of *Limacina retroversa*, which does not now live in the Mediterranean. Transitional climates between Glacial and Postglacial are marked by an increase in relative numbers of *Clio pyramidata*, a eurythermal euthecosome. Alternating bands of low-latitude, warm-water species with high-latitude, temperate species indicate the climatic oscillations that have occurred in this area during the Pleistocene. Similar climatic fluctuations can be traced from pteropods preserved in outcroppings on the island of Sicily (Buccheri, 1984).

The Red Sea also is an area that contains well-preserved pteropods in sediments, some dating back to about 150,000 years before present. Among the numerous studies of palaeoceanographic changes in this area, those of Almogi-Labin (1982) and Ivanova (1985) are particularly notable in their use of pteropod assemblages to indicate past environmental changes that included fluctuations in general productivity as well as in temperature and salinity ranges.

In some instances, morphological changes within a species can be detected in sediments. Rampal (1975), for example, made biometric studies of shells of *Cavolinia inflexa* collected from Mediterranean sediments and plankton. This euthecosome exhibits considerable variation in shell form, and three subspecific varieties are recognized. In addition to the extant forms, Rampal found distinctly different, Postglacial fossil forms of this species in sediments deposited about 1,000 to 1,500 years ago. She was able to link fluctuations in abundance of *C. inflexa* as well as biometric changes in the shells to periods of temperature change.

Changes in hydrologic environmental conditions over time affect fish populations as well as zooplankton. Undisturbed, anaerobic sediments that have accumulated in the Santa Barbara Basin off California have been examined for evidence of such changes (Soutar and Isaacs, 1969; Soutar, 1971). Because the sediments are strongly stratified, they yield a good time-record of pteropod shells as well as of the scales of such fish as the Pacific sardine, northern anchovy, and Pacific hake. The presence or absence (and relative abundance) of shells of *Limacina helicina*, a cold-water thecosome, was used to show the extent and timing of intrusions of subarctic waters into a subtropical zone, and such events were correlated with changes in fish populations over a period spanning about 1,500 years. Such historical information becomes increasingly useful in predictions of fishery harvests that may be affected by natural, long-term changes in climate as well as by such human activities as pollution and overfishing.

Some attempts have been made to use certain pteropod-shell deposits as indications of past changes in sea level. Herman and Rosenberg (1969), for example, examined the ratio of numbers of specimens of *Creseis acicula* and *C. virgula* to *Limacina inflata* in sediments from depths ranging from 40 to 310 m off the western coast of India. The *Creseis*: *L. inflata* ratio was high in water shallower than 100 m and decreased rapidly with increasing depth. They took this change to show that the species are restricted to different depth ranges, *Creseis* species being considered as epipelagic and *L. inflata* as mesopelagic. They theorized that the vertical depth change in the relative abundance of these pteropod species in sediments, if coupled with absolute age determinations of the sediments, might yield information on former water depths over continental shelves. Several points limit the application of this theory, however. One is that *Creseis* species are found living over deep ocean waters as well as in shallow inshore zones, and the ratio of *Creseis*: *L. inflata* can be high in populations living offshore. The major difference between these pteropods lies in their different patterns of vertical migratory behavior; *L. inflata* performs extensive migrations compared with *Creseis*, and this difference can lead to separation of the species as populations are carried into shallow areas. Finally, it would be of considerable interest to know whether rates of dissolution vary in these species, since this could also lead to a shift in the species composition with increasing depth.

Sedimentary records can provide valuable insights into past hydrologic and climatic events. It is essential, however, to be aware of changes that can alter shell records and lead to misinterpretations. Although not yet determined experimentally, dissolution rates probably vary for different sizes of pteropod shells, just as they are known to do for foraminiferan tests. In areas where dissolution rates are high, thin shells would be expected to dissolve faster than larger, thicker shells. Thus species composition and relative abundance may change through the loss of shells of small species (e.g. *Limacina inflata*) compared with large forms (e.g. *Cavolinia tridentata*), and the size-frequency distribution of the original population may shift through faster removal of the shells of young individuals compared with those of adults. Similarly, bottom currents may displace lighter shells farther relative to heavier ones. Geologists are well aware, also, of the effects of benthic organisms on pelagic sediments. Their burrowing activities may displace calcareous shells and other particles, resulting in misinterpretations of geologic dates related to biological events recorded at different sediment core depths. Furnestin (1979) has also rightly pointed out that it is not possible to transpose the unmodified ecological requirements of Recent species into the past. Tolerance ranges, particularly to temperature, may have altered over time just as the behavior of a species may change.

The total abundance of shells present in sediments depends not only on the original population density but on the balance between sedimentation rates, dissolution rates, and burial by detrital material. Although high abundance in sediments is generally taken to be indicative of favorable environmental conditions prevailing in the water column at the time of deposition, this is not necessarily the correct explanation. In shallow waters, slow dissolution rates over long time periods may permit large accumulations of aragonite shells from modest-sized populations. Near-shore deposits of shells may also reflect exceptionally high mortality of oceanic species that are continually carried into unfavorable neritic conditions by prevailing currents. In this regard, the suggestion that red-tide poisoning may cause high pteropod mortality and high accumulations of shells in inshore areas (Diester-Haass and van der Spoel, 1978) is probably not valid, in light of White's (1977) work showing that *Limacina retroversa* is not adversely affected by dinoflagellate toxins and, indeed, may act as a vector that transmits the toxin to fish. In fact, Diester-Haass and van der Spoel acknowledge another explanation advanced by Berger (1977). This attributes the high numbers of pteropod shells in late Pleistocene sediments off northwest Africa and Portugal to decreased dissolution rates due to changes in water chemistry. This is only one of many suggestions that past chemical changes have enhanced the preservation of aragonite shells at certain times, and it is a topic considered in more detail in a later section. However, this further strengthens the argument presented here that abundance of shells in sediments, particularly in older deposits, is not necessarily a direct reflection of the densities of the original living populations.

Finally, it should be pointed out that the sedimented shells of pteropods may be utilized by benthic organisms. Sipunculid worms, gnathiid isopods, and corophiid amphipods have all been found living inside empty, intact shells of *Cuvierina columnella* (Just, 1977), and it is possible that the shells of other species also provide refuge for small animals. Thiel (1983) found that pteropod shells, collected from Red Sea sediments at depths between 450 and 2,000 m, were riddled by microborings probably made by fungi. This suggests that microorganisms can use the organic matrix of the shells as a food source.

Pteropods and the Carbonate Cycle

In recent years, recognition of the increasing amounts of carbon dioxide being released into the atmosphere by human activities has led to mounting interest and concern about its ultimate fate and the effects of major additions of this gas in the oceans (e.g. Andersen and Malahoff, 1977). The quantity of carbon dioxide that can be absorbed in surface waters depends largely on the existing concentration of dissolved carbon dioxide, and this is regulated, in part, by biological activities. These include photosynthesis, plant and animal respiration, and the formation and dissolution of calcium carbonate skeletons by certain marine organisms. In regard to the latter process, considerable attention has been given to planktonic foraminifera and coccolithophorids, both of which secrete calcium carbonate exoskeletons. Because the tests of these organisms are composed of calcite, which is a relatively insoluble form of calcium carbonate, large amounts of bound carbon are removed from the cycle in sedimentary deposits that cover extensive areas of ocean bottom.

Pteropods and shelled heteropods also secrete calcium carbonate shells, but these are composed primarily of aragonite, which has a greater attraction for water molecules and is thus more soluble than calcite. These mollusks, like the calcite-secreting microzooplankton, utilize bicarbonate ions dissolved in near-surface waters (<200 m) for shell formation. In the case of pelagic snails, the bicarbonate ions are bound with calcium to produce aragonitic shells, and dissolved carbon dioxide is released. Following the death of the mollusks, the shells sink through the water column and either form sea-floor sediments in relatively shallow waters or are transported into water below the aragonite compensation depth, in which case the shells dissolve. Dissolution is the reverse of shell secretion and results in the release of bicarbonate ions that eventually are returned to near-surface waters by mass transfer. Thus the carbon dioxide concentration and carbonate alkalinity of marine waters is regulated, in part, by shell secretion and carbon dioxide production in shallow water, by the removal of bound carbon dioxide from the cycle through the burial of undissolved carbonate on seafloor areas above the aragonite compensation depth, and by the dissolution and production of bicarbonate ions in deeper waters.

Until very recently, little attention was directed toward the role played by pteropods in the carbonate cycle, because these mollusks are often less abundant in the plankton than foraminifera and coccolithophorids, they have longer generation times, and sedimentary deposits of their shells are relatively rare. Shell deposits only accumulate in rather shallow, confined areas of tropical and subtropical oceans that are saturated with respect to calcium carbonate (Fig. 49). Berger (1978) estimated that 2.4 percent of the sea floor of the Atlantic Ocean is covered by pteropod ooze. There the ooze mostly occurs in the vicinity of oceanic islands and submarine elevations at average maximum depths of 2,800 meters (Berner, 1977). In contrast, in the older, more stratified and more acidic waters of the Pacific and Indian oceans, pteropod ooze is present on less than 0.2 percent of the sea bed, and these sediments generally do not occur below 500 meters. Although the total area occupied by pteropod ooze seems small, it covers about 1.5 million square kilometers of ocean floor (Sverdrup et al., 1942). It is apparent, however, that there is an underrepresentation of pteropods in deep-sea sediments compared with their widespread distribution and abundance in surface oceanic waters. This discrepancy results from the dissolution of shells during settlement through the water column and after deposition on the sea floor.

It has now become clear that calcium carbonate budgets that consider aragonite sedimentation as negligible or insignificant compared with calcite sedimentation are not in agreement with measured $CaCO_3$ accumulation rates in sediments of the present-day oceans (e.g. Broecker, 1971; Berner, 1977). Omission of aragonite has led to an underestimate of the amount of $CaCO_3$ secreted in shallow water and dissolved in deep water, and this has led to a reconsideration of the role played by planktonic mollusks, particularly the more abundant pteropods, in the carbonate cycle.

Berner (1977) discussed several approaches that could be, or have been, taken to estimate the relative importance of pteropod aragonite sedimentation in the carbonate cycle. One approach has been to compare numbers of living pteropods to calcite-secreting organisms in plankton tows and to compare these with the pteropod:foram + coccolithophorid ratio in sediments; this is based on the premise that the difference will provide an approximation of aragonite dissolution. There are very few data, however, which provide accurate assessments of the relative numbers of these plankton, since pteropods, forams, and coccolithophorids comprise distinctly different size fractions requiring different sampling techniques. Comparisons of the numbers of different-sized organisms also lead to inaccuracy; ideally, the relative mass (numbers × weight) of aragonite and calcite should be compared, but there are no simple conversions for geometrically complex shapes, and few samples have been subjected to mineralogical analysis by X-ray diffraction. This approach also does not recognize the very different generation times and turnover rates of pteropods and of calcitic microzooplankton. Finally, there is no assessment of the actual numbers of dead pteropods and empty shells that are exposed to dissolution during sinking, since some proportion of pteropods are certainly removed from near-surface populations by predators that crush the shells or partially or wholly dissolve the shells during passage through the acidic environment of the predator's gut (e.g. Richter, 1977).

Berner (1977) approximated the relative importance of aragonite versus calcite sedimentation by examining the relative abundance of pteropods, foraminifera, and coccoliths in sediments that had undergone little or no dissolution. This second approach requires that the sediments be obtained from relatively shallow areas and that the sediments have not been disturbed by currents, slumping, or other causes. Using the few available data and a rough conversion of numbers to weight, Berner concluded that about 50 percent by weight of the calcium carbonate sedimenting to the sea floor consists of aragonite in the form of pteropod shells. There has been some criticism that Berner's sediment data included samples in which pteropod shells were overrepresented by winnowing effects and that the value of 50 percent is too high (Berger, 1982).

A third and more direct approach to the study of aragonite sedimentation was suggested by Berner (1977) and has subsequently been employed by a number of investigators. This involves the deployment of traps designed to capture sinking pteropod shells (as well as other sinking detritus) at various depths in oceanic areas. Using this technique, researchers have estimated that aragonite may constitute at least 12 percent (Berner and Honjo, 1981) to more than 50 percent of the total $CaCO_3$ flux in some areas (Betzer et al., 1984). This wide range of values is due partly to the use of different methods. Studies that have suspended traps for several months before collection or that have placed traps in undersaturated water may have underestimated the aragonite contribution because of dissolution of the shells (e.g. Berner and Honjo, 1981). On the other hand, the work of Betzer et al.

(1984) and the conclusions of Whitfield (1984) have been criticized by Harbison and Gilmer (1986) as overestimating aragonite owing to the placement of traps at depths with high pteropod densities and the inadvertent trapping of a substantial number of live pteropods along with dead animals and empty shells. The latter authors compared relative numbers of *Limacina* spp. and copepods collected in plankton tows and in sediment traps deployed in the same area for 4 to 8 hours. In all cases, *Limacina* spp. were preferentially collected in traps, probably because the normal downward escape behavior by swimming or sinking does not allow living pteropods to escape from traps.

The depth at which aragonite undergoes significant dissolution is of interest, since, as Berner and Honjo (1981) and others have pointed out, dissolution in relatively shallow waters would neutralize some of the excess fossil-fuel CO_2 that may have penetrated to those depths. The importance of aragonite in this respect is underscored by its greater solubility at all depths compared with calcite.

The rate at which dead pteropods sink through the water column will determine the amount of dissolution that occurs during settling and that which occurs at the sea floor-water interface. In this regard, laboratory studies of the settling rates of pteropods give an approximation of the time it may take uneroded shells to reach the sea bottom.

The rates at which dead pteropods and empty shells sink in experimental water columns have been compiled in Table 19. Although the methods varied and only a limited number of specimens and species have been studied, the settling velocities lie within a narrow range. Smaller species (*Limacina* and *Creseis* spp.) sink at rates ranging between about 1.0 and 3.0 cm/second. A larger species with a heavy adult shell, such as *Cuvierina columnella*, may sink at 5.0 cm/second. Heteropod sinking rates are not included in Table 19, but they are similar and range from 1.2 to 2.4 cm/second for the empty shells of several species of *Atlanta* (Vinogradov, 1961; Richter, 1973). As expected, sinking velocities vary with the mass and geometry of the shell. These laboratory values are only approximations, since in natural conditions dead pteropods and empty shells would be slowed by encountering temperature and density boundaries and by a reduction in weight owing to dissolution effects in deep water. Assuming no slowing with descent and taking 3,800 m as the average depth of the oceans (Sverdrup et al., 1942), the experimental sinking velocities indicate that shells would reach the sea bottom in less than 5 days. This would tend to support Berner's (1977) contention that most aragonite dissolution takes place on the sea floor.

There is disagreement with this position, however, in the work of Betzer et al. (1984) and Byrne et al. (1984). These authors suggest that earlier predictions of aragonite dissolution rates (e.g. Milliman, 1975; Honjo and

TABLE 19

Sinking rates of euthecosomes under experimental conditions

Species	Sinking rate (cm/sec)	Method	Reference
Limacina bulimoides	1.4–2.6	dead animals	a
	2.5	empty shells	b
	3.1	empty shells	c
L. helicina	1.0–1.4	empty shells	b
L. inflata	1.0 (juveniles)	empty shells	b
	1.3 (maximum)	empty shells	d
	2.2–2.5	dead animals	a
	3.2	empty shells	c
L. lesueuri	2.9	empty shells	c
L. trochiformis	1.4–1.7	dead animals	a
Creseis acicula	1.0–1.4	dead animals	a
	1.1–1.6	anesthetized animals	e
C. spp. (*acicula* and *virgula*)	0.95	empty shells	d
Cuvierina columnella	5.0	empty shells	b
Cavolinia longirostris	1.9	empty shells	d

REFERENCES: a, Vinogradov, 1961; b, Byrne et al., 1984; c, Richter, 1973; d, Diester-Haass, 1973; e, Kornicker, 1959.

Erez, 1978) were too low, because the experimental conditions created unusually high saturation levels of aragonite, owing to lack of water exchange in containers. Byrne et al. (1984) conducted dissolution experiments by subjecting pteropod shells to about 375 atm and 5° C seawater collected from 2,000 m depth. The results were used to calculate the cumulative percent dissolution of pteropods settling at a velocity of 1.4 cm/second (from Vinogradov, 1961). Depending on latitude, they predicted that, at a minimum, 20 percent (at 35° N) to 40 percent (at 50° N) of aragonite shells would be dissolved at 3,000 m in the Pacific Ocean. When values were corrected for mass loss through dissolution and subsequent reduced settling rates, then pteropods of less than 500 μm diameter would dissolve completely by depths of 2,700 to 3,500 m in the highly undersaturated water at 50° N in the Pacific. Only large species would undergo substantial dissolution after reaching the sea floor. Similarly, Betzer et al. (1984) predicted that only about 10 percent of aragonite sinking from the upper 100 m of the North Pacific would settle deeper than 2,200 m.

It is obvious that more work is needed to resolve the importance of aragonite in the present-day global carbonate cycle. It would appear that some of the conclusions regarding the ability of aragonite to buffer excess amounts of anthropogenic CO_2 may be premature and overly optimistic. On the other hand, it is apparent that shelled pteropods are not insignificant players in this cycle. Resolution of the details requires cooperative efforts between geochemists, who contribute knowledge of the kinetics of dissolution in the water column and in sediments, and biologists, who can contribute understanding of thecosome behavior patterns, population densities, and generation times.

The average maximum depths at which appreciable pteropod shells can accumulate presently lie at about 2,800 m in the open Atlantic Ocean and 500 m in the central Pacific Ocean, with these values becoming shallower toward high latitudes and on continental slopes (Berner, 1977). However, certain sea-floor deposits of pteropod shells suggest that the aragonite compensation depth itself has changed in the past. Chen (1968b), for example, noted that pteropod shells were abundant in Caribbean Sea sediments from water depths greater than 2,000 m during the last glacial period but were rare or absent in recent sediments at those depths. He interpreted this as an indication that the present compensation depth is shallower than during the last glacial period, when it was located between 3,000 and 4,000 meters. In 1977, Berger noted that past changes in aragonite and calcite compensation depths had occurred worldwide in open oceans. He agreed with Chen that the shallowest compensation depths are found in the recent Postglacial period (about 12,000 years ago to present). Prior to this period, deeper compensation depths were present for a narrow time period lasting about 1,000 years. This "Deglacial" period is marked by strikingly higher numbers of both aragonitic pteropod shells and calcitic foraminiferan tests in sediments; this maximum accumulation of pelagic skeletons is referred to as a preservation peak. During the last Glacial (ca. 20,000 years B.P.), compensation levels were at intermediate depths compared to the later two periods. A preservation peak has been noted in other areas, such as in South China Sea sediments, where the aragonite compensation depth apparently has shallowed to about 1,000 m from more than 1,700 m in the past (Rottman, 1979). Although it has been suggested that increased preservation of carbonates has occurred synchronously on a global basis during glacial intervals and thus can be useful for correlation of geological dates between sites (Berger, 1977; Rottman, 1979), there is also evidence that the timing of preservation peaks may differ in the Atlantic and Pacific (Price et al., 1985). Further, in some areas, deposits of pteropod shells ranging in thickness from a few millimeters to 1.2 meters seem to have resulted not only from a downward shift in the compensation depth but from concomitant disturbances of sediments during periods of increased strength of bottom currents (Bellaiche and Thiriot-Quiévreux, 1982; Price et al., 1985).

Although the factors responsible for changes in the compensation depths of aragonite and calcite are not yet understood, several explanations have been advanced to account for global changes in water chemistry. These include an increase in the temperature of deep ocean water, a decrease in the chlorinity of deep waters through the addition of large amounts of glacial melt water, and an increase in the availability of calcium carbonate in deep water owing to large-scale redeposition of shallow-water sediments (Berger, 1977). The preservation peak of pelagic carbonate skeletons also has been linked with the advent of a proliferation of tropical rainforests during the late Pleistocene (Shackleton, 1977). It is suggested that the great increase in terrestrial photosynthesis would have removed large amounts of carbon dioxide from the atmosphere, thus increasing carbonate concentrations in the sea and shifting the compensation depths of aragonite and calcite to deeper water.

Evolution

Deposits of sedimented shells provide a good fossil record for this group of planktonic mollusks. Affinities with ancient Palaeozoic organisms, such as *Conularia* and *Hyolithes*, have been largely dismissed (Moore et al., 1952), as have claims of Cretaceous species (Curry and Morris, 1967). The earliest thecosome species appear with certainty in the Lower Eocene (about 55–60 million years ago) in a number of widely distributed rock formations (Collins, 1934; Curry, 1965; Curry and Morris, 1967). The fossil record does not lend support to the general idea that the spirally coiled forms (*Limacina* or *Peraclis*) gave rise to the bilaterally symmetrical forms of the Cavoliniidae. Cavoliniids are represented by conical *Creseis* and *Clio*-like forms in Eocene rocks as old as those containing fossil spiral thecosomes. *Limacina inflata* is the oldest Recent limacinid species positively identified as abundant in rocks of the Middle Miocene (Collins, 1934); it does not appear in formations prior to the appearance of cavoliniids. It is possible that the survival of deposits of the larger cavoliniids has been enhanced, or that the radiation of thecosomes occurred within the bounds of detection of the fossil record. Recognizable forms of *Cavolinia* and *Diacria* are later cavoliniid arrivals that do not appear before the Miocene (ca. 25 million years ago). The fossil cavoliniids in general had thicker shells and thicker shell apertures than the modern species, suggesting that an overall reduction in shell weight has occurred. There are no known fossils for the pseudothecosome families Peraclididae, Cymbuliidae, and Desmopteridae.

There has been considerable speculation about which ancestral groups of mollusks might have given rise to the thecosomes and which modern groups are most closely related. Thecosomes have been associated with most of the major lineages of opisthobranchs, illustrating not only the remarkable similarities often displayed by different gastropods but also the difficulties of separating characters that reflect true homology from those that derive from convergent evolution.

A number of recent workers (e.g. Morton, 1954a; Boettger, 1955) have tended to favor the position first expressed by Boas (1886b) and expanded by Pelseneer (1888b) that the thecosomes are closely related to the bulloid cephalaspideans. Similarities between these two opisthobranch orders are found in the similarly shaped radular teeth and small numbers of lateral teeth; in certain features of the gut, including the presence of gizzard teeth in both groups; in the relatively unconcentrated nervous systems; and in the similarity of the opercula of *Limacina* and *Actaeon*. Lemche (1948) suggested that the thecosomes arose by neoteny from free-swimming larvae of benthic bulloids. This hypothesis provides a mechanism of reducing body size and of gaining access to the pelagic environment. It would also explain why the shells of some thecosomes are sinistrally coiled, since this character is shared by the larvae of many opisthobranch groups. Although this hypothesis is theoretically attractive, it does not conclusively link thecosomes with the cephalaspideans. This idea also has been criticized by Jägersten (1972), who pointed out that it is unlikely thecosomes arose from veliger larvae that became sexually mature, since there is no evidence for retention of the velum as a swimming organ in adults. Instead, the foot, by transforming from a creeping organ to swimming fins, seems to have played the decisive role in the transition to a holopelagic existence. This argues for a gradual evolution in an adult stage of a benthic ancestor, with transition from a temporary to a permanent swimming habit.

Bandel et al. (1984) also proposed a neotenic origin for thecosomes, but derived the group from benthic prosobranchs. This idea was based primarily on superficial similarities between the shells of *Limacina* and *Peraclis* and those of the large larvae of certain prosobranchs (Architectonicacea, Tonnacea) that have either sinistral or dextral coiling and an extended pelagic life. They avoided the issue of the velum as a larval character only and proposed that the frontal and lateral portions of the developing larval foot could have been used as wings, with the foot cilia substituting for the velar food groove. This hypothesis ignores the opisthobranch anatomical features that are thought to prevail in adult thecosomes (cf. Pelseneer, 1888a, 1888b).

Although most recent workers have followed Boas's (1886b) early lead in placing the thecosomes and gymnosomes in separate orders, there is not unanimity for this position. Ghiselin (1965) reexamined the reproductive systems of all opisthobranch groups and used this information as a basis for critically assessing phylogenetic studies. He concluded that similar reproductive tract anatomy in thecosomes and gymnosomes hints at a common ancestry, with present differences in the two groups resulting from an early divergence in feeding habits. These ideas are considered further in a discussion of gymnosome evolution in Chapter 5.

The planktonic specializations of this group, as well as morphological simplifications relating to size and weight reduction, are such that the thecosomes cannot be linked with assurance to any close relatives. It may be that additional comparative studies of living thecosomes (and gymnosomes) and more detailed anatomical studies of a greater number of species will provide additional clues to the affinities of these animals.

There is general agreement, however, that the spirally coiled genus *Limacina* includes the most primitive members of the euthecosomes. All species of the families Limacinidae and Cavoliniidae show close relationships in internal organization, feeding methods, and swimming behavior. The cavoliniids are set apart by having undergone a secondary detorsion of the mantle cavity and viscera relative to the limacinids (cf. Boas, 1886a, 1886b), and this has resulted in bisymmetry of the shell and body (cf. Ghiselin, 1966). Boas (1886a) hypothesized that primitive, spirally coiled thecosomes led directly to straight, conical species (i.e. *Creseis*) by an uncoiling of the shell during detorsion.

Many phylogenetic studies (e.g. Boas, 1886a, 1886b; Pelseneer, 1888b, 1906; Meisenheimer, 1905b, 1906) assumed a close affinity of *Peraclis* with *Limacina* because of the spiral character of the shells and the supposed similar positions of the footlobes and wings. This apparent similarity in orientation, however, is an artifact of preservation caused by the manner in which *Peraclis* retracts into the shell (Gilmer and Harbison, 1986). All pseudothecosomes (excluding *Desmopterus*) share a common body plan. The orientation of the sinistrally coiled shell of *Peraclis* with respect to soft body parts is the same as the relative orientation of the pseudoconch in adult cymbuliids, and it is also apparent in cymbuliid larvae that have a spirally coiled, external, calcareous shell (Krohn, 1860; and above, Fig. 45). The cymbuliids could easily be derived from a primitive *Peraclis*-like form by replacing the calcareous shell with a gelatinous pseudoconch and leaving the orientation of the soft parts unchanged. Interestingly, the pseudothecosome orientation is reminiscent of an untorted form (termed exogastric), such as is present in *Nautilus*; this is considered to be the most advantageous configuration for a swimming gastropod with a spiral shell (Naef, 1911; Ghiselin, 1966). In this configuration, the apex of the shell lies in front of the mouth and foot and also leads through the water during swimming. This constitutes a fundamental difference between the pseudothecosomes, which usually employ this mode of swimming, and the euthecosomes, which pull the shell behind the wings as they swim. It is also noteworthy that the supposed "gill" of *Peraclis*, which is simply retracted and folded mantle tissue, has been a major criterion for regarding this genus as primitive to all other thecosomes (e.g. Pelseneer, 1906; Tesch, 1946).

The absence of a gill in *Peraclis*, the evidence that certain cavoliniids produce temporary pseudoconchs, and the discovery of the commonality of a mucous-web feeding mechanism in all thecosomes should lead to a re-evaluation of phylogenetic relationships within this group. We believe that more detailed studies of the biology of living animals, particularly of reproductive biology, are necessary for the development of a mature appraisal. Similarly, we suggest that studies of feeding and reproduction in *Desmopterus* may resolve the phylogenetic position and taxonomic affinities of this unusual mollusk that presently occupies an uneasy position in a separate family of pseudothecosomes.

Some workers have attempted to base phylogenetic studies on the basic differences in shell microstructure that have been noted between the spirally coiled species of *Limacina* and *Peraclis*, which have a crossed-lamellar and/or prismatic pattern, and the symmetrical cavoliniids with a helical arrangement in the adult shell (e.g. Rampal, 1977). The larval shell of *Cavolinia* species also has a spiral crystalline pattern, but the protoconch in other cavoliniid genera is composed of the crossed-lamellar or prismatic structure that is regarded by Rampal (1973, 1974, 1977) as reflecting the origin of cavoliniids from spirally coiled ancestral limacinids via detorsion of the shell. Curry and Rampal (1979) have investigated the shell microstructure of fossil thecosomes from the Middle and Lower Eocene and the Lower Miocene. They found that a spirally coiled fossil species had a prismatic crystalline arrangement resembling that of modern *Limacina* species. Similarly, the two fossil species with straight shells had a helical pattern like that of the modern cavoliniids. The authors interpreted these results as indicating that the evolution from spirally coiled to straight shells was reflected at a very early stage in the shell microstructure of fossil forms. The implications of using shell microstructure as a phylogenetic character are not clear, however, and it has been shown that the helical crystalline pattern is not unique to thecosomes, as once thought (Bé et al., 1972), but is present in the larval shells of several other mollusks (Richter, 1976).

Several molluscan organs have undergone major changes in thecosomes. The foot has not only become modified as a swimming organ, but all thecosomes share a common modification of the footlobes associated with the development of a feeding mechanism that can filter a variety of sizes of small organisms and suspended organic particles from large volumes of water. The development of external feeding webs has probably been of considerable importance in the spread of thecosomes throughout oligotrophic oceans, where large quantities of water must be cleared by suspension feeders just to account for simple body metabolism (Jørgensen, 1966). Although all species have similar feeding habits, the footlobes do exhibit some variation in morphology with a tendency toward reduction and concentration of the area of the ciliated tracts used in moving the mucous feeding web to the mouth. In the pseudothecosomes, the development and gradual lengthening of a proboscis has resulted in the separation of the feeding web from the body and wings. This is exemplified in *Gleba*, which can position its body as much as two body widths away from its web. The common feeding mechanism is also apparently reflected in the well-developed, mucus-secreting, pallial gland that is present in all thecosomes except *Desmopterus*. The gut of thecosomes has several features that are considered to be relics of a macrophagous ancestry, including the muscular gizzard and a style sac that is probably functionally vestigial (Morton, 1960); since these features are found in a number of mollusks, they are unlikely to indicate close phylogenetic ties.

The development of buoyancy mechanisms also marks the evolution of this planktonic group. Members of the families Limacinidae, Cavoliniidae, Peraclididae, and Cymbuliidae all have the capacity to remain neutrally buoyant, or nearly so, for apparently longer periods than they spend swimming. Buoyancy control not only has resulted in a saving of energy normally used in swimming, but it has been intimately linked with the evolution of the feeding mechanism, since it has permitted the cessation of swimming during feeding. The common ability of all thecosomes to float indicates that neither shell shape nor the extent of gelatinous tissue (i.e. Cymbuliidae) is the only factor regulating buoyancy control. Although there is no obvious reduction of shell size in euthecosomes, as there is in the evolution of heteropods, there is a tendency for the shell to become internal. Functionally, the shells of *Cavolinia* and *Peraclis* spp. have become secondarily internal, since they are normally completely covered with mantle tissue. Similarly, *Cavolinia* has developed a dorsal, temporary pseudoconch that is neutrally buoyant and covered by folds of extruded mantle. In cymbuliids, the shell has been completely replaced by a permanent pseudoconch that is covered by integument.

Reproductive anatomy generally appears to be conservative within this group, with few differences between species; the major exceptional species is the epipelagic brooder, *Limacina inflata*. Some euthecosomes are now known to accomplish fertilization via spermatophore exchange, and it is likely that more species will be discovered to do so. Spermatophore production should be regarded as a common phenomenon in many pelagic animals and not indicative of close relationship to other molluscan groups with this type of reproduction. The asexual divisions of *Clio pyramidata* and *C. polita* have not been demonstrated to produce viable offspring, and the adaptive function of this type of reproduction is not clear; in any event, it is difficult to accept van der Spoel's theory (1979) that this adds evidence for linking mollusks with a distant coelenterate origin.

Only a few thecosomes are known to have partially or totally suppressed a free-swimming veliger stage, and all but one are mesopelagic or bathypelagic. These deep-water species may have developed brood protection or ovoviviparity in response to an environment with low quantities of food suitable for ciliary feeding larvae. Indeed, many deep-sea species of other phyla also show suppression of free larval stages. The one epiplanktonic brooder, *Limacina inflata*, is the most successful limacinid in terms of abundance and widespread distribution; in fact, Lalli and Wells (1978) have suggested that it may be the most abundant marine gastropod, including benthic species. Its success is probably linked with its unusual reproductive strategy, but this remains obscure. It may be that the production of few young and the concomitant loss of accessory reproductive glands allows for a smaller species size; this is of obvious adaptive value for flotation in a planktonic animal living in warm water of low density, and it places a lower demand for food in oligotrophic tropical waters.

The mantle and mantle cavity have been involved in most of the major adaptational changes that have taken place in this group. The mantle not only is involved in shell formation in the euthecosomes, but also presumably secretes the temporary pseudoconchs of *Cavolinia* species and the permanent pseudoconchs of cymbuliids. Extrusions of mantle folds from the shell play an important role in buoyancy in a variety of cavoliniids. The pallial or mantle gland is an outgrowth of the mantle lining, and this gland is suspected of providing the major amounts of mucus for feeding webs; in *Peraclis*, prominent protrusions on the edge of the mantle also appear to be involved in mucus production. The mantle cavity itself not only is the site of water transport for respiration, but functions as a brood chamber in *Limacina inflata* and in at least two species of *Clio*. The extraordinary plasticity of this uniquely molluscan organ has been pointed out before (e.g. Yonge and Thompson, 1976) as being of overwhelming significance in the radiation of mollusks.

List of Recognized Species

(Synonymy is given in van der Spoel, 1967 and 1976)

Suborder Euthecosomata

Family Limacinidae

Limacina helicina (Phipps, 1774)

L. retroversa (Fleming, 1823)

L. bulimoides (d'Orbigny, 1836)
L. inflata (d'Orbigny, 1836)
L. lesueuri (d'Orbigny, 1836)
L. trochiformis (d'Orbigny, 1836)
L. helicoides Jeffreys, 1877

Family Cavoliniidae

Subfamily Clioinae

Creseis virgula (Rang, 1828)
C. acicula (Rang, 1828)
C. chierchiae (Boas, 1886)
Styliola subula (Quoy and Gaimard, 1827)
Hyalocylis striata (Rang, 1828)
Clio pyramidata Linnaeus, 1767
C. cuspidata (Bosc, 1802)
C. recurva (Childern, 1823)
C. chaptali Gray, 1850
C. sulcata (Pfeffer, 1879)
C. andreae (Boas, 1886)
C. convexa (Boas, 1886)
C. scheelei (Munthe, 1888)
C. antarctica Dall, 1908
C. campylura (Tesch, 1948)
C. orthotheca (Tesch, 1948)

Subfamily Cuvierininae

Cuvierina columnella (Rang, 1827)

Subfamily Cavoliniinae

Diacria trispinosa (de Blainville, 1821)
D. quadridentata (de Blainville, 1821)
(Considered by van Leyen and van der Spoel (1982) to consist of five species, four subspecies, and four formae.)
D. major (Boas, 1886)
D. rampali Dupont, 1979
Cavolinia tridentata (Niebuhr, 1775)
C. inflexa (Lesueur, 1813)
C. longirostris (de Blainville, 1821)
C. uncinata (Rang, 1829)
C. gibbosa (d'Orbigny, 1836)
C. globulosa (Gray, 1850)

Suborder Pseudothecosomata

Family Peraclididae

Peraclis reticulata (d'Orbigny, 1836)

P. triacantha (Fischer, 1882)

P. bispinosa Pelseneer, 1888

P. moluccensis Tesch, 1903

P. valdiviae (Meisenheimer, 1905)

P. apicifulva Meisenheimer, 1906

P. depressa Meisenheimer, 1906

Family Cymbuliidae

Cymbulia peroni de Blainville, 1818

C. parvidentata Pelseneer, 1888

?*C. sibogae* Tesch, 1903 (a species of uncertain validity)

?*C. tricavernosa* Zhang, 1964 (a species of uncertain validity)

Gleba cordata Niebuhr, 1776

?*G. chrysostricta* (Troschel, 1854) (a species of uncertain validity)

Corolla spectabilis Dall, 1871

?*C. ovata* (Quoy and Gaimard, 1832) (a species of uncertain validity)

C. calceola (Verrill, 1880)

?*C. intermedia* (Tesch, 1903) (a species of uncertain validity)

Family Desmopteridae

Desmopterus papilio Chun, 1889

D. gardineri Tesch, 1910

?*D. pacificus* Essenberg, 1919 (a species of uncertain validity)

References Cited

Works with recommended keys or aids for the identification of species are indicated by an asterisk.

Ackman, R. G., J. Hingley, and K. T. MacKay. 1972. Dimethyl sulfide as an odor component in Nova Scotia fall mackerel. *J. Fish. Res. Bd Can.* 29: 1085–88.

Agassiz, A. 1866. On the habits of a species of pteropod (*Spirialis flemingii?*). *Proc. Boston Soc. nat. Hist.* 10: 14–15.

Almogi-Labin, A. 1982. Stratigraphic and paleoceanographic significance of Late Quaternary pteropods from deep-sea cores in the Gulf of Aqaba (Elat) and northernmost Red Sea. *Mar. Micropaleont.* 7: 53–72.

Andersen, N. R., and A. Malahoff, eds. 1977. *The Fate of Fossil Fuel CO_2 in the Oceans*. New York: Plenum Press. 749 pp.

Arnaud, P. M. 1973. Le genre *Lepas* Linné, 1758, dans les terres australes et Antarctiques Françaises (Cirripedia). *Crustaceana* 24: 157–62.

Bainbridge, V., and B. J. McKay. 1968. The feeding of cod and redfish larvae. *Int. Comm. Northwest Atl. Fisheries, Spec. Publ.* No. 7. Environmental surveys—Norwestlant 1–3, 1963. Part I. Pp. 187–217.

Bandel, K., A. Almogi-Labin, C. Hembleben, and W. G. Deuser. 1984. The conch of *Limacina* and *Peraclis* (Pteropoda) and a model for the evolution of planktonic gastropods. *Neues Jb. Geol. Paläont. Abh. 168*: 87–107.

*Bé, A. W. H., and R. W. Gilmer. 1977. A zoogeographic and taxonomic review of euthecosomatous pteropoda. In: *Oceanic Micropaleontology*, vol. 1, chap. 6. A. T. S. Ramsay, ed. London: Academic Press. Pp. 733–808.

Bé, A. W. H., C. MacClintock, and D. C. Currie. 1972. Helical shell structure and growth of the pteropod *Cuvierina columnella* (Rang) (Mollusca, Gastropoda). *Biomineralization Res. Reps.* 4: 47–79.

Beers, J. R. 1966. Studies on the chemical composition of the major zooplankton groups in the Sargasso Sea off Bermuda. *Limnol. Oceanogr. 11*: 520–28.

Bellaiche, G., and C. Thiriot-Quiévreux. 1982. The origin and significance of a thick deposit of pteropod shells in the Rhône deep-sea fan. *Palaeogeogr., Palaeoclim., Palaeoecol.* 39: 129–37.

Berger, W. H. 1977. Deep-sea carbonate and the deglaciation preservation spike in pteropods and foraminifera. *Nature, Lond.* 269: 301–4.

———. 1978. Deep-sea carbonate: pteropod distribution and the aragonite compensation depth. *Deep-Sea Res.* 25: 447–52.

———. 1982. Processes in the deep ocean. *Science 218*: 465–66.

Berner, R. A. 1977. Sedimentation and dissolution of pteropods in the ocean. In: *The Fate of Fossil Fuel CO_2 in the Oceans*. N. R. Andersen and A. Malahoff, eds. New York: Plenum Press. Pp. 243–60.

Berner, R. A., and S. Honjo. 1981. Pelagic sedimentation of aragonite: Its geochemical significance. *Science 211*: 940–42.

Betzer, P. R., R. H. Byrne, J. G. Acker, C. S. Lewis, R. R. Jolley, and R. A. Feely. 1984. The oceanic carbonate system: A reassessment of biogenic controls. *Science* 226: 1074–77.

Biggs, D. C. 1977. Respiration and ammonium excretion by open ocean gelatinous zooplankton. *Limnol. Oceanogr.* 22: 108–17.

de Blainville, H. M. D. 1824. *Dictionnaire des sciences naturelles*, vol. 32. Paris: Levrault. 567 pp.

Blanc-Vernet, L., H. Chamley, and C. Froget. 1969. Analyse paléoclimatique d'une carotte de Méditerranée nord-occidentale. Comparaison entre les résultats de trois études: Foraminifères, Ptéropodes, fraction sédimentaire issue du continent. *Palaeogeogr., Palaeoclim., Palaeoecol.* 6: 215–35.

Boas, J. E. V. 1886a. Spolia Atlantica. Bidrag til Pteropodernes. Morfologi og systematik samt til Kundskaben om deres geografiske udbredelse. *K. danske Vidensk. Selsk. Skr.*, 6 Raekke, naturv. mat. Afd., 4: 1–231.

———. 1886b. Zur Systematik und Biologie der Pteropoden. *Zool. Jb.* *1*: 311–40.

Boettger, C. R. 1955. Die Systematik der Euthyneuren Schnecken. *Zool. Anz.* (*Suppl.* 18). *Verh. dt. Zool. Ges.* (1954): 253–80.

Boltovskoy, D. 1971. Pteropodos thecosomados del Atlantico suboccidental. *Malacologia* *11*: 121–40.

———. 1975. Ecological aspects of zooplankton (Foraminifera, Pteropoda and Chaetognatha) of the southwestern Atlantic Ocean. *Veliger* *18*: 203–16.

Bonnevie, K. 1913. Pteropoda. *Rep. "Michael Sars" N. Atl. Deep Sea Exped.* *3*: 1–85.

———. 1916. Mitteilungen über Pteropoden. I. Beobachtungen über den Geschlectsapparat von *Cuvierina columnella* Rang. *Jena. Z. Naturw.* *54*: 245–76.

Broecker, W. S. 1971. Calcite accumulation rates and glacial to interglacial changes in oceanic mixings. In: *Late Cenozoic Glacial Ages*. K. K. Turekian, ed. New Haven: Yale Univ. Press. Pp. 239–65.

Bruland, K. W., and M. W. Silver. 1981. Sinking rates of fecal pellets from gelatinous zooplankton (salps, pteropods, doliolids). *Mar. Biol.* *63*: 295–300.

Buccheri, G. 1984. Pteropods as climatic indicators in Quaternary sequences: A Lower-Middle Pleistocene sequence outcropping in Cava Puleo (Ficarazzi, Palermo, Italy). *Palaeogeogr., Palaeoclim., Palaeoecol.* *45*: 75–86.

Byrne, R. H., J. G. Acker, P. R. Betzer, R. A. Feely, and M. H. Cates. 1984. Water column dissolution of aragonite in the Pacific Ocean. *Nature, Lond.* *312*: 321–26.

Chen, C. 1964. Pteropod ooze from Bermuda Pedestal. *Science* *144*: 60–62.

———. 1966. Calcareous zooplankton in the Scotia Sea and Drake Passage. *Nature, Lond.* *212*: 678–81.

———. 1968a. The distribution of thecosomatous pteropods in relation to the Antarctic Convergence. *Antarct. J. U.S.* Pp. 155–57.

———. 1968b. Pleistocene pteropods in pelagic sediments. *Nature, Lond.* *219*: 1145–49.

Chen, C., and A. W. H. Bé. 1964. Seasonal distributions of euthecosomatous pteropods in the surface waters of five stations in the western North Atlantic. *Bull. mar. Sci. Gulf Caribb.* *14*: 185–220.

Chen, C., and N. S. Hillman. 1970. Shell-bearing pteropods as indicators of water masses off Cape Hatteras, North Carolina. *Bull. mar. Sci.* *20*: 350–67.

Chun, C. 1889. Bericht über eine nach den Canarischen Inseln im Winter 1887/1888 ausgeführte Reise. *Sber. preuss. Akad. Wiss.* 2: 519–53.

Collins, R. L. 1934. A monograph of the American Tertiary pteropod molluscs. *Johns Hopk. Univ. Stud. Geol.* *11*: 137–234.

Conover, R. J. 1978. Transformation of organic matter. In: *Marine Ecology*, vol. 4: Dynamics, chap. 5. O. Kinne, ed. New York: Wiley. Pp. 221–499.

Conover, R. J., and C. M. Lalli. 1972. Feeding and growth in *Clione limacina* (Phipps), a pteropod mollusc. *J. expl mar. Biol. Ecol.* *9*: 279–302.

Conover, R. J., and M. A. Paranjape. 1977. Comments on the use of a deep tank in planktological research. *Helgoländer wiss. Meeresunters.* *30*: 105–17.

Cornelius, P. F. S. 1982. Hydroids and medusae of the family Campanulariidae recorded from the eastern North Atlantic, with a world synopsis of genera. *Bull. Br. Mus. nat. Hist.* (*Zool.*) 42: 37–148.

Costa, A. 1867. Illustrazione della *Spirialis recurvirostra*. *Annls Mus. Zool. Univ. Napoli* 4(1864): 56–58; 102–3.

Curl, H. 1962. Standing crops of carbon, nitrogen, and phosphorus and transfer between trophic levels, in continental shelf waters south of New York. *Rapp. P.-v. Réun. Cons. perm. int. Explor. Mer.* *153*: 183–9.

Curry, D. 1965. The English Palaeogene pteropods. *Proc. malac. Soc. Lond.* *36*: 357–71.

Curry, D., and N. J. Morris. 1967. Mollusca: Amphineura, Monoplacophora and Gastropoda. In: *The Fossil Record: A Symposium with Documentation*. W. B. Harland et al., eds. Geol. Soc. London. Pp. 423–30.

Curry, D., and J. Rampal. 1979. Shell microstructure in fossil thecosome pteropods. *Malacologia* *18*: 23–25.

Cuvier, G. 1804. Mémoire concernant l'animal de l'*Hyale*, un nouveau genre de mollusques nus, intermédiaire entre l'*Hyale* et le *Clio* et l'établissement d'un nouvel ordre dans la classe des mollusques. *Annls Mus. Hist. nat. Paris* 4: 223–34.

———. 1817. *Le Règne animal distribué d'après son organisation, pour servir de base à l'histoire naturelle des animaux*, vols. 1–4. Paris: Déterville.

Denton, E. J., and T. I. P. Shaw. 1961. The buoyancy of gelatinous marine animals. *J. Physiol., London* *161*: 14P-15P (Proceedings).

Diester-Haass, L. 1973. Holocene climate in the Persian Gulf as deduced from grain-size and pteropod distribution. *Mar. Geol. 14*: 207–23.
Diester-Haass, L., and S. van der Spoel. 1978. Late Pleistocene pteropod-rich sediment layer in the Northeast Atlantic and protoconch variation of *Clio pyramidata* Linné 1767. *Palaeogeogr., Palaeoclim., Palaeoecol.* 24: 85–109.
Dunbar, M. J. 1942. Marine macroplankton from the Canadian Eastern Arctic. II. Medusae, Siphonophora, Ctenophora, Pteropoda, and Chaetognatha. *Can. J. Res. 20D*: 71–77.
Dupont, L. 1979. Note on variation in *Diacria* Gray, 1847, with descriptions of a species new to science, *Diacria rampali* nov. spec., and a forma new to science, *Diacria trispinosa* forma *atlantica* nov. forma. *Malacologia 18*: 37–52.
Essenberg, C. 1919. The pteropod *Desmopterus pacificus* sp. nov. *Univ. Calif. Publs Zool. 19*: 85–88.
Fager, E. W., and J. A. McGowan. 1963. Zooplankton species groups in the North Pacific. *Science 140*: 453–60.
Fol, H. 1875. Etudes sur le développement des Mollusques. Premier mémoire: Sur le développement des Ptéropodes. *Archs Zool. exp. gén.* 4: 1–214.
Fraser, J. H. 1962. *Nature Adrift: The Story of Marine Plankton*. London: Foulis. 178 pp.
———. 1970. The ecology of the ctenophore *Pleurobrachia pileus* in Scottish waters. *J. Cons. perm. int. Explor. Mer* 33: 149–68.
Froget, C. 1967. Les ptéropodes dans les sédiments sous-marins du quaternaire: Caractérisation du régime "nord-atlantique" au cours des périodes glaciaires en Méditerranée par le ptéropode *Spiratella retroversa* Fleming. *C. r. hebd. Séanc. Acad. Sci., Paris* 264: 2968–69.
Furnestin, M.-L. 1979. Planktonic molluscs as hydrological and ecological indicators. In: *Pathways in Malacology*. S. van der Spoel, A. C. van Bruggen, and J. Lever, eds. Utrecht: Bohn, Scheltema, and Holkema. Pp. 175–94.
Gaudy, R. 1974. Feeding four species of pelagic copepods under experimental conditions. *Mar. Biol. 25*: 125–41.
Ghiselin, M. T. 1965. Reproductive function and the phylogeny of opisthobranch gastropods. *Malacologia* 3: 327–78.
———. 1966. The adaptive significance of gastropod torsion. *Evolution* 20: 337–48.
Gilmer, R. W. 1972. Free-floating mucus webs: A novel feeding adaptation for the open ocean. *Science 176*: 1239–40.
———. 1974. Some aspects of feeding in the thecosomatous pteropod molluscs. *J. expl mar. Biol. Ecol. 15*: 127–44.

———. 1986. Preservation artifacts and their effects on the study of euthecosomatous pteropod mollusks. *Veliger* 29: 48–52.

Gilmer, R. W., and G. R. Harbison. 1986. Morphology and field behavior of pteropod molluscs: feeding methods in the families Cavoliniidae, Limacinidae and Peraclididae (Gastropoda: Thecosomata). *Mar. Biol.* 91: 47–57.

Grossman, E. L., P. R. Betzer, W. C. Dudley, and R. B. Dunbar. 1986. Stable isotopic variation in pteropods and atlantids from North Pacific sediment traps. *Mar. Micropaleont.* 10: 9–22.

Haagensen, D. A. 1976. *Caribbean Zooplankton.* Part II—Thecosomata. Office of Naval Research, Dept. of the Navy: Washington, D.C. Pp. 551–712.

Hamner, W. M. 1974. Blue-water plankton. *Natn. geogr.* 146: 530–45.

Hamner, W. M., L. P. Madin, A. L. Alldredge, R. W. Gilmer, and P. P. Hamner. 1975. Underwater observations of gelatinous zooplankton: Sampling problems, feeding biology, and behavior. *Limnol. Oceanogr.* 20: 907–17.

Harbison, G. R., D. C. Biggs, and L. P. Madin. 1977. The associations of Amphipoda Hyperiidea with gelatinous zooplankton—II. Associations with Cnidaria, Ctenophora and Radiolaria. *Deep-Sea Res.* 24: 465–88.

Harbison, G. R., and R. W. Gilmer. 1986. Effects of animal behavior on sediment trap collections: implications for the calculation of aragonite fluxes. *Deep-Sea Res.* 33: 1017–24.

Hardy, A. C. 1924. The herring in relation to its animate environment. Part I. The food and feeding habits of the herring with special reference to the East Coast of England. *Fishery Invest., London,* series II (3): 1–53.

Hartmann, J., and H. Weikert. 1969. Tagesgang eines Myctophiden (Pisces) und zweier von ihm gefressener Mollusken des Neustons. *Kiel. Meeresforsch.* 25: 328–30.

Heath, H., and M. H. Spaulding. 1904. The anatomy of a pteropod, *Corolla (Cymbuliopsis) spectabilis* Dall. *Zool. Jb. Anat. Ontog. Tiere* 20: 67–80.

Herman, Y. 1971. Vertical and horizontal distribution of pteropods in Quaternary sequences. In: *The Micropalaeontology of Oceans.* B. M. Funnell and W. R. Riedel, eds. Cambridge: Univ. Press. Pp. 463–86.

———. 1981. Paleoclimatic and paleohydrologic record of Mediterranean deep-sea cores based on pteropods, planktonic and benthonic Foraminifera. *Revista esp. Micropaleont.* 13: 171–200.

Herman, Y., and P. E. Rosenberg. 1969. Pteropods as bathymetric indicators. *Mar. Geol.* 7: 169–73.

Hirohito. 1977. Five hydroid species from the Gulf of Aqaba, Red Sea. Biological Laboratory, Imperial Household, Tokyo. 26 pp.

Ho, J.-S. 1966. Larval stages of *Cardiodectes* sp. (Caligoida: Lernaeoceriformes), a copepod parasitic on fishes. *Bull. mar. Sci. 16*: 159–99.

Hochberg, F. G., and R. R. Seapy. 1985. Parasites of holopelagic molluscs. Intl. Symp. Mar. Plankton, July, 1984, Tokai University, Shimizu, Japan. *Bull. mar. Sci.* 37: 767 (Abstract).

Honjo, S., and J. Erez. 1978. Dissolution rates of calcium carbonate in the deep ocean: An in-situ experiment in the North Atlantic Ocean. *Earth Planet. Sci. Lett.* 40: 287–300.

Howells, H. H. 1936. The anatomy and histology of the gut of *Cymbulia peronii* (Blainville). *Proc. malac. Soc. Lond.* 22: 62–72.

Hsiao, S. C. T. 1939a. The reproductive system and spermatogenesis of *Limacina* (*Spiratella*) *retroversa* (Flem.). *Biol. Bull. mar. biol. Lab., Woods Hole* 76: 7–25.

———. 1939b. The reproduction of *Limacina retroversa* (Flem.). *Biol. Bull. mar. biol. Lab., Woods Hole* 76: 280–303.

Ikeda, T. 1974. Nutritional ecology of marine zooplankton. *Mem. Fac. Fish. Hokkaido Univ.* 22: 1–97.

Ikeda, T., and E. H. Fay. 1981. Metabolic activity of zooplankton from the Antarctic Ocean. *Aust. J. mar. freshw. Res.* 32: 921–30.

Ito, J. 1964. Food and feeding habit of Pacific salmon (genus *Oncorhynchus*) in their oceanic life. *Bull. Hokkaido reg. Fish. Res. Lab.* 29: 85–97.

Ivanova, E. V. 1985. Late Quaternary biostratigraphy and paleotemperatures of the Red Sea and the Gulf of Aden based on planktonic Foraminifera and pteropods. *Mar. Micropaleont.* 9: 335–64.

Jägersten, G. 1972. *Evolution of the Metazoan Life Cycle*. New York: Academic Press. 282 pp.

Jørgensen, C. B. 1966. *Biology of Suspension Feeding*. Oxford: Pergamon. 357 pp.

Just, J. 1977. A new genus and species of corophiid Amphipoda from pteropod shells of the bathyal western Atlantic, with notes on related genera (Crustacea). *Steenstrupia* 4: 131–38.

Keferstein, W. 1862–66. Kopf-Weichthiere. Malacozoa Cephalota. In: *Klassen und Ordnungen des Thierreichs*, vol. 3. H. G. Bronn, ed. Leipzig: Winter. Pp. 523–1500.

Kobayashi, H. A. 1974. Growth cycle and related vertical distribution of the thecosomatous pteropod *Spiratella* ("*Limacina*") *helicina* in the central Arctic Ocean. *Mar. Biol.* 26: 295–301.

Kölliker, A., and H. Müller. 1853. Bericht über einige im Herbste 1852 in Messina angestellte vergleichend-anatomische Untersuchungen. Chromatophoren bei *Cymbulia*. *Z. wiss. Zool.* 4: 332–33.

Kornicker, L. S. 1959. Observations on the behavior of the pteropod *Creseis acicula* Rang. *Bull. mar. Sci. Gulf Caribb.* 9: 331–36.

Kramp, P. L. 1922. *Kinetocodium danae* n. g., n. sp., a new gymnoblastic hydroid, parasitic on a pteropod. *Vidensk. Medd. dansk naturh. Foren.* 74: 1–21.

Krohn, A. 1860. *Beiträge zur Entwickelungsgeschichte der Pteropoden und Heteropoden.* Leipzig: W. Engelmann. 45 pp.

Kubota, T., and T. Uyeno. 1970. Food habits of lancetfish, *Alepisaurus ferox* (Order Myctophiformes) in Suruga Bay, Japan. *Jap. J. Ichthyol.* 17: 22–28.

Lalli, C. M., and F. E. Wells. 1973. Brood protection in an epipelagic thecosomatous pteropod, *Spiratella* ("*Limacina*") *inflata* (d'Orbigny). *Bull. mar. Sci.* 23: 933–41.

———. 1978. Reproduction in the genus *Limacina* (Opisthobranchia: Thecosomata). *J. Zool., Proc. zool. Soc. Lond.* 186: 95–108.

Lebour, M. V. 1932. *Limacina retroversa* in Plymouth waters. *J. mar. biol. Ass. U.K.* 18: 123–29.

LeBrasseur, R. J. 1966. Stomach contents of salmon and steelhead trout in the northeastern Pacific Ocean. *J. Fish. Res. Bd Can.* 23: 85–100.

Lemche, H. 1948. Northern and Arctic tectibranch gastropods. I/II. *K. danske Vidensk. Selsk. Skr.* 5: 1–136.

van Leyen, A., and S. van der Spoel. 1982. A new taxonomic and zoogeographic interpretation of the *Diacria quadridentata* group (Mollusca, Pteropoda). *Bull. zool. Mus. Univ. Amst.* 8: 101–19.

Lippa, E. J. R. 1965. Unusual contamination of groundfish caught in northern Hecate Strait. *J. Fish. Res. Bd Can.* 22: 1311–12.

MacDonald, A. G., I. Gilchrist, and J. M. Teal. 1972. Some observations on the tolerance of oceanic plankton to high hydrostatic pressure. *J. mar. biol. Ass. U.K.* 52: 213–23.

MacDonald, J. D. 1885. On the general characters of the genus *Cymbulia. Proc. R. Soc.* 38: 251–53.

Martens, F. 1675. *Spitzbergische oder grönlandische Reise Beschreibung gethan im Jahr 1671.* Hamburg: Schultzen.

Massy, A. L. 1920. Mollusca. III. Eupteropoda (Pteropoda Thecosomata) and Pterota (Pteropoda Gymnosomata). *Br. Antarct. Terra Nova Exped. 1910, Nat. Hist. Rep. Zool.* 2: 203–32.

———. 1932. Mollusca: Gastropoda Thecosomata and Gymnosomata. *"Discovery" Rep.* 3: 267–96.

McGowan, J. A. 1968. The Thecosomata and Gymnosomata of California. *Veliger* 3 (supplement): 103–25.

———. 1971. Oceanic biogeography of the Pacific. In: *The Micropalaeon-*

tology of Oceans. B. M. Funnell and W. R. Riedel, eds. Cambridge: Univ. Press. Pp. 3–74.

———. 1974. The nature of oceanic ecosystems. In: *The Biology of the Oceanic Pacific.* C. B. Miller, ed. Corvallis: Oregon State Univ. Press. Pp. 9–28.

McLaren, I. A. 1958. The biology of the ringed seal (*Phoca hispida* Schreber) in the Eastern Canadian Arctic. *Bull. Fish. Res. Bd Can. 118*: 1–97.

———. 1963. Effects of temperature on growth of zooplankton, and the adaptive value of vertical migration. *J. Fish. Res. Bd Can.* 20: 685–727.

Meisenheimer, J. 1905a. Die arktischen Pteropoden. *Fauna arct.* 4: 408–30.

———. 1905b. Pteropoda. *Wiss. Ergebn. dt. Tiefsee-Exped. "Valdivia"* 9: 1–314.

———. 1906. Die Pteropoden der deutschen Südpolar-Expedition 1901–1903. *Dt. Südpolar-Exped.* 9 (Zool.). *1*: 94–153.

Mileikovsky, S. A. 1962. [Pelagic larvae of Gastropoda from the region of the White Sea Biological Station of Moscow State University.] In: *The Biology of the White Sea.* L. A. Zenkevitch, ed. Moscow: Moscow State Univ. Press. Pp. 171–200. [In Russian.]

Milliman, J. D. 1975. Dissolution of aragonite, Mg-calcite and calcite in the North Atlantic Ocean. *Geol., geol. Soc. Amer.* 3: 461–62.

Mironov, G. N. 1977. [On the composition of the stomach contents in the young stages of Thecosomata (Mollusca, Pteropoda) from the tropical ocean areas.] *Biol. Morya, Kiev.* 42: 33–39. [In Russian.]

Moore, M., and P. W. Valentine. 1984. Pteropods invade beaches. *Washington Post*, August 24, final edition, section C.

Moore, R. C., C. G. Lalicker, and A. G. Fischer. 1952. *Invertebrate Fossils.* New York: McGraw Hill. 766 pp.

Morton, J. E. 1954a. The biology of *Limacina retroversa. J. mar. biol. Ass. U.K.* 33: 297–312.

———. 1954b. The pelagic Mollusca of the Benguela Current. Part I. First survey, R.R.S. "William Scoresby," March 1950, with an account of the reproductive system and sexual succession of *Limacina bulimoides. "Discovery" Rep.* 27: 163–99.

———. 1960. The functions of the gut in ciliary feeders. *Biol. Rev.* 35: 92–140.

———. 1964. Locomotion. In: *Physiology of Mollusca*, vol. 1, chap. 12. K. M. Wilbur and C. M. Yonge, eds. New York: Academic Press. Pp. 383–423.

Murray, J., and J. Hjort. 1912. *The Depths of the Ocean.* London: Macmillan. 821 pp.

*Myers, T. D. 1968. Horizontal and vertical distribution of thecosomatous pteropods off Cape Hatteras. Unpubl. Ph.D. diss., Duke Univ. 224 pp.
Naef, A. 1911. Studien zur generellen Morphologie der Mollusken. *Ergeb. Fort. Zool.* 3: 73–164.
Niebuhr, C. 1772. *Beschreibung von Arabien.* Kopenhagen: Môller. 432 pp.
Nishimura, S. 1965. Droplets from the plankton net. XX. "Sea stings" caused by *Creseis acicula* Rang (Mollusca: Pteropoda) in Japan. *Publs Seto mar. biol. Lab.* 8: 287–90.
Nival, P., S. Nival, and I. Palazzoli. 1972. Données sur la respiration de différents organismes communs dans le plancton de Villefranche-sur-Mer. *Mar. Biol.* 17: 63–76.
Okutani, T. 1960. *Argonauta boettgeri* preys on *Cavolinia tridentata. Venus* 21: 39–41.
Omori, M. 1969. Weight and chemical composition of some important zooplankton in the North Pacific Ocean. *Mar. Biol.* 3: 4–10.
Pafort-van Iersel, T., and S. van der Spoel. 1979. The structure of the columellar muscle system in *Clio pyramidata* and *Cymbulia peroni* (Thecosomata, Gastropoda) with a note on the phylogeny of both species. *Bijdr. Dierk.* 48: 111–26.
———. 1986. Schizogamy in the planktonic opisthobranch *Clio*—A previously undescribed mode of reproduction in the Mollusca. *Intl. J. invert. Reprod. Develop.* 10: 43–50.
Paranjape, M. A. 1968. The egg mass and veligers of *Limacina helicina* Phipps. *Veliger* 10: 322–26.
Pearcy, W. G., and L. F. Small. 1968. Effects of pressure on the respiration of vertically migrating crustaceans. *J. Fish. Res. Bd Can.* 25: 1311–16.
Pelseneer, P. 1887. Report on the Pteropoda collected by H.M.S. Challenger during the years 1873–76. I. The Gymnosomata. *Scient. Rep. "Challenger,"* Zoology 19: 1–74.
———. 1888a. Report on the Pteropoda collected by H.M.S. Challenger during the years 1873–76. II. The Thecosomata. *Scient. Rep. "Challenger,"* Zoology 23: 1–132.
———. 1888b. Report on the Pteropoda collected by H.M.S. Challenger during the years 1873–76. III. Anatomy. *Scient. Rep. "Challenger," Zoology* 23: 1–97.
———. 1906. Biscayan plankton. VII. Mollusca (excluding Cephalopoda). *Trans. Linn. Soc. Lond.*, 2nd ser. 10: 137–57.
Perkins, P. S. 1983. The life history of *Cardiodectes medusaeus* (Wilson), a copepod parasite of lanternfishes (Myctophidae). *J. crustacean Biol.* 3: 70–87.

Péron, F., and C. A. Lesueur. 1810. Histoire de la famille des Mollusques Ptéropodes: Caractères des dix genres qui doivent la composer. *Annls Mus. Hist. nat. Paris 15*: 57–69.
Phipps, C. J. 1774. *A Voyage Towards the North Pole.* London: Bowyer and Nichols. 275 pp.
Picard, J. 1956. Le premier stade de l'Hydroméduse *Pandea conica*, issu de l'Hydropolype *Campaniclava cleodorae. Bull. Inst. océanogr., Monaco 1086*: 1–11.
Price, B. A., J. S. Killingley, and W. H. Berger. 1985. On the pteropod pavement of the eastern Rio Grande Rise. *Mar. Geol.* 64: 217–35.
Purcell, J. E. 1981. Selective predation and caloric consumption by the siphonophore *Rosacea cymbiformis* in nature. *Mar. Biol.* 63: 283–94.
Rampal, J. 1973. Phylogenie des ptéropodes thécosomes d'après la structure de la coquille et la morphologie du manteau. *C. r. hebd. Séanc. Acad. Sci., Paris* 277: 1345–48.
———. 1974. Structure de la coquille des Ptéropodes au microscope à balayage. *Rapp. Comm. int. Mer Médit.* 22: 133–34.
———. 1975. Les thécosomes (Mollusques pélagiques). Systématique et évolution—Écologie et biogéographie Méditerranéennes. Unpubl. D.Sc. diss., Univ. de Provence, Marseille. 485 pp.
———. 1977. Diversité de structure de la coquille chez les thécosomes. *Atti Soc. ital. Sci. nat. Museo civ. Stor. nat. Milano 118*: 207–11.
Rang, P. C. A. L., and L. F. A. Souleyet. 1852. *Histoire naturelle des Mollusques Ptéropodes. Monographie comprenant la description de toutes les espèces de ce groupe de Mollusques.* Paris: J. B. Baillière. 86 pp.
Redfield, A. C. 1939. The history of a population of *Limacina retroversa* during its drift across the Gulf of Maine. *Biol. Bull. mar. biol. Lab., Woods Hole* 76: 26–47.
Rees, W. J. 1956. A revision of the hydroid genus *Perigonimus* M. Sars, 1846. *Bull. Br. Mus. nat. Hist.* (*Zool.*) 3: 337–50.
Richter, G. 1973. Zur Stammesgeschichte pelagisher Gastropoden. *Natur Mus., Frankf. 103*: 265–75.
———. 1976. Zur Frage der Verwandtschaftsbeziehungen von Limacinidae und Cavoliniidae. *Arch. Mollusken. 107*: 137–44.
———. 1977. Jäger, Fallensteller und Sammler (Zur Ernährung planktischer Schnecken). *Natur Mus., Frankf. 107*: 221–34.
*———. 1979. Die thecosomen Pteropoden der "Meteor"-Expedition in den Indischen Ozean 1964/65. "*Meteor*" *Forsch.-Ergebnisse* 29(D): 1–29.
———. 1983. Lebensformen und Nahrungsketten der Hochsee. Teil I and II. *Natur Mus., Frankf. 113*: 131–38 and 166–77.
Rottman, M. L. 1976. Euthecosomatous pteropods (Mollusca) in the Gulf

of Thailand and the South China Sea: Seasonal distribution and species associations. *Naga Rep.* 4: 1–117.

———. 1978. Ecology of recurrent groups of pteropods, euphausiids, and chaetognaths in the Gulf of Thailand and the South China Sea. *Mar. Biol.* 48: 63–78.

———. 1979. Dissolution of planktonic Foraminifera and pteropods in South China Sea sediments. *J. foram. Res.* 9: 41–49.

———. 1980. Net tow and surface sediment distributions of pteropods in the South China Sea region: Comparison and oceanographic implications. *Mar. Micropaleont.* 5: 71–110.

Russell, H. D. 1960. Heteropods and pteropods as food of the fish genera, *Thunnus* and *Alepisaurus*. *Nautilus* 74: 46–56.

Sakthivel, M. 1972a. Studies on *Desmopterus* Chun 1889 species in the Indian Ocean. *"Meteor" Forsch.-Ergebnisse* 10(D): 46–57.

———. 1972b. Swarming of a pteropod *Cavolinia uncinata pulsata* (Rang, 1829; Spoel, 1969) in the inshore waters off Cochin. *Indian J. mar. Sci.* 1: 148.

Shackleton, N. J. 1977. Carbon-13 in *Uvigerina*: Tropical rainforest history and the equatorial Pacific carbonate dissolution cycles. In: *The Fate of Fossil Fuel CO_2 in the Oceans*. N. R. Andersen and A. Malahoff, eds. New York: Plenum Press. Pp. 401–22.

Silver, M. W., and K. W. Bruland. 1981. Differential feeding and fecal pellet composition of salps and pteropods, and the possible origin of the deep-water flora and olive-green "cells." *Mar. Biol.* 62: 263–73.

Sipos, J. C., and R. G . Ackman. 1964. Association of dimethyl sulphide with the "blackberry" problem in cod from the Labrador area. *J. Fish. Res. Bd Can.* 21: 423–25.

Smith, K. L., Jr., and J. M. Teal. 1973. Temperature and pressure effects on respiration of thecosomatous pteropods. *Deep-Sea Res.* 20: 853–58.

Souleyet, L. F. A. 1843. Observations anatomiques, physiologiques et zoologiques sur les mollusques ptéropodes. *C. r. hebd. Séanc. Acad. Sci., Paris* 17: 662–75.

Soutar, A. 1971. Micropalaeontology of anaerobic sediments and the California Current. In: *The Micropalaeontology of Oceans*. B. M. Funnell and W. R. Riedel, eds. Cambridge: Univ. Press. Pp. 223–30.

Soutar, A., and J. D. Isaacs. 1969. History of fish populations inferred from fish scales in anaerobic sediments off California. *Calif. Coop. Oceanic Fish. Invest. Rep.* 13: 63–70.

van der Spoel, S. 1962. Aberrant forms of the genus *Clio* Linnaeus, 1767, with a review of the genus *Proclio* Hubendick, 1951 (Gastropoda, Pteropoda). *Beaufortia* 9: 173–200.

———. 1964. Notes on some pteropods from the North Atlantic. *Beaufortia 10*: 167–76.
*———. 1967. *Euthecosomata, a Group with Remarkable Developmental Stages (Gastropoda, Pteropoda)*. Gorinchem: J. Noorduijn. 375 pp.
———. 1968. The shell and its shape in Cavoliniidae (Pteropoda, Gastropoda). *Beaufortia 206*: 185–89.
———. 1969. The shell of *Clio pyramidata* L., 1767 forma *lanceolata* (Lesueur, 1813) and forma *convexa* (Boas, 1886) (Gastropoda, Pteropoda). *Vidensk. Medd. dansk naturh. Foren. 132*: 95–114.
———. 1970. The pelagic Mollusca from the "Atlantide" and "Galathea" Expeditions collected in the East Atlantic. *Atlantide Rept. 11*: 99–139.
———. 1973a. Growth, reproduction and vertical migration in *Clio pyramidata* Linné, 1767 forma *lanceolata* (Lesueur, 1813), with notes on some other Cavoliniidae (Mollusca, Pteropoda). *Beaufortia 21*: 117–34.
———. 1973b. Strobilation in a mollusc; the development of aberrant stages in *Clio pyramidata* Linnaeus, 1767 (Gastropoda, Pteropoda). *Bijdr. Dierk.* 43: 202–14.
———. 1974. Notes on the adult and young stages in *Diacria* (Gastropoda, Pteropoda). *Basteria 38*: 19–26.
*———. 1976. *Pseudothecosomata, Gymnosomata and Heteropoda (Gastropoda)*. Utrecht: Bohn, Scheltema, and Holkema. 484 pp.
———. 1979. Strobilation in a pteropod (Gastropoda, Opisthobranchia). *Malacologia 18*: 27–30.
———. 1982. Are pteropods really ptero-pods? (Mollusca, Gastropoda, Pteropoda). *Bull. zool. Mus. Univ. Amst.* 9: 1–6.
*van der Spoel, S., and D. Boltovskoy. 1981. Pteropoda. In: *Atlas del zooplancton del Atlántico sudoccidental y métodos de trabajo con el zooplancton marino*. D. Boltovskoy, ed. Publicación especial del INIDEP: Mar del Plata, Argentina. Pp. 493–531.
Stepien, J. C. 1980. The occurrence of chaetognaths, pteropods and euphausiids in relation to deep flow reversals in the Straits of Florida. *Deep-Sea Res.* 27: 987-1011.
Stock, J. H. 1973. *Nannallecto fusii* n. gen., n. sp., a copepod parasitic on the pteropod, *Pneumodermopsis*. *Bull. zool. Mus. Univ. Amst. 3*: 21–24.
Sverdrup, H. U., M. W. Johnson, and R. H. Fleming. 1942. *The Oceans. Their Physics, Chemistry, and General Biology*. New York: Prentice-Hall. 1,087 pp.
Takeuchi, I. 1972. Food animals collected from the stomachs of three salmonid fishes (*Oncorhynchus*) and their distribution in the natural environments in the northern North Pacific. *Bull. Hokkaido reg. Fish. Res. Lab. 38*: 1–119.

Tesch, J. J. 1913. Pteropoda. In: *Das Tierreich*. F. E. Schulze, ed. Berlin: R. Friedländer & Sohn. *36*: 1–154.

*———. 1946. The thecosomatous pteropods. I. The Atlantic. *Dana Rep.* No. *28*. 82 pp.

———. 1948. The thecosomatous pteropods. II. The Indo-Pacific. *Dana Rep.* No. *30*. 45 pp.

Thiel, H. 1983. Pteropod shells: Another food source for deep-sea organisms. *Senckenberg. marit.* *15*: 147–55.

Thiriot-Quiévreux, C. 1970. Transformations histologiques lors de la métamorphose chez *Cymbulia peroni* de Blainville (Mollusca, Opisthobranchia). *Z. Morph. Tiere* *67*: 106–17.

Troost, D. G., and S. van der Spoel. 1972. Juveniles of *Cavolinia inflexa* (Lesueur, 1813) and *Cavolinia longirostris* (de Blainville, 1821), their discrimination and development (Gastropoda, Pteropoda). *Bull. zool. Mus. Univ. Amst.* 2: 221–35.

Vande Vusse, F. J. 1980. A pelagic gastropod first intermediate host for a hemiuroid trematode. *J. Parasit.* *66*: 167–68.

Vayssière, A. 1915. Mollusques euptéropodes (Ptéropodes thécosomes). *Résult. Camp. scient. Prince Albert I Monaco* 47: 1–174.

Vecchione, M., and G. C. Grant. 1983. A multivariate analysis of planktonic molluscan distribution in the Middle Atlantic Bight. *Contin. Shelf Res.* *1*: 405–24.

Vinogradov, M. Ye. 1961. Food sources of the deep-water fauna. Speed of decomposition of dead Pteropoda. *Doklady akad. nauk SSSR*. *138*: 1439–42.

Wells, F. E. 1975. Comparison of euthecosomatous pteropods in the plankton and sediments off Barbados, West Indies. *Proc. malac. Soc. Lond.* *41*: 503–9.

———. 1976a. Growth rate of four species of euthecosomatous pteropods occurring off Barbados, West Indies. *Nautilus* *90*: 114–16.

———. 1976b. Seasonal patterns of abundance and reproduction of euthecosomatous pteropods off Barbados, West Indies. *Veliger* *18*: 241–48.

———. 1978. Subgeneric relationships in the euthecosomatous pteropod genus *Limacina* Bosc, 1817. *J. malac. Soc. Aust.* 4: 1–5.

White, A. W. 1977. Dinoflagellate toxins as probable cause of an Atlantic herring (*Clupea harengus harengus*) kill, and pteropods as apparent vector. *J. Fish. Res. Bd Can.* *34*: 2421–24.

———. 1981. Marine zooplankton can accumulate and retain dinoflagellate toxins and cause fish kills. *Limnol. Oceanogr.* *26*: 103–9.

Whitfield, M. 1984. Surprise from the shallows. *Nature, Lond.* *312*: 310.

Wolff, T. 1968. The Danish Expedition to "Arabia Felix" (1761–1767).

Bull. Inst. océanogr., Monaco, No. spécial 2 [Congr. int. Hist. Océanogr., 1]: 581–601.
Wormelle, R. L. 1962. A survey of the standing crop of plankton of the Florida Current. VI. A study of the distribution of the pteropods of the Florida Current. *Bull. mar. Sci. Gulf Caribb.* 12: 95–136.
Wormuth, J. H. 1981. Vertical distributions and diel migrations of Euthecosomata in the northwest Sargasso Sea. *Deep-Sea Res.* 28: 1493–1515.
———. 1985. The role of cold-core Gulf Stream rings in the temporal and spatial patterns of euthecosomatous pteropods. *Deep-Sea Res.* 32: 773–88.
Yonge, C. M. 1926. Ciliary feeding mechanisms in the thecosomatous pteropods. *J. Linn. Soc. (Zool.)* 36: 417–29.
Yonge, C. M., and T. E. Thompson. 1976. *Living Marine Molluscs*. London: Collins. 288 pp.
Zarnik, B. 1911. Über den Chromosomencyclus bei Pteropoden. *Verh. Dt. zool. Ges.* 20: 205–15.
Zhang, Fu-sui. 1964. The pelagic molluscs off the China coast. I. A systematic study of Pteropoda (Opisthobranchia), Heteropoda (Prosobranchia) and Janthinidae (Ptenoglossa, Prosobranchia). *Studia mar. Sinica* 5: 125–226. [In Chinese.]

5

The Gymnosomes
Shell-less Pteropods

Class Gastropoda
  Subclass Opisthobranchia
    Order Gymnosomata
      Suborder Gymnosomata
        Family Pneumodermatidae
        Family Notobranchaeidae
        Family Cliopsidae
        Family Clionidae
      Suborder Gymnoptera
        Family Hydromylidae
        Family Laginiopsidae

Although gymnosomatous pteropods form a sizable group of about 45 to 50 species, they are for a variety of reasons perhaps the most poorly known of the holoplanktonic gastropods. First, most gymnosomes are oceanic and thus generally inaccessible to capture except by the use of plankton nets. Only a few specimens have been hand-collected by scuba divers or from inshore localities (e.g. Rosewater, 1959; Dexter, 1962). Most specimens have been obtained from oceanographic zooplankton samples preserved in formaldehyde. Second, adult gymnosomes do not have a shell, and the body of these animals is highly contractile. Specimens placed directly into a preservative, without prior anaesthetic, contract into indistinguishable forms in which all taxonomic features are hidden. This problem was referred to by Tesch (1950), who described such specimens as "unsightly, shapeless, deformed balls, contracted to the utmost." Though this is an apt description of preserved specimens, it is a most inappropriate description of living gymnosomes. The formidable problems that arise when species identifications must be based on the dissection of minute, contracted animals have discouraged many potential researchers. Finally, many gymnosome species are rare, probably as a result of their highly specialized carnivorous habits. It is also possible that some fast-swimming species may be able to avoid net capture.

The major anatomical modifications of gymnosomes for a planktonic life include paired swimming wings or parapodia (a feature shared with thecosomatous pteropods), the loss of the shell and consequent lightening and streamlining of the body, and small body size. Despite the obstacles encountered when working with these relatively rare, oceanic animals, progress has been made in discovering something of their habitats and habits. Gymnosomes have been found in all the major oceans, including the Arctic and Antarctic oceans, where one abundant bipolar species (*Clione limacina*) forms part of the diet of baleen whales. The geographic distributions of all the species have been compiled by van der Spoel (1976). Most gymnosomes are confined to epipelagic or mesopelagic zones, with only a few species, such as *Schizobrachium polycotylum*, living below these depths. Diel vertical migrations probably are carried out by many species, but this has not been well documented in the group (e.g. Mackie, 1985).

The following discussion applies to the majority of gymnosome species. The two species that comprise the Suborder Gymnoptera are considered in a later section, since they differ in several anatomical respects.

Swimming, External Anatomy, and Feeding Structures

The external anatomy of gymnosomes gives few clues to their molluscan affinities. The adults do not have a shell; the name "gymnosome" refers to a naked (gymno-) body (-some). They also lack the mantle and mantle cavity that figure so importantly in most other mollusks.

The shell-less gymnosome body is bilaterally symmetrical and either streamlined or baglike in shape (Color Figs. 11, 12; Figs. 50, 52–54). Maximal body length ranges from less than 2 mm in several species to 85 mm in *Clione limacina* living in subarctic Atlantic regions (Conover and Lalli, 1972; Gilmer, pers. obs.). The most conspicuous distinguishing features of adult gymnosomes are the paired, muscular, swimming wings projecting from the sides of the body and derived from part of the foot. Compared with the thecosomatous pteropods, gymnosome wings are shorter and broader, and provide a smaller wing surface relative to body size. The

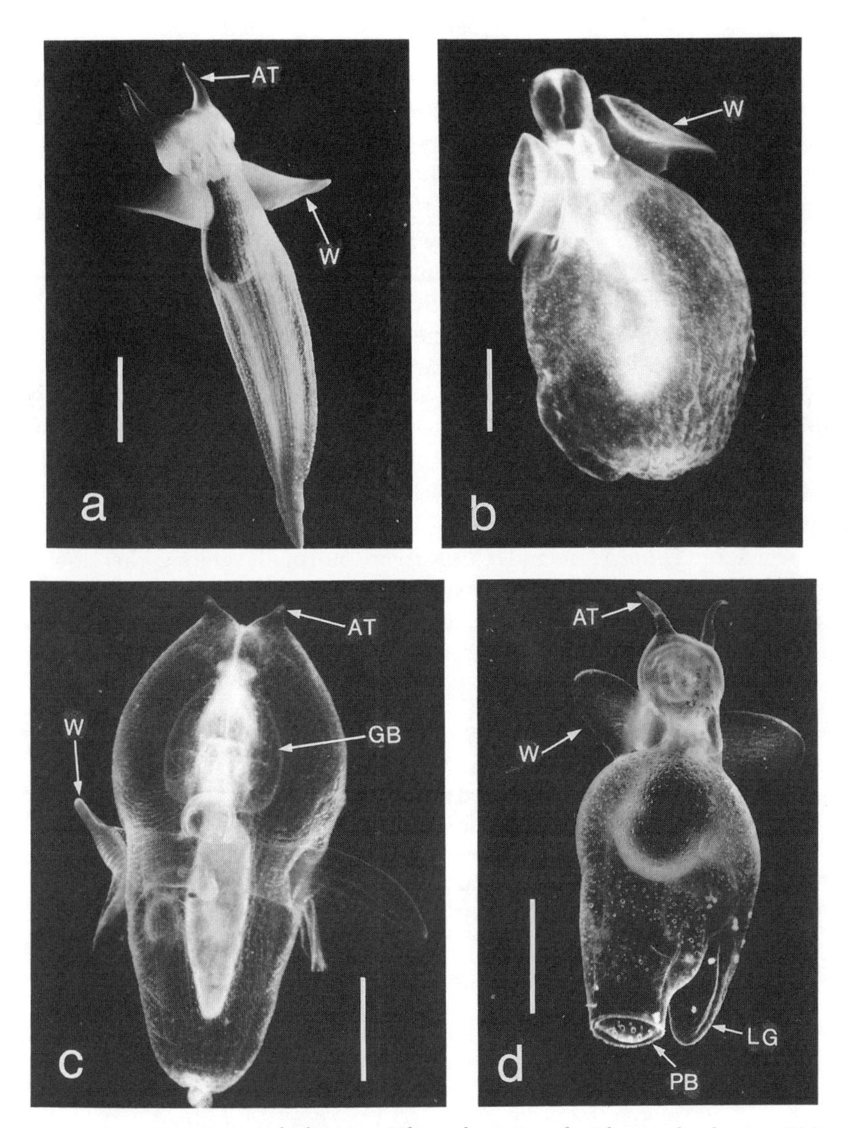

FIG. 50. Gymnosome morphology: *a, Clione limacina*; *b, Cliopsis krohni*; *c, Thliptodon* sp.; *d, Pneumodermopsis canephora*. AT, anterior tentacle; GB, gullet bladder; LG, lateral gill; PB, posterior ciliary band; W, wing. Scale lines represent 5 mm.

wings do not attain full development until late in larval development and, until that time, gymnosome larvae swim by means of three ciliary bands that encircle the body. Neural control of ciliary swimming in larvae of *Pneumoderma atlanticum* has been investigated by Mackie et al. (1976).

Morton (1958) and van der Spoel (1976) have described the swimming of adult *Clione limacina* as a type of sculling action (Fig. 51). After each stroke, the wing twists slightly at the point where it attaches to the body. This action results in the leading anterior edge being directed relatively more strongly downward in a downstroke and upward on the return stroke. Thus the ventral wing surface is directed posteriorly after a downstroke and the dorsal surface is inclined backward after the upstroke. While moving ventrally, the wing gives an upward as well as a forward thrust, and on the upstroke the wing provides both downward and forward thrust. Other gymnosome species employ a similar type of swimming, and this ability to utilize both downward and upward strokes in a sculling action can provide fast movement and rapid maneuverability.

Species of *Thliptodon* and *Cliopsis* are normally among the slowest-moving gymnosomes. They often swim at speeds slower than 1 cm/sec by moving their wings at less than 2 beats/second. Their avoidance swimming speeds, however, approach 35 cm/second (Gilmer, pers. obs.). Species of

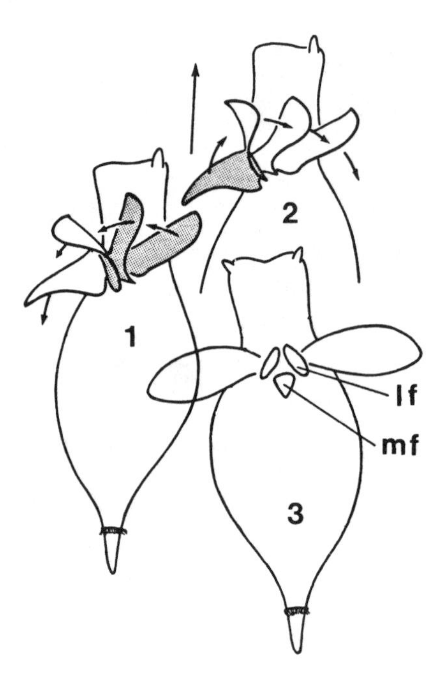

FIG. 51. Swimming movements in *Clione limacina,* showing the successive positions of the wings in downward (1) and upward (2) strokes (left-side view) and in ventral (3) view (modified after Morton, 1958). lf, lateral foot lobe; mf, median foot lobe.

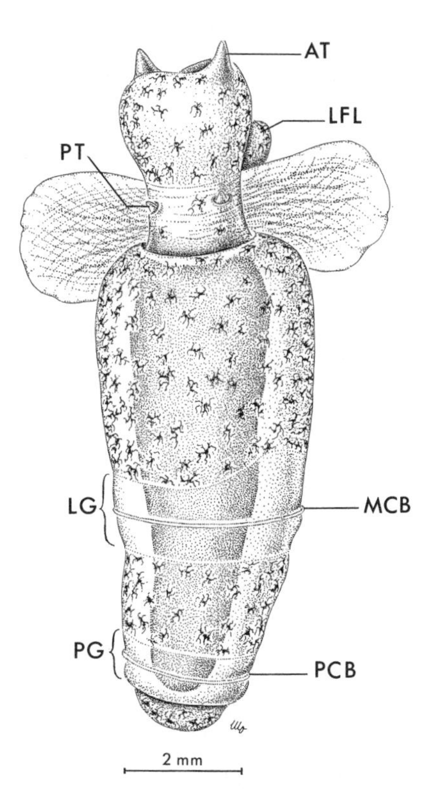

FIG. 52. Dorsal view of a juvenile *Spongiobranchaea australis,* a common species in the South Atlantic and Subantarctic waters. AT, anterior tentacle; LFL, lateral foot lobe; LG, lateral gill; MCB, median ciliary band; PCB, posterior ciliary band; PG, posterior gill; PT, posterior tentacle.

Pneumodermopsis and *Notobranchaea* are capable of swimming at speeds in excess of 1 m/sec for at least short time intervals (Gilmer in Hamner et al., 1975). Unlike the thecosomes, gymnosomes rarely seem to cease swimming completely, although *Clione limacina* is an exception and has often been observed hanging motionless and upside down. This behavior has not been seen in other gymnosomes. Active swimming does cease, however, during mating or spawning; at these times, the animals remain motionless, with only slow rhythmic beating of the wings. Gymnosomes are not known to have any special buoyancy mechanisms such as ionic regulation or the accessory buoyancy structures produced by thecosomes. In the polytrochous larvae of *Clione limacina* and other species, however, large subintegumentary oil droplets may provide buoyancy during this stage, as well as serving as food reserves.

The gymnosome body usually consists of a well-defined head separated from the trunk by a narrowed neck area. The head bears two pairs of tentacles; the anterior set (Figs. 52–54) is larger and retractile, and the posterior tentacles (Fig. 54b) are positioned on the dorsal neck region and are

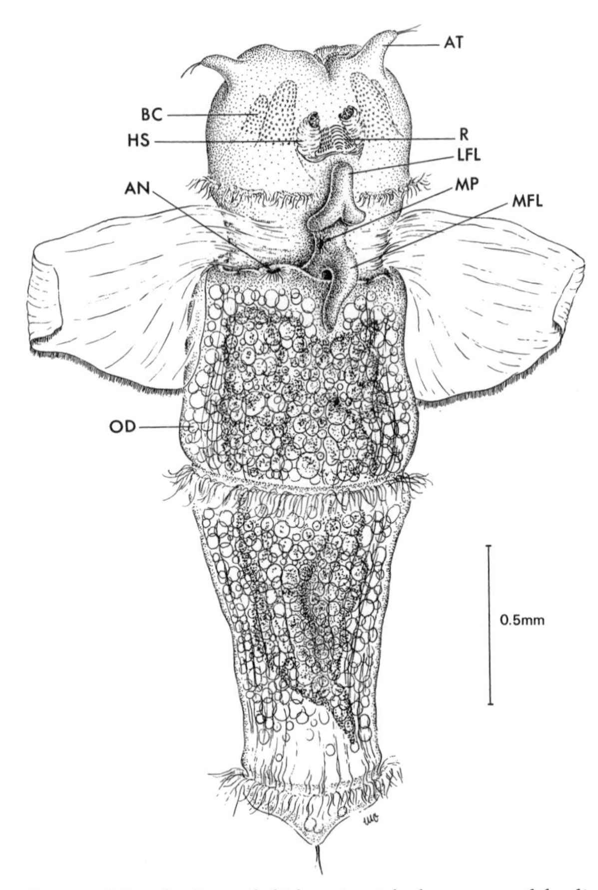

FIG. 53. Ventral view of *Paedoclione doliiformis* with the retracted feeding structures as they appear in histologically cleared specimens (from Lalli, 1972). This neotenous species retains the three ciliary bands, which are characteristic of gymnosome larvae, throughout its adult life. AN, anus; AT, anterior tentacle; BC, buccal cones; HS, hook sac; LFL, lateral foot lobes; MFL, median foot lobe; MP, male genital pore; OD, integumentary oil droplet; R, radula.

usually set into depressions. The anterior tentacles seem to be tactile organs, whereas the histology of the posterior pair suggests that they are light-sensitive. A mouth or an opening to the proboscis sac, depending on the species, is located terminally on the head. The remnant of the gastropod foot is located ventrally, separated from but between the wings, and consists of two lateral lobes and one unpaired medial lobe (Figs. 53, 54a). The anus and single osphradium (Figs. 53, 54a) are located on the right ventral side beside the foot lobes.

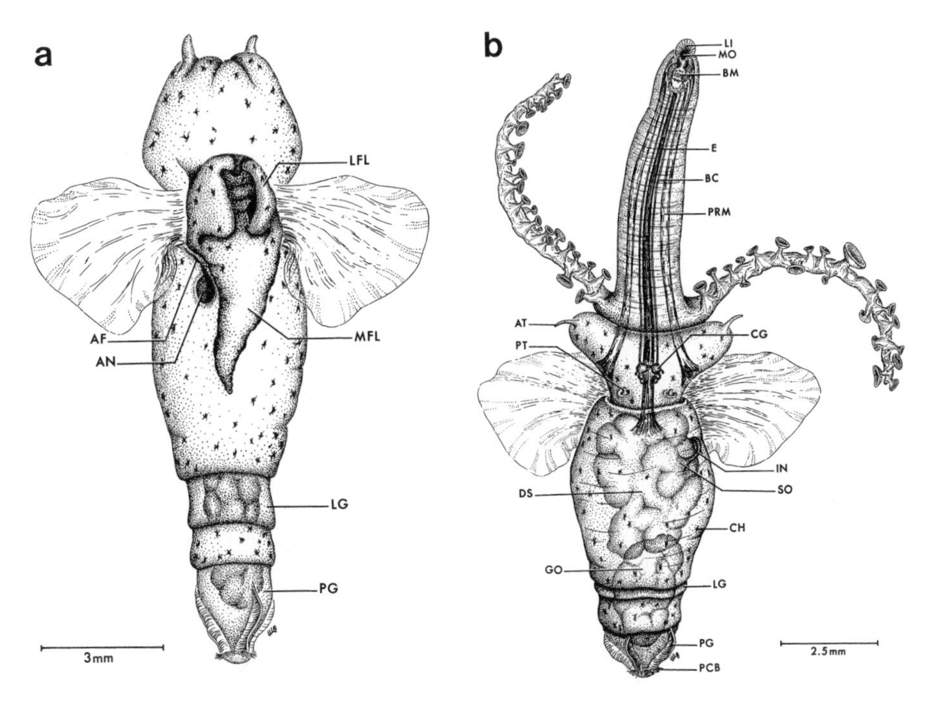

FIG. 54. *Crucibranchaea* sp. (from Lalli, 1970a): *a*, ventral view of a specimen in the resting position, with proboscis and buccal tentacles withdrawn; *b*, dorsal view of an animal in the feeding position, with the feeding structures everted. AF, anal field; AN, anus; AT, anterior tentacle; BC, buccal-cerebral ganglia connective; BM, buccal mass; CG, cerebral ganglion of nerve ring; CH, chromatophore; DS, digestive sac; E, esophagus; GO, gonad; IN, intestine; LFL, lateral foot lobe; LG, lateral gill; LI, lips; MFL, median foot lobe; MO, mouth; PCB, larval posterior ciliary band; PG, lateral fold of posterior gill; PRM, proboscis retractor muscle; PT, posterior tentacle; SO, stomach area of digestive sac.

The integumentary covering of the body is tough and elastic, and may be either transparent or opaque. In some species, there are stellate chromatophores (Figs. 52, 54) scattered over the body. These pigment-filled sacs, which are morphologically similar to those of cephalopods, do not, however, seem to respond rapidly to changes in illumination or to mechanical stimulation. Changes in chromatophore shape and consequent change in body color have been reported to occur only very slowly, when the animals were changed experimentally from light to dark conditions, or vice versa (Lalli, 1970a). Some of the gymnosomes also have integumentary gills that are secondarily evolved structures and not true molluscan ctenidia. The gills

TABLE 20

A comparison of feeding structures in gymnosome genera, arranged subjectively in order of decreasing complexity

Genus	Median arm (number of suckers)	Lateral arms (number of suckers/arm)	Buccal cones (total number)	Proboscis	Radula	Jaw	Hooks (number/sac)
Pneumodermopsis	3–5	arms reduced, 1–16		+	+	+	5–50
Abranchaea	4	no arms, 12–18		+	+	+	6
Crucibranchaea	no arm, 3–5	8–55		+	+	+	16–50
Schizobrachium		branched arms, >1,000		+	+	+	large sacs, ca. 20?
Pneumoderma		7–150		+	+	+	20–100
Spongiobranchaea		7–20 or more		+	+	+	20–34
Clione			6		+		15–30
Paraclione			4		+	+, reduced	10–60
Notobranchaea			4 or absent		+	+	9–20
Paedoclione			4 (1 reduced)		+		reduced sacs, 7
Thalassopterus			4		+		reduced sacs, ?
Fowlerina			2		+	+	13–15
Cliopsis				+, long	+	+	40–60
Pruvotella				+	+	+	ca. 40
Massya					+, large	+	16–20
Thliptodon					+		large sacs, 17–50
Cephalobrachia					+		large sacs, 16–40
Hydromyles					+	+, reduced	
Laginiopsis				long, nonretractile			

NOTE: + signifies presence.

of gymnosomes (Figs. 50d, 52, 54) are simply thin, unpigmented, nonglandular areas of integument that may encircle the lower trunk area or project from it on one side (a lateral gill), or the gill may form the posterior tip of the body (a posterior gill).

The identification of gymnosome species depends largely on the anatomical features of the feeding apparatus. Since these organs usually are not visible in preserved, contracted specimens or in nonfeeding, live individuals, taxonomic identification often involves dissection. Figures 54a and 54b illustrate, by way of example in one species, the dramatic change in external appearance that results when a gymnosome everts its feeding apparatus. Figure 54b also emphasizes the position taken by early taxonomists who erroneously placed gymnosomes in the Class Cephalopoda. It has now been established that the anatomical similarities of sucker-bearing tentacles and stellate chromatophores on the body, both of which are present in some gymnosomes as well as in squids and octopuses, are the results of convergent evolution and are not indicative of close phylogenetic relationships.

Evolution within the Order Gymnosomata has resulted in an almost bewildering array of different feeding organs. These are listed, for each genus, in Table 20. The radula (Fig. 55) is the one feeding structure that is common to all gymnosomes but one. (The single exception is *Laginiopsis trilobata*, a deep-sea species belonging to the Suborder Gymnoptera; the only feeding organ described for this animal is a long, nonretractile proboscis provided with three terminal lobes.) There is no common radular formula, but most species have a single, serrate, median tooth bordered on each side by one to 15 long, curved, lateral teeth in each row. The lateral teeth are adapted for grasping and pulling prey into the gut. Most gymnosomes also have paired hook sacs (Figs. 56–58) located on either side of the radula. These sacs, not present in any other mollusks, are unique to gymnosomes. Each sac contains from five to 100 curved, chitinous hooks (Fig. 55), the size and number of the hooks depending on the species. The eversible hooks are used to obtain a purchase on the soft body tissues of prey and to pull the prey into the buccal cavity. A jaw (Fig. 56), composed of clusters of small, chitinous spines, is present in many species. The radula, hook sacs, and jaw make up the buccal mass proper, and this complement of organs can be protruded from the buccal cavity during feeding. In some species (of the Families Pneumodermatidae and Cliopsidae), the buccal mass proper is located at the tip of a long proboscis that is everted when feeding (Fig. 54b). In these species, the opening on the head of a nonfeeding animal marks the opening of the proboscis sac and is not the true mouth. The mouth is always located terminally on the proboscis (Fig. 54b) and is seen only upon the eversion of this structure.

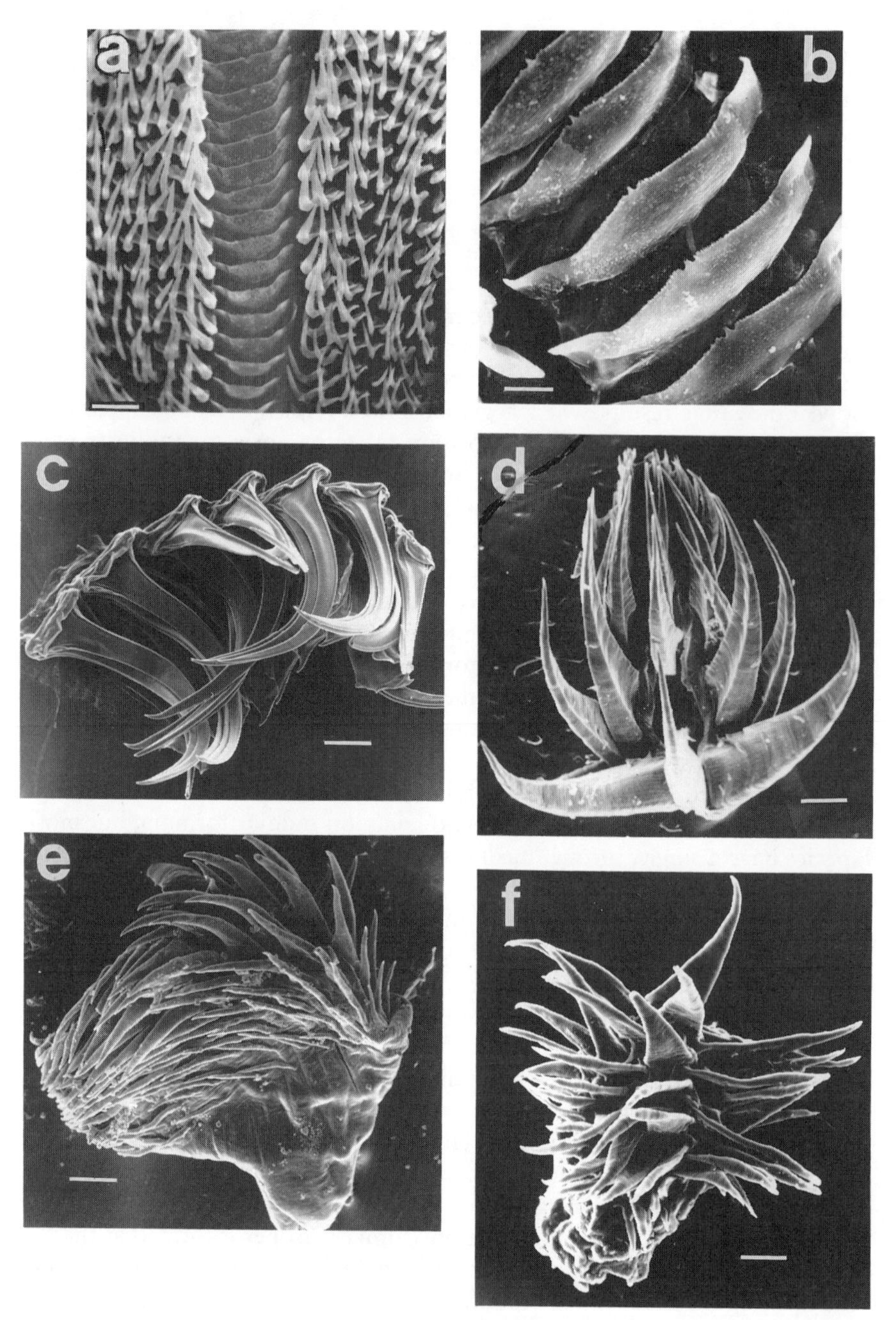

FIG. 55. Scanning electron micrographs of gymnosome radulae and hooks: *a*, *Clione limacina* radula; *b*, median radular teeth of *C. limacina*; *c*, lateral radular teeth of *Thliptodon diaphanus*; *d*, hooks from one hook sac of *Clione limacina*; *e*, a small portion of the hooks from one hook sac of *Pneumoderma atlanticum*; *f*, hooks from one hook sac of *Cliopsis krohni*. Scale lines represent 35 μm in *a*, 10 μm in *b*, 120 μm in *c*, 25 μm in *d*, 60 μm in *e*, and 30 μm in *f*.

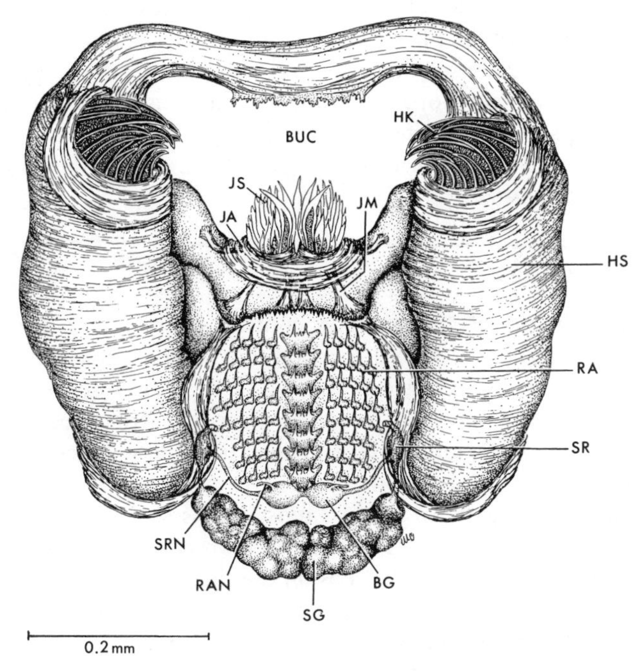

FIG. 56. Dorsal view of the buccal mass of *Crucibranchaea* sp.; the ventrally placed jaw has been displaced anteriorly to show its structure (from Lalli, 1970a). BG, buccal ganglion; BUC, buccal cavity; HK, hooks; HS, hook sac; JA, jaw papilla; JM, jaw muscle; JS, spines of jaw; RA, lateral teeth of radula; RAN, radular nerve; SG, salivary gland; SR, salivary reservoir; SRN, salivary reservoir nerve.

Prehensile organs, used to capture prey, are found in some gymnosomes. The Family Pneumodermatidae contains all of the gymnosomes that have sucker-bearing arms similar to the tentacles of cephalopods (Fig. 54b). Normally the sucker-bearing tentacles are retracted into the sides of the buccal cavity or into the spacious sac formed by the invaginated proboscis, and they are everted only during prey capture. Within the pneumodermatids, the relative development and number of arms and the number of suckers vary; these features are used to distinguish genera and species. The suckers may be stalked, in which case they are capable of independent movement, or they may be sessile. Prehensile organs of a different type are present in six genera (Table 20). These are buccal cones or cephaloconi—eversible tentacles without suckers but with a papillate surface (Figs. 57, 61). There is one pair of buccal cones in *Fowlerina* and three pairs in *Clione*; the remaining four genera have two pairs. Although different in structure, the buccal cones serve the same functions of capturing and holding prey as do the sucker-bearing arms of the pneumodermatids.

Because of the structural diversity and complexity of the gymnosome feeding organs, their functional anatomy will be considered in more detail in the following section. The few studies of prey capture by living gymnosomes have revealed the intricacies and specialized uses of the components of the feeding apparatus.

Food and Feeding

The feeding habits of gymnosomes contrast sharply with those of the thecosomatous pteropods, whose food rarely exceeds a few millimeters in size. The present evidence suggests that all of the gymnosomes are highly specialized carnivores, and that their diverse feeding structures are adapted for the capture and ingestion of different species of thecosomes and possibly other relatively large zooplankton.

Feeding behavior has been studied most intensively in the well known "sea angel," *Clione limacina*. It has long been known that *Clione* is a bipolar species that coexists with and feeds on the thecosome *Limacina helicina* in Arctic and Antarctic waters (Boas, 1886b; Meisenheimer, 1905a; Manteufel, 1937; McGowan, 1968) and preys on *L. retroversa* in temperate areas of both hemispheres (Lebour, 1931). The following account of prey capture by *Clione* is from observations by Lalli (1970b) and Conover and Lalli (1972).

The capture of either *Limacina helicina* or *L. retroversa* by *Clione* is initiated by direct contact with the prey. *Clione* immediately everts its six buccal cones to seize the shell of the prey (Fig. 57). The eversion and elongation of these prehensile, tentacle-like structures is accomplished by changes in blood pressure that force blood from the open hemocoel into the central cavities of the cones. The flexible buccal cones grip the prey shell and adhere to the shell surface through the interaction of a secretion produced by epithelial glands and disklike terminations of modified epithelial cells of the cones. The cones rapidly manipulate the prey so that the shell aperture is pressed against the mouth of *Clione*. At this time, chitinous hooks are evaginated from the paired hook sacs and are inserted into the soft body tissues of the prey. The evagination and erection of the hooks results from a complex interplay of muscles in the hook sacs that forces fluid into the anterior areas of the sacs (Fig. 58). When *Clione* has a secure hold on the prey, the hooks and radula are used to pull the body of the *Limacina* from its shell and the prey is swallowed whole. The total time from capture to ingestion of the prey is 2 to 45 minutes. The intact, empty prey shell is immediately

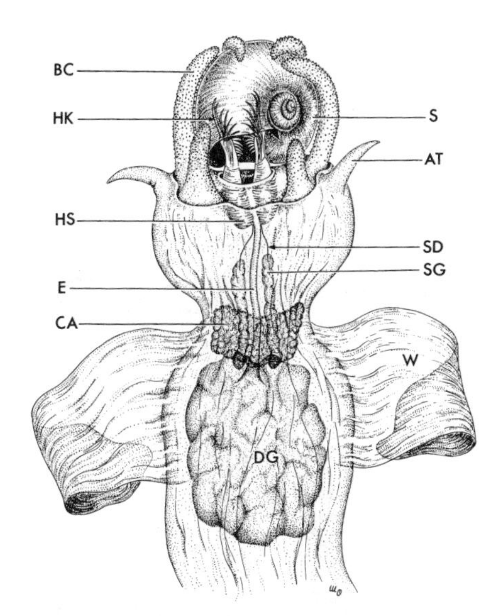

FIG. 57. *Clione limacina* feeding on *Limacina helicina* (from Lalli, 1970b). AT, anterior tentacle; BC, buccal cone; CA, copulatory apparatus; DG, digestive gland; E, esophagus; HK, hooks; HS, hook sac; S, shell of the prey; SD, salivary duct; SG, salivary gland; W, wing.

discarded after ingestion, and a second *Limacina* can be captured by the same predator within 2 minutes.

Occasionally, *Clione* is unsuccessful in extracting a captured *Limacina* from its shell. This occurs if the prey retracts rapidly enough and far enough into its shell so that the predator is unable to secure a hold with its hooks. In such cases, a starved *Clione* may hold the shell for several hours. If no other prey are available, other *Clione* may attempt to wrest the captured prey from the original predator, either by grasping the prey shell with the buccal cones and pulling at it or by attempting to push the unsuccessful *Clione* from the shell aperture. These aggressive encounters end either when the prey is finally extracted and eaten by one of the competitors, or when the prey dies; in the latter event, the *Limacina* is dropped and ignored.

Clione does not exhibit a feeding response to dead *Limacina* or to empty prey shells. To determine the stimuli involved in the recognition of prey by *Clione*, Conover and Lalli (1972) devised a series of laboratory experiments using live *Limacina* from which the shells were carefully removed. These shell-less prey, which were able to make feeble wing movements but were unable to swim, were presented to starved *Clione*. Some of the *Clione* captured and ingested shell-less *Limacina* within a few minutes, but in all such instances the buccal cones remained retracted during feeding; the prey were captured and ingested only by the use of the hooks and radula and the swallowing movements of the pharynx. In most cases, however, *Clione* responded to shell-less prey by everting its buccal cones and then immediately

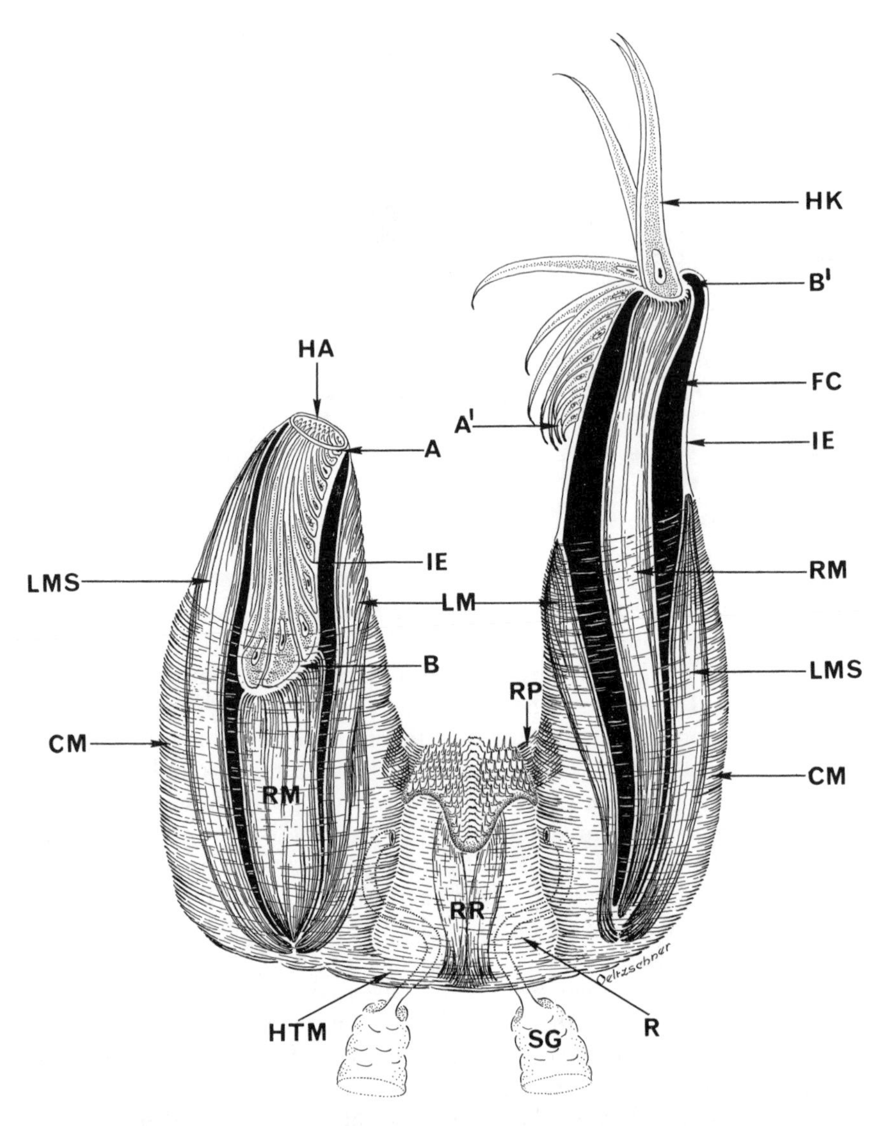

FIG. 58. The hook sacs and radula of *Clione limacina,* showing the position of the hooks and associated musculature in the invaginated or resting state (left) and the evaginated or feeding position (right) (from Lalli, 1970b). Points A–B and A′–B′ show changes in orientation during evagination; A and A′ mark the positions of the smallest hooks, B and B′ those of the attachment of the retractor muscle to the largest hooks. CM, outer circular muscle; FC, fluid-filled cavity; HA, hook sac aperture; HK, hooks; HTM, hook sac transverse muscle; IE, secretory epithelium of hooks; LM, median longitudinal muscle; LMS, lateral longitudinal muscle; R, radula; RM, retractor muscle of hooks; RP, radula protractor muscle; RR, retractor muscle of radular membrane; SG, salivary glands and ducts.

retracting them after touching the *Limacina*. This was followed by prolonged testing by the predator, during which time the cones were repeatedly everted and retracted after contact. The typical feeding response was blocked for some time until, eventually, *Clione* seized the prey with its hooks and swallowed it.

In a second experiment, *Clione* was presented simultaneously with a shell-less *Limacina* and an empty prey shell at close proximity to each other. The predator responded by first turning toward the shell-less prey and everting its buccal cones. If the cones touched both the shell-less prey and the empty shell, the predator immediately selected and picked up the shell, which it carried for up to 35 minutes before dropping it. This is the only time *Clione* will respond to an empty shell. These experiments indicate that the initial stimulus triggering the recognition and capture of prey emanates from live prey, possibly involving a chemical recognition or a response to the wing movements of the prey. The feeding response, however, is reinforced by the tactile recognition of the *Limacina* shell by the buccal cones of *Clione*.

Experiments also have been performed to determine whether *Clione* shows a feeding preference for either *Limacina helicina* or *L. retroversa* (Table 21). Six *Clione*, which previously had been fed only *L. retroversa* in the laboratory, were offered equal numbers of both prey species for an ensuing 3-week period. Although more *L. retroversa* were eaten, the difference was not statistically significant.

Clione does, however, show a significant selection of prey by size (Fig. 59). In these experiments of Conover and Lalli (1972), individual *Clione* were classified according to size (dry weight) and provided with a large range of *Limacina* of different sizes. When the size frequencies of uneaten

TABLE 21

Comparison of numbers of Limacina retroversa *and* L. helicina *of comparable sizes that were eaten by* Clione limacina *over a 3-week period*

Clione code name	Total eaten	
	L. retroversa	*L. helicina*
EW6	2	6
EW2	10	6
EC1	7	2
EW8	1	0
EC3	1	5
EC2	10	3
TOTAL	31	22

SOURCE: Conover and Lalli, 1972.

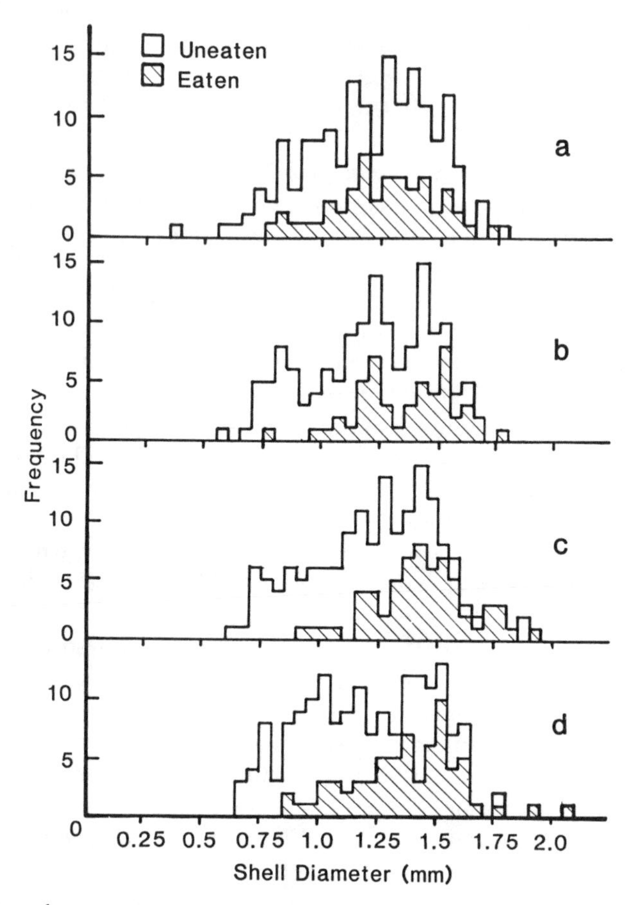

FIG. 59. Size selection of prey by *Clione limacina* of different sizes, when offered a range of sizes of *Limacina retroversa* (from Conover and Lalli, 1972): *a*, small *Clione* (<1 mg dry wt); *b*, medium small (1–4 mg); *c*, medium large (4–10 mg); *d*, large (>10 mg).

and eaten prey are compared, it is clear that the three largest size categories of *Clione* selectively fed on larger prey and left uneaten large numbers of smaller *Limacina*. The smallest size category of *Clione* (polytrochous larvae weighing less than 1 mg dry weight) did not show this selection, probably because small predators cannot effectively capture and manipulate prey of larger sizes.

Because of its food specificity, *Clione limacina* is an ideal subject for the study of the ecological energetics involved in a linear food link between a suspension-feeding prey and a carnivorous predator. Further, the discarded intact prey shells serve as natural tags of feeding rates and of food quantity and quality, since shell size can be converted to units of ingested dry weight, carbon, nitrogen, or calories. Following these premises, Conover and Lalli (1972, 1974) studied feeding rates, assimilation, metabolism, and the growth efficiency of this species.

When food is not limited, the feeding rate of *Clione* is directly related to the environmental temperature, to the size of the predator, and to the size of the prey available. The feeding rate increases with temperature to a limit of ca. 17° C (Fig. 60a), which is the approximate maximal temperature encountered within the species' geographic range. Feeding rates also increase with the size of the predator (Fig. 60b), from about 0.2 mg dry weight of

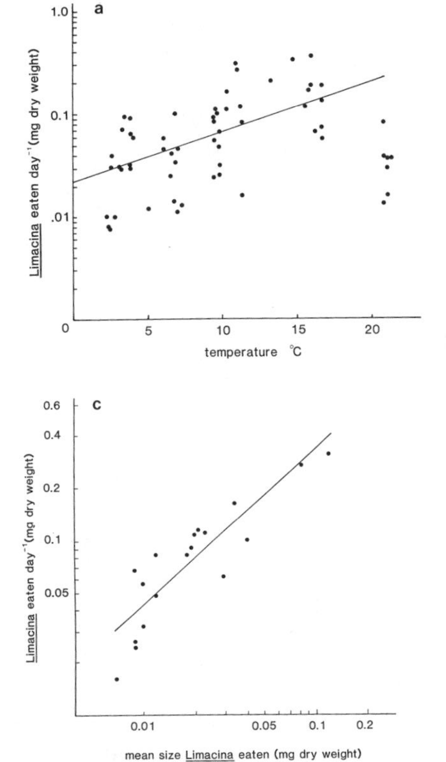

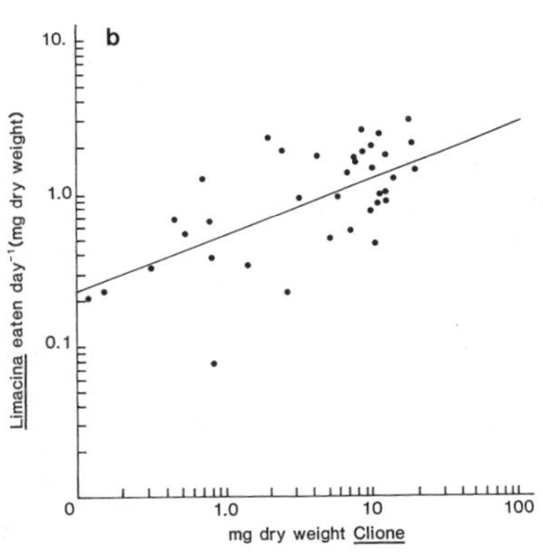

FIG. 60. Feeding rates of *Clione limacina* under different experimental conditions (from Conover and Lalli, 1972). *a*, Semi-log plot of the feeding rate (mg dry weight of *Limacina* eaten per day) against temperature for *Clione* weighing between 0.1 and 0.399 mg dry weight; the equation for the line is $\log \hat{y} = 0.048x - 1.653$ (data from temperatures > 20° C have been omitted from the regression analysis). *b*, Log-log plot of the feeding rate against dry weight (mg) of *Clione* at 13° to 15° C; the equation for the line is $\log \hat{y} = 0.365 \log x - 0.260$. *c*, Log-log plot of the feeding rate against size of *Limacina* eaten (mg dry weight) for *Clione* of the same sizes as in *a* at temperatures between 9.8° and 11.3° C; the equation for the line is $\log \hat{y} = 0.888 \log x + 0.409$.

TABLE 22

Comparison of growth and feeding rates for Clione limacina *fed on either large or small* Limacina retroversa *for 20 days*

Prey size (dry weight)	Number of *Clione* feeding	Average dry weight of *Clione* at start (mg)	Total number of prey eaten	Total dry weight of prey eaten (mg)	Average dry weight of prey eaten (mg)	Net growth of *Clione* (mg)
Large prey (>4.4 μg)	9	0.086	167	1.342	0.0080	0.500
Small prey (<4.4 μg)	8	0.067	92	0.251	0.0027	−0.023

SOURCE: Conover and Lalli, 1972.

Limacina eaten per day, by small *Clione* of less than 0.1 mg dry weight, to ca. 1.1 mg eaten per day, by *Clione* weighing more than 10.0 mg dry weight. The size of the prey eaten, however, is the variable that most influences feeding rate (Fig. 60c) and largely determines the growth rate of *Clione*. Table 22 shows the result of one experiment in which small *Clione* (<0.2 mg dry weight) were fed on either small (<0.5 mm shell diameter = 4.4 μg dry weight) or large (>0.5 mm) *Limacina*. Only those *Clione* fed large *Limacina* showed positive growth, whereas the replicate group of *Clione* lost weight, despite the fact that any given individual feeding for a given period ate approximately the same number of prey regardless of size. This result is strengthened by observations of the natural populations of *Clione* and *Limacina* in which the largest predators are present when the mean size of the prey is greatest but prey numbers are low. It thus appears that the *size* of the prey, not necessarily the numbers of prey, governs the growth rate and ultimate size of *Clione* in nature. Further, the fact that *Clione* actively selects larger and larger prey as it grows results in the maintainence of high growth efficiency throughout the predator's life cycle (Conover and Lalli, 1974).

The ecological disadvantages of extreme food specialization are obvious, the advantages more subtle. Conover and Lalli (1972) suggested that a major advantage of monophagy is increased ecological efficiency. When feeding is restricted to two very similar species, as it is in *Clione*, then all aspects of feeding can be refined to maximize energy yield with respect to energy spent in such activities. One such feeding aspect is the percentage of food actually assimilated by the predator from ingested prey. Later experiments (Conover and Lalli, 1974) demonstrated that *Clione* assimilates carbon from its prey with greater than 90-percent efficiency and nitrogen with nearly 100-percent efficiency. These extremely high values confirm that ecological efficiency is indeed high in this specialized prey-predator food link.

Paedoclione doliiformis (Figs. 53, 61) is a small, neotenous gymnosome that retains a small body size and external larval features throughout its life. This species also feeds exclusively on *Limacina retroversa* or *L. helicina* (Lalli, 1972). Because of its small size (<2.5 mm), the species is additionally restricted to feeding upon young, small prey of less than 1.0 mm shell diameter. Prey capture is similar to that employed by *Clione* (Fig. 57), but *Paedoclione* has only three functional buccal cones (the fourth cone is reduced in size and does not play a role in feeding). A short proboscis is thrust into the shell aperture of the captured prey, and the relatively large radular teeth are used to grip the body of the prey and pull it from the shell. The hook sacs are poorly developed and each contains only seven hooks; they

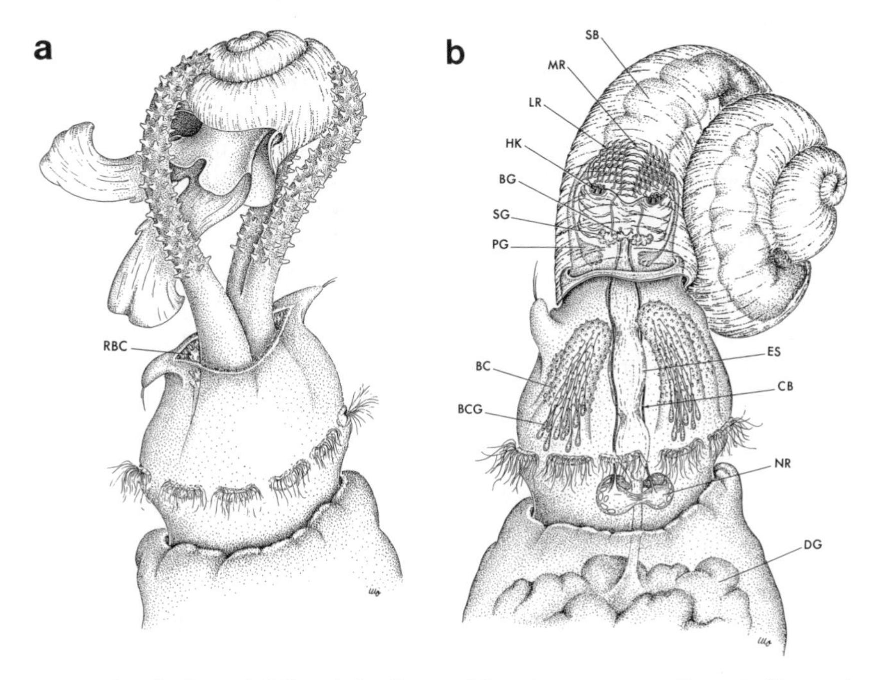

FIG. 61. *Paedoclione doliiformis* feeding on *Limacina retroversa* (from Lalli, 1972). *a,* Prey capture: three buccal cones are everted to capture the prey; the fourth, rudimentary buccal cone (RBC) remains retracted. *b,* Extraction and ingestion of the prey: the small proboscis is everted into the shell aperture, the radula is protruded from the mouth, and the buccal cones are retracted; the feeding organs within the proboscis are slightly enlarged to show structural detail. BC, buccal cone; BCG, buccal cone gland; BG, buccal ganglion; CB, cerebro-buccal connective; DG, digestive gland; ES, anterior area of esophagus distended with ingested food; HK, hooks; LR, lateral radular teeth; MR, median radular tooth; NR, nerve ring; PG, labial gland; SB, body of *Limacina*; SG, salivary gland.

appear to play only a minor role in the extraction of the prey. Upon the completion of ingestion, *Paedoclione* discards the intact, empty *Limacina* shell.

Paedoclione coexists with *Clione limacina* in the Gulf of Maine and the shelf waters of southern Nova Scotia, and both predators feed exclusively on *Limacina*. This coexistence of two feeding specialists competing for the same food has led Lalli (1972) to suggest that neoteny in *Paedoclione* lessens competition and permits continued cohabitation. Since *Paedoclione* is capable of feeding only on prey of less than 1.0 mm diameter, the species in effect competes only with larval *Clione*, which also take prey in this size range. Juvenile and adult *Clione* preferentially select *Limacina* larger than 1.0 mm, and in fact become physically incapable of eating small prey as they increase in size. Thus neoteny in *Paedoclione* limits food competition to the small-sized and more abundant prey.

Gymnosomes of families other than the Clionidae also feed on thecosomatous pteropods. In 1965, Sentz-Braconnot described the feeding of young *Pneumodermopsis paucidens*. The buccal armature of this species consists of a radula, a well-developed spinous jaw, poorly developed hook sacs, a long evaginable proboscis, a crown of short suckers representing reduced lateral arms, and a long median arm bearing five suckers on very long stalks. In the laboratory, *Pneumodermopsis* ate only *Creseis virgula* and *C. acicula*, and ignored *Cavolinia inflexa* and *Styliola subula*. On contact with *Creseis*, *Pneumodermopsis* immediately evaginated its suckers and proboscis. The prey shell was grasped by the suckers, and the proboscis was introduced into the shell aperture of the retracted prey. During ingestion, which took approximately 10 minutes, the rosette of lateral suckers held the inside of the prey shell while the body of the prey was gradually engulfed by the proboscis. When the prey was completely swallowed, the retractor muscle of the prey was severed by the predator (presumably by the cutting actions of the radula, jaw, and hooks) and the proboscis was withdrawn from the prey shell and invaginated. The empty, intact *Creseis* shells were left behind. *P. paucidens*, however, extends farther north than any species of *Creseis* (van der Spoel, 1967, 1976), and has been observed feeding on *Diacria trispinosa* in the field (Fig. 62a). This thecosome extends via the Irminger Current into Icelandic waters, where *P. paucidens* is common (Cooper and Forsyth, 1963). Van der Spoel (1976) also lists *Limacina bulimoides* as a prey species of this gymnosome.

Other gymnosomes are also reported to prey on thecosomes (Table 23). *Pneumodermopsis canephora* has been observed attacking a feeding *Cavolinia longirostris* (Figs. 62c, d). When the gymnosome contacted the extended feeding web of the euthecosome, it immediately ceased swimming and remained motionless on the surface of the web. The predator's probos-

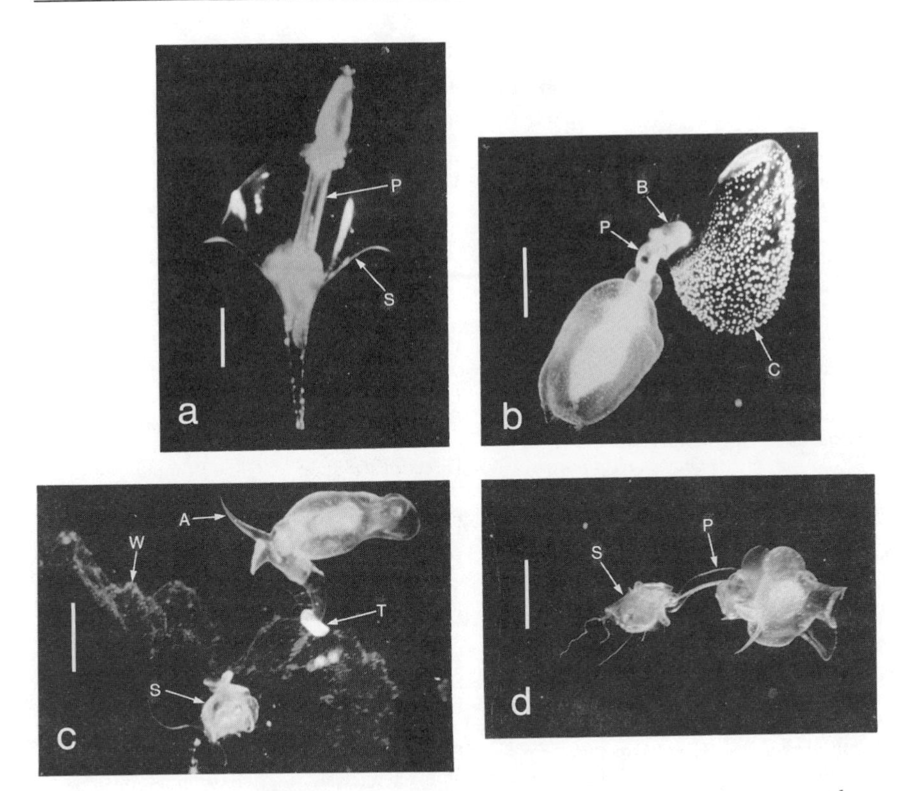

FIG. 62. Gymnosome predation on thecosomes: *a, Pneumodermopsis paucidens* feeding on a young *Diacria trispinosa* (by G. Dietzmann). *b, Cliopsis krohni* feeding on the pseudothecosome *Corolla* sp. (by J. Carlton). *c, Pneumodermopsis canephora* attacking a *Cavolinia longirostris*; the large terminal sucker of the gymnosome has attached to the mucous feeding web of the prey. *d,* The same animals as in *c,* but the *Cavolinia* has ingested its web, drawing the motionless predator closer; at this stage, the gymnosome has grasped the body of the prey with its radula and will pull the animal from its shell. A, anterior tentacle of gymnosome; B, body of prey; C, pseudoconch of prey; P, proboscis of gymnosome; S, shell of prey; T, terminal sucker of gymnosome; W, mucous feeding web of prey. Scale lines represent 3 mm in *a,* and 10 mm in *b, c,* and *d.*

cis was slowly everted over the next 30 seconds, and the animal was gradually pulled toward its prey as the *Cavolinia* began to ingest its web. As soon as contact was made between the predator and its prey, *Pneumodermopsis* attached a single large sucker to the dorsal surface of the prey's shell, and in the next 10 minutes proceeded to completely extract and ingest the body of the thecosome. *Cliopsis krohni* captures and feeds on the pseudothecosome

TABLE 23
Prey species of gymnosomatous pteropods

Gymnosome species	Prey species	Method	Reference
Clione limacina	*Limacina helicina* *Limacina retroversa*	Direct observations and stomach analyses	Boas, 1886b; Meisenheimer, 1905a; Lebour, 1931; Manteufel, 1937; McGowan, 1968; Lalli, 1970b; Conover & Lalli, 1972, 1974
Paedoclione doliiformis	*Limacina retroversa* *Limacina helicina*	Direct observation	Lalli, 1972
Paraclione sp.	*Limacina inflata*	Direct observation	Lalli (unpubl. obs.)
Pneumodermopsis paucidens	*Creseis virgula* *Creseis acicula*	Direct observation	Sentz-Braconnot, 1965
	Diacria trispinosa	Direct observation	Gilmer (unpubl. obs.)
	Limacina bulimoides	?	van der Spoel, 1976
Pneumodermopsis ciliata	Unidentified thecosomes	Stomach analysis	Massy, 1917
Pneumodermopsis canephora	*Cavolinia longirostris*	Direct observation	Gilmer (unpubl. obs.)
	Unidentified mollusks	Stomach analysis	Pruvot-Fol, 1924
Pneumoderma atlanticum	Chaetognaths?	Preserved specimen observation	Pruvot-Fol, 1924, 1954
	Unidentified thecosomes	?	van der Spoel, 1976
Pneumoderma sp.	*Cavolinia tridentata*	Stomach analysis	Boas, 1886b
Spongiobranchaea australis	*Clio pyramidata*	Feeding experiments	Lalli (unpubl. obs.)
Cliopsis krohni	*Corolla* sp.	Direct observation	Gilmer (unpubl. obs.)
Hydromyles gaudichaudii	Unidentified thecosomes	Stomach analysis	Lalli (unpubl. obs.)

Corolla in a similar way (Fig. 62b). *Cliopsis* too will cease swimming on contact with the feeding web of its prey, and the predator is drawn closer to its prey during the ingestion of the web. The long proboscis of the predator is eventually manipulated down to the visceral mass and the columellar muscle that is severed. Finally, the entire wingplate and viscera of *Corolla* are engulfed and ingested by the gymnosome, even though the pseudothecosome may be more than three times the size of the predator. Only the empty pseudoconch remains floating in the water. *Spongiobranchaea australis* feeds on *Clio pyramidata* in Subantarctic waters (Lalli, pers. obs.), and an unidentified species of *Paraclione* (3.6 mm long) has been seen feeding on *Limacina inflata*, eating five individuals in about 5 hours (Lalli, pers. obs.). The remains of unidentified thecosome prey were found in the alimentary tracts of *Pneumodermopsis ciliata* (Massy, 1917), and the remains of *Cavolinia tridentata* in the guts of a small number of *Pneumoderma* sp. (Boas, 1886b). Finally, Schiemenz (1906) reported seeing unidentified gymnosome larvae capturing thecosomes by entangling the prey in mucous strings emitted by integumentary glands, then extracting the prey from the shell by the use of the proboscis and hooks.

It has been suggested that some gymnosomes feed on zooplankton other than thecosomes, but these reports have not been verified with living animals. Pruvot-Fol (1924) found molluscan remains, including a turbinate-shaped shell (heteropod larva?), in the stomach of a *Pneumodermopsis canephora*. She also (1924, 1954) provided figures of *Clione limacina* holding a larval fish and *Pneumoderma atlanticum* ingesting a chaetognath. These were preserved animals, however, and the capture of the presumed prey may have occurred incidentally during collection or rapid fixation, when gymnosomes frequently evert their mouthparts.

Present evidence (Table 23) suggests that gymnosomes prey primarily, or exclusively, on thecosomatous pteropods, and that the diverse feeding structures of gymnosomes are adapted to the capture and extraction of specific shelled pteropods. Only three gymnosome species, however, have been well studied in this respect, leaving more than 40 species for which there is little or no food-habit information.

There is no doubt, however, that all gymnosomes are carnivores that occupy the niche of secondary consumers in the marine food web. It is not surprising that many predators of gymnosomes are the same as those of thecosomes. Known zooplankton predators of *Clione limacina* include medusae (Fraser, pers. comm.) and ctenophores (Gilmer, pers. obs.). It is interesting, however, that *Clione* is one of the few zooplanktonic species that is not eaten by the amphipod *Parathemisto gaudichaudi*; individuals captured by this hyperiid are rapidly released (Sheader and Evans, 1975). Fish predators include herring (Lebour, 1931; Glover, 1957), mackerel (Lebour, 1931; Massy, 1932), and salmon (Synkova, 1951; Ito, 1964), and *Pneumoderma atlanticum* has been found in the stomachs of lancetfish (Russell, 1960). Since the initial descriptions of *Clione* in the seventeenth and eighteenth centuries, baleen whales in the Arctic have been known to include this gymnosome in their diets. Various species of sea birds also prey on epipelagic gymnosomes (Meisenheimer, 1905a; Massy, 1932).

Reproduction and Development

The general anatomy of the hermaphroditic reproductive tract of gymnosomes has been described by Meisenheimer (1905b and references cited therein), Tesch (1950), and Morton (1958), but many species have not yet been studied in detail. Present knowledge suggests that the reproductive anatomy is very similar to that of the thecosomatous pteropods (Ghiselin, 1965). The ovotestis is situated within the trunk, and a hermaphrodite duct leads forward to a common genital pore on the right ventral side, anterior to the anus. The duct provides a storage area for mature sperm, and in the female stage differentiates into the albumen and mucous accessory glands. The male organs—penis, accessory copulatory organ, and prostate gland—are located within the head, with the male pore opening externally on the right side of the foot lobes. The penis, normally retracted, is armed with small spines in only a few species (e.g. *Paraclione longicaudata*). The accessory copulatory organ is a long tentacle-like structure attached to the base of the penis; it may bear a terminal sucker in some species. An open ciliated sperm groove connects the two genital pores. Often-quoted accounts of gymnosomes with external reproductive organs are not considered here, since these are certainly descriptions of animals damaged during collection and preservation (Pruvot-Fol, 1938; Lalli, pers. obs.).

In contrast to the thecosomes, some gymnosome species frequently exhibit neoteny, with sexual maturity and spawning occurring at a small size, before the loss of all external larval features. Further, it is not clear that gymnosomes are protandrous hermaphrodites in the same sense as the thecosomes. Although it is true that spermatogenesis and the development of the male reproductive organs precede oogenesis and female development, it has not been clearly established that degeneration of male organs occurs in mature female gymnosomes. There is no evidence that mating occurs before full attainment of female maturity, as it does in thecosomes; rather, the available information suggests that gymnosomes function as simultaneous hermaphrodites at maturity.

The behavioral aspects of reproduction and the details of larval development are best known in the common and widely distributed species, *Clione limacina*, and in *Paedoclione doliiformis*, an abundant species in New England and eastern Canadian waters. Considerably more information, preferably attained from observations on living animals, is needed on other gymnosome species before generalities can be established for this group of mollusks.

Copulation in *Clione limacina* was first described by Boas (1886a) from specimens preserved *in copulo*. The animals were united ventrally, and fertilization was reciprocal, with the penis of each inserted into the common genital pore of the mate. In these specimens, the large accessory copulatory organ was free and extended posteriorly. More recently, copulation in living *Clione* (Fig. 63a) has been observed by scientists working in submersibles (Szabo, pers. comm.) and by scuba divers (Pratt, 1982; Gilmer, pers. obs.). Szabo observed 15 mating pairs of *C. limacina* off Vancouver Island in late May at depths of 140 to 150 m (bottom depth = 155 m). All the pairs consisted of individuals of similar size that were joined ventrally with their heads directed toward the surface. Gilmer's observations were made in Au-

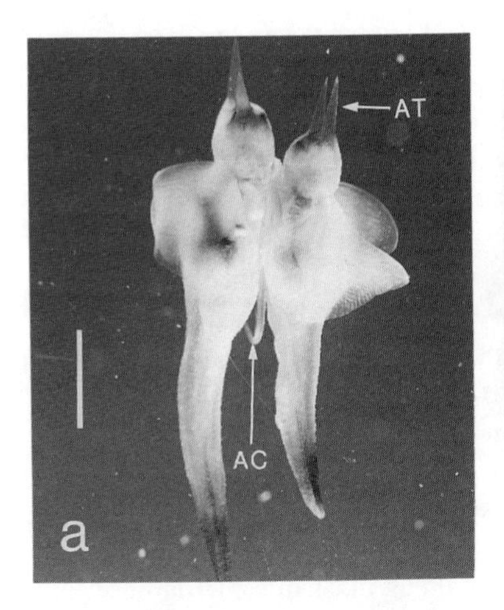

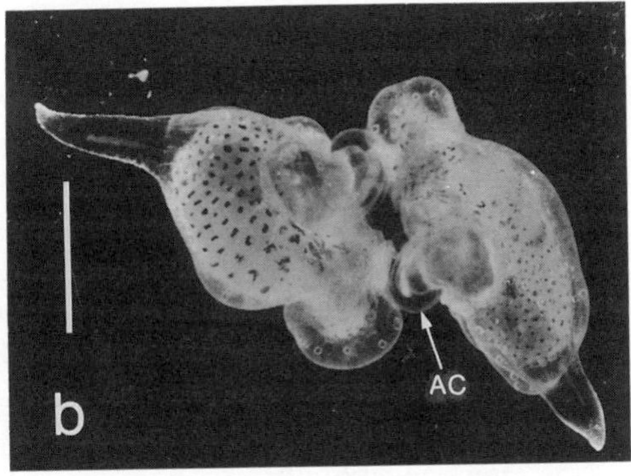

FIG. 63. Reproduction in gymnosomes: *a*, copulating *Clione limacina* (by G. Dietzmann); *b*, mating *Crucibranchaea macrochira* exchanging masses of sperm or spermatophores. AC, accessory copulatory organ; AT, anterior tentacles. Scale lines represent 5 mm.

gust off Greenland and in the Labrador Sea, in the upper 30 m. Most copulating individuals were of similar size, although in one instance an 82-mm individual was mating with a 34-mm *Clione*, and another pair consisted of a 77-mm and a 38-mm individual. In all cases, the terminal sucker on the accessory copulatory organ of each mate was attached to the body wall of the partner. This attachment leaves a small scar; some of the larger individuals bore as many as four such scars, indicating that mating occurs numerous times throughout their life. These scars are rarely evident once an animal has been preserved. Fertilization is reported to require about 4 hours in this species (Knipowitsch, 1891). The mating animals exhibit slow wing beats during this period, and they may continue to feed. In several of the pairs observed, one of the partners was feeding on *Limacina helicina*. Separation involves either rapid swimming by both partners in a spiralling direction or rhythmic swimming by both in opposite directions.

At least one gymnosome species, *Crucibranchaea macrochira*, appears to produce spermatophores that are exchanged during copulation (Fig. 63b). It is probable, upon closer examination of living specimens, that other species will be found to transfer sperm in this manner.

The fact that reciprocal copulation may be a lengthy process in gymnosomes is supported by observations of *Paedoclione doliiformis*, which pairs for at least 3 hours. Mating individuals of *Paedoclione* are always simultaneous hermaphrodites, with both male and female organs fully developed and with mature sperm in the hermaphrodite duct and large eggs in the ovotestis (Lalli and Conover, 1973).

Paedoclione doliiformis spawns from September through November, the period when its prey (*Limacina retroversa*) is most abundant, and spawning behavior has been observed in the laboratory (Lalli and Conover, 1973). Gravid females that are about to deposit eggs swim to or near the water surface and then hover or swim slowly at this level. The three foot lobes become distended and elevated from the body, with the median lobe directed upward (Fig. 64). Fusiform encapsulated eggs emerge singly in a stream of mucus from the common genital pore, at intervals of 7 to 10 seconds. The eggs swell and become ovoid within 10 to 20 seconds. The emerging egg ribbon passes through the foot lobes, which direct the flow. As more eggs emerge, the female begins to swim backward, using short wing strokes and frequently changing direction. These directional changes mold the continuous ribbon into a roughly spherical egg mass measuring between 1.7 and 3.3 mm in diameter. After the last egg is deposited, the female continues to swim around the surface of the egg mass for up to 20 minutes, adding additional mucus. Spawning occupies 20 to 30 minutes from the release of the first egg to the time the female swims away from the floating egg mass. The foot lobes of the spawned female return to their usual size and position within seconds of leaving the egg mass. Females may spawn

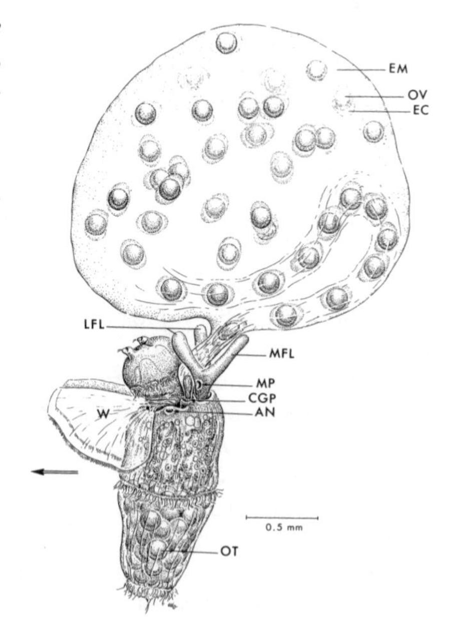

FIG. 64. A spawning *Paedoclione doliiformis* (from Lalli and Conover, 1973). The arrow indicates direction of movement produced by the wings. AN, anus; CGP, common genital pore; EC, egg capsule; EM, egg mass; LFL, lateral foot lobes; MFL, median foot lobe; MP, male pore; OT, ova in ovotestis; OV, ovum; W, wings.

TABLE 24

Fecundity of Paedoclione doliiformis *at different temperatures and food concentrations*

Datum	Temperature (°C)					
	6°	10.2°	10.2°	16.4°	16.4°	17–19°
Number of specimens	6	6	6	5	5	6
Prey concentration	unlimited	unlimited	starved	unlimited	starved	unlimited
Total number of *Limacina retroversa* eaten	21	120	—	312	—	658
Average number of *L. retroversa* eaten	3.5	20.0	—	62.4	—	109.7
Total number of egg masses produced	0	18	5	50	12	34
Average number of egg masses per specimen	0	3.0	0.8[a]	10.0	2.4	5.7[b]
Average number of eggs per egg mass	0	25.9	30.4	46.4	36.4	77.5
Average number of eggs per specimen	0	77.8	25.3	463.8	87.4	439.0

SOURCE: Lalli and Conover, 1973.
[a]Two specimens did not spawn.
[b]One specimen died after producing a single egg mass.

successively over a period of several days, with each egg mass containing from 30 to 165 eggs.

The influence of temperature and food availability on the fecundity of *Paedoclione doliiformis* has been examined under controlled laboratory conditions by Lalli and Conover (1973), and some of the results are presented in Table 24. It is apparent that a cold temperature (6° C) inhibits feeding and prevents maturation and spawning, and that lack of food depresses egg production at both 10° and 16° C. Fecundity is highest in specimens provided with unlimited food at temperatures of 16° to 19° C, with each individual producing an average of between 439 and 464 eggs. It is of interest to note that the individuals used in these experiments initially showed no visual evidence of mature eggs in the ovotestis and were kept isolated. This suggests that either *Paedoclione* is capable of self-fertilization or that it copulates before maturity and stores exogenous sperm until the ova are mature; the latter suggestion, however, is contrary to observations of mating pairs.

Eggs of *Paedoclione doliiformis* (Fig. 65) measure about 0.15 mm in diameter, with the surrounding ellipsoidal egg capsule averaging 0.16 by 0.25 mm in size. Development proceeds by typical molluscan spiral cleavage, and the embryos develop a larval shell and bilobed velum within 48 hours at a temperature of 17° to 19° C. Free-swimming veligers (Fig. 66) hatch in less than 3 days after spawning. The veliger shell of newly hatched

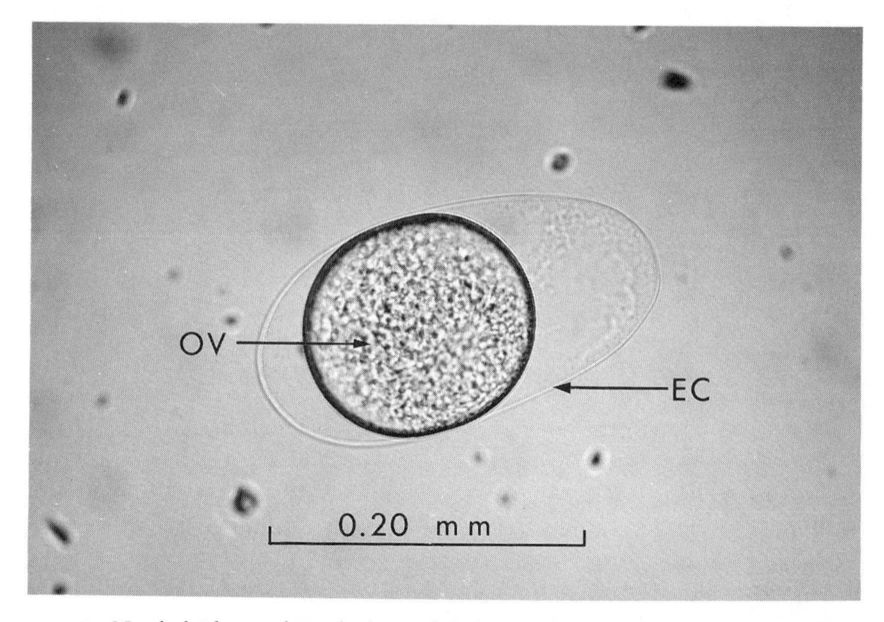

FIG. 65. Newly laid egg of *Paedoclione doliiformis* (from Lalli and Conover, 1973). EC, egg capsule; OV, ovum.

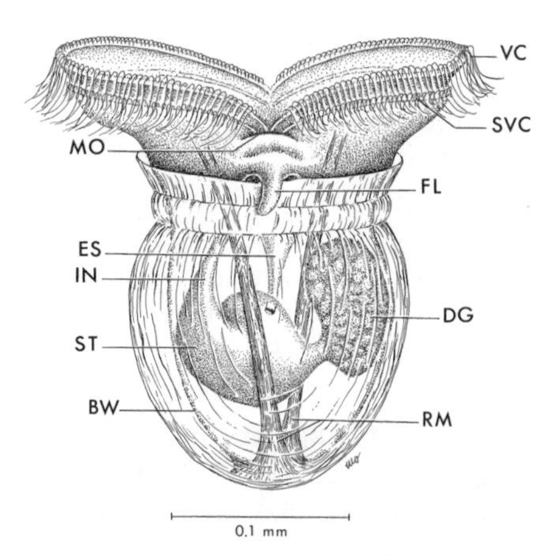

FIG. 66. A free-swimming veliger of *Paedoclione doliiformis* with the retractor muscles slightly displaced from their normal midventral and middorsal positions for clarity (from Lalli and Conover, 1973). BW, larval body wall; DG, digestive gland; ES, esophagus; FL, undifferentiated foot lobe; IN, intestine; MO, mouth; RM, retractor muscle; ST, stomach; SVC, postoral cilia of subvelum; VC, preoral cilia of velum.

young is thimble-shaped and 0.16–0.18 mm in maximum width by 0.17–0.19 mm in length (Fig. 67a). Veligers feed on phytoplankton (*Isochrysis* sp.) in the laboratory and grow rapidly, adding to the dimensions of the shell. At 17° to 19° C, the larval shells are cast off 11 days after hatching, when they measure 0.21 to 0.28 mm in total length (Fig. 67b). Cast shells show a sharp demarcation between the smooth embryonic area, which is laid down before hatching, and the thicker postembryonic area, which is deposited during the free-swimming stage. The postembryonic shell is sculptured and exhibits conspicuous growth rings (Lalli and Conover, 1976). Successive metamorphic changes, which occur in less than 12 hours after shell loss, are shown in Figure 68. The body of the larva elongates, the velum gradually disintegrates, and three ciliary bands develop, encircling the posterior tip of the body, the trunk, and the head. Locomotion is accomplished solely by ciliary action, since the wings are not yet completely developed. The larvae, now referred to as polytrochs, begin feeding in 2 days on veligers of *Limacina retroversa*. In other gymnosome species, there is a second metamorphosis in which the ciliary bands gradually disappear as the body size increases and the wings take over the role of locomotion. *Paedoclione doliiformis*, however, is a neotenous species that retains larval characteristics and small size (2.5 mm length) throughout sexual maturity.

Peak periods of breeding and spawning of *Clione limacina* coincide with local periods of maximal abundance of the phytoplankton species that serve as food for the veliger stage (Mileikovsky, 1970). Animals may lay eggs within 20 to 24 hours after fertilization (Knipowitsch, 1891), and egg masses of this species have been described by the same author, by Lebour (1931), and by Lalli and Conover (1973). Eggs are encapsulated and laid in a free-floating, oblong or spherical, gelatinous mass. The sizes of eggs and egg masses vary directly with the maximal size of the adult. Two size races of *C. limacina* are generally recognized: the northern, cold-water race in

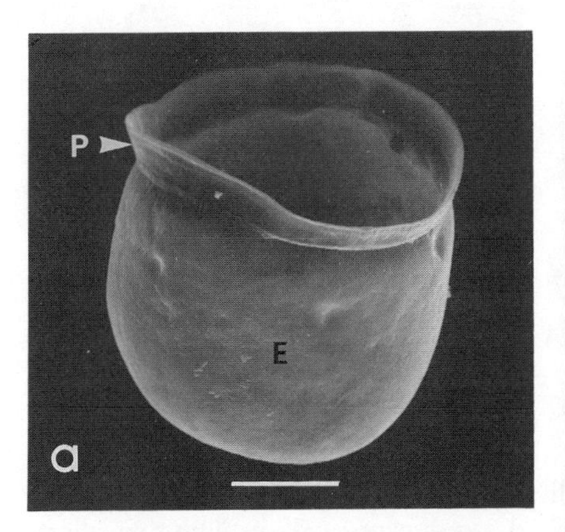

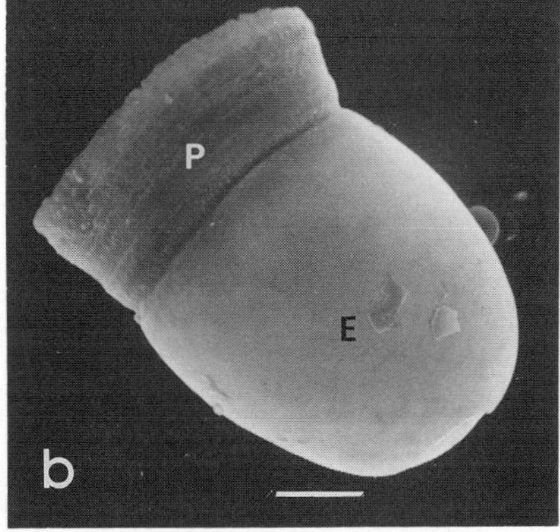

FIG. 67. *Paedoclione doliiformis* veliger shells (from Lalli and Conover, 1976): *a*, shell of a 1-day-old veliger, showing the embryonic shell (E) and the initial formation of the postembryonic shell (P); *b*, discarded veliger shell collected from the plankton, showing an increase in size of the postembryonic shell (P). Scale lines represent 50 μm.

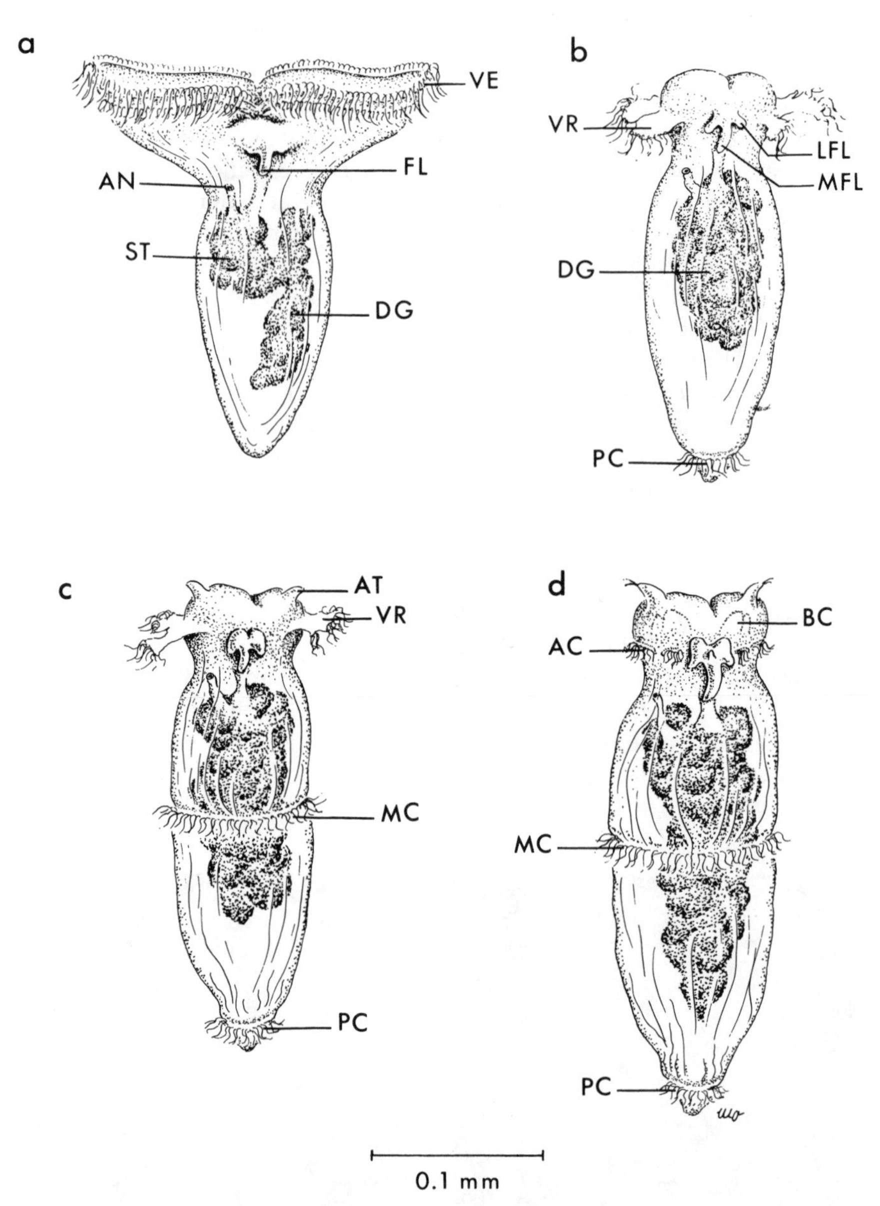

FIG. 68. *Paedoclione doliiformis,* successive stages in metamorphosis from shell-less veliger to polytrochous larva (from Lalli and Conover, 1973): *a,* shell-less veliger larva; *b,* larva with one ciliary band and velar rudiments; *c,* larva with two ciliary bands and velar rudiments; *d,* polytrochous larva with three ciliary bands and completely developed head structures. AC, anterior ciliary band; AN, anus; AT, anterior tentacle; BC, retracted buccal cones; DG, digestive gland; FL, undifferentiated foot lobe; LFL, lateral foot lobe; MC, middle ciliary band; MFL, median foot lobe; PC, posterior ciliary band; ST, stomach; VE, velum; VR, velar rudiments.

TABLE 25

Number of eggs produced by Clione limacina *collected from the Denmark Strait and Labrador Sea*

Clione body length (mm)	Egg mass diameter (mm)	Approximate number of eggs
35	40	1,200–1,500
45	70	1,700–2,000
52	70	1,800
55	82	2,500–2,800
77	85	2,800

SOURCE: Gilmer, pers. obs.

NOTE: Measurements were taken either from collected egg masses or from field photographs made while scuba diving.

which maximal length is 70 to 85 mm; and the southern dwarf form, which matures at 4 to 5 mm and attains a maximal length of 12 mm (Lebour, 1931). Knipowitsch (1891) described the egg masses of large *Clione* from the White Sea as measuring up to 4 cm in diameter and containing eggs of 0.12 mm diameter enclosed in ovoid egg capsules of 0.21 by 0.16 mm. Dwarf specimens from the English Channel lay smaller egg masses of 1.0 to 1.2 mm in diameter containing eggs of 0.09 by 0.08 mm (Lebour, 1931). Two small specimens of *C. limacina* collected in Nova Scotian waters each deposited approximately 150 eggs per egg mass (Lalli and Conover, 1973). Larger *Clione* from the Labrador Sea produce considerably more eggs (Table 25). Both the size of the egg masses and the number of eggs per mass increase directly with the size of the spawning individual.

The early embryology of *Clione limacina*, from the first cleavage through gastrulation, has been described by Knipowitsch (1891), and Lebour (1931), Mileikovsky (1962), and Lalli and Conover (1973) have described development after hatching. At 16° C, veligers hatch within 3 to 4 days. The thin, transparent, thimble-shaped shell of newly hatched veligers measures 0.11 mm in diameter by 0.12 to 0.16 mm in length. The bilobed velum, ciliated foot lobe, and paired statocysts are prominent features of these larvae. Growth is rapid, with the shell doubling in length in approximately 1 week. About 2 weeks after hatching, the veligers cast their shells, which then measure 0.30 to 0.36 mm in length (Fig. 69). Examination by scanning electron microscopy reveals the abrupt transition between the embryonic and postembryonic areas of the discarded shells (Lalli and Conover, 1976). The embryonic portion is without sculpture, whereas the postembryonic flared extension shows conspicuous growth bands on both inner and outer surfaces. The loss of the shell is followed within 24 hours by the loss of the

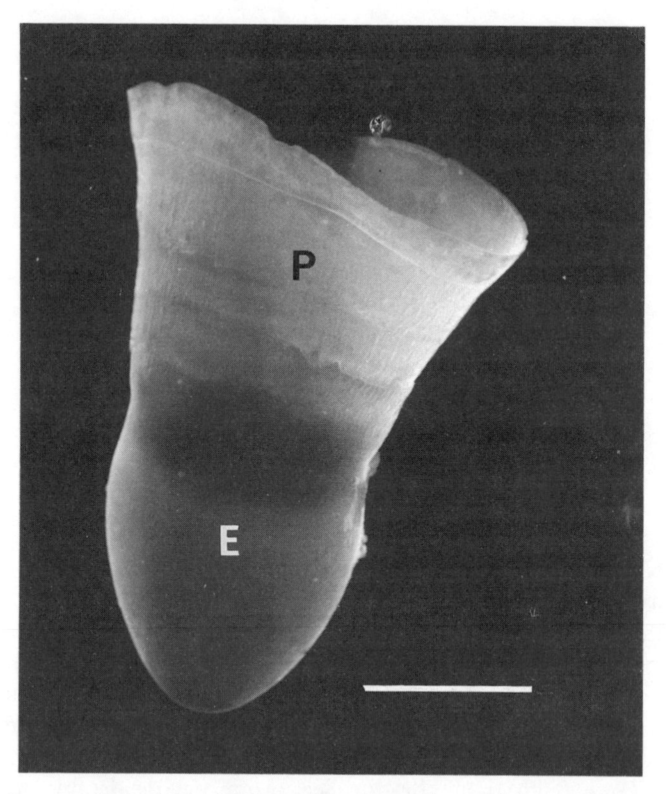

FIG. 69. A cast veliger shell of *Clione limacina* collected from a plankton sample, showing the embryonic shell (E) and the postembryonic shell (P) (from Lalli and Conover, 1976). The scale line represents 100 μm.

velum and the appearance of the three ciliary bands that mark the polytrochous larval stage. The feeding apparatus rapidly completes development at this time, and polytrochous larvae may begin feeding on *Limacina retroversa* veligers within 2 to 3 days after the loss of the velum (Conover and Lalli, 1972). In contrast to *Paedoclione doliiformis*, *Clione limacina* undergoes a second, slow metamorphosis in which the body elongates, the wings grow in size, and the ciliary bands gradually disappear, beginning with the anterior ring. Larvae of *Clione* from the English Channel complete most of these metamorphic changes and show sexual development at less than 3 mm length (Lebour, 1931; Morton, 1958). Polytrochous larvae from northern areas, however, may attain a length of 15 mm without showing any sign of metamorphosis to the adult form (Pelseneer, 1887; Lalli, pers. obs.). Records of seasonal abundance and vertical distribution of polytrochous larvae of *C. limacina* have been compiled by Mileikovsky (1970) throughout the northern range of this species.

The discarded shells of gymnosome veligers are sometimes found in plankton and sediment samples and can be recognized easily, since they are unique in shape and size among molluscan larval shells. Shells of unknown gymnosome species have been found in mixed pteropod sediments of late Pleistocene origin from off Portugal and northwest Africa (van der Spoel and Diester-Haass, 1976), but as Lalli and Conover (1976) have pointed out, gymnosome shells will be relatively rare in deep-water sediments, since predation on veligers decreases the number of shells sedimenting intact and the remaining will be subject to dissolution in deep water. It is not known, though, whether the carbonate shells are composed of aragonite, as are euthecosome shells, or of calcite, as is the case in many other mollusks. Thus the depth at which dissolution occurs will vary with the respective compensation depth and with location.

Parasites

Gymnosomes are remarkably free of epifauna, and to date only three organisms are known to parasitize gymnosomes. Stock (1971, 1973) described two new genera and species of ectoparasitic copepods from gymnosomes—*Micrallecto uncinata* from a single specimen of *Pneumoderma atlanticum* and *Nannallecto fusii* from a single *Pneumodermopsis paucidens*. These parasites appear to belong to the Family Splanchnotrophidae, which includes copepods known to parasitize nudibranchs. More recently, Stock and van der Spoel (1976) reported finding an endoparasitic organism from a single specimen of *Notobranchaea macdonaldi*; the parasite has been provisionally assigned to the Subclass Copepoda.

Suborder Gymnoptera

The Suborder Gymnoptera contains only two species, in separate families. One of the species, *Hydromyles gaudichaudii** (Color Fig. 13; Fig. 70), is of considerable interest, since not only is it very abundant, but it differs in many respects from all other gymnosomes. Some of the anatomical differ-

*We have followed the taxonomic opinions of Pruvot-Fol (1942), Tesch (1950), and van der Spoel (1976), regarding the generic name of this species. We have not adopted van der Spoel's suggestion, however, that the species name *globulosa* be applied as a *nomen conservandum*. To do so would mean taking the name of a species that was collected only in the Atlantic Ocean and was originally described by Rang as *Psyche globulosa*. The description of *Euribia* (*Psyche*) *globulosa* given in Rang and Souleyet (1852) clearly does not apply to the animal we describe here. The species discussed herein corresponds in all essential features to *Euribia gaudichaudii* as described and illustrated in the same publication.

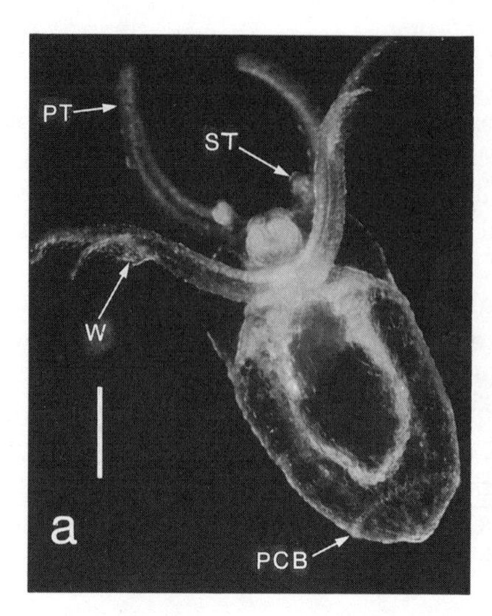

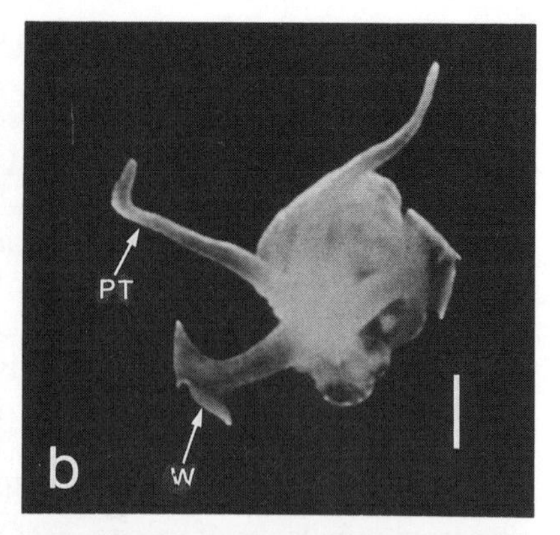

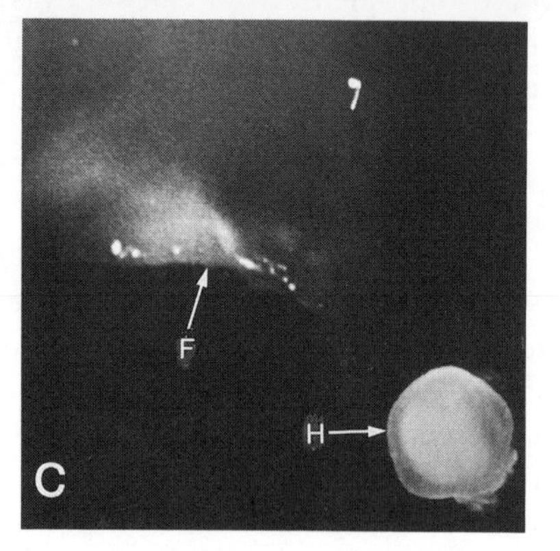

FIG. 70. *Hydromyles gaudichaudii*: *a*, an adult (by G. R. Harbison); *b*, an *in situ* photograph of an individual swimming toward the viewer, showing the positions of the wings and tentacles (by M. Jones); *c*, a disturbed individual emitting mucus and brownish fluid (by M. Jones). F, ejected fluid; H, *Hydromyles*; PCB, posterior ciliary band; PT, primary tentacle; ST, secondary tentacle; W, wing. Scale lines represent 3 mm.

ences are such that there has been reasonable doubt about whether this species should be included with the gymnosomes.

Hydromyles occurs in both the Indian and Pacific oceans, extending at times to 50° N and 50° S, but the highest concentrations are in the Malaysian Archipelago and off eastern Australia (van der Spoel, 1976). According to Tesch (1950), *Hydromyles* constituted 98.5 percent of all gymnosomes collected by the *Dana* Expedition in the tropical Indo-Pacific. More than 40,000 specimens of *Hydromyles* were collected, compared with only 650 specimens of all other gymnosomes; as many as 3,000 individuals were sometimes collected in a single plankton tow. These high population densities suggest that *Hydromyles* may differ biologically and ecologically, as well as anatomically, from other gymnosomes.

The anatomy of *Hydromyles gaudichaudii* has been described previously by several authors, with the reports of MacDonald (1858), Pelseneer (1888), and Meisenheimer (1905b) being notable for their detail. We have been able to add to these descriptions by the examination of preserved specimens that were collected by scuba divers in waters off eastern Australia.

The expanded integument of *Hydromyles gaudichaudii* (Fig. 70a) forms an ovoid, tough, transparent covering that is flattened ventrally and rounded dorsally. It contains a sizable opening on the anteroventral surface, into which the head and wings can be completely withdrawn. The body itself is attached by fine muscle strands to the integument, but the connection appears to be tenuous, since preserved animals are sometimes detached from their integumentary covering. A large space separates the viscera and the outer covering. The integument can be regarded as a functional equivalent of the external molluscan shell, since both are enveloping structures that protect the soft body parts. The capacious space between the integumentary covering and the inner viscera recalls a mantle cavity and serves as a brood chamber for this species. The integument is composed of large hexagonal cells that form a faintly visible pattern over the surface. Small, scattered oil droplets, which are embedded in the integument, may contribute to the buoyancy of the animal. Two ciliary bands may be present, encircling the outer surface of the integument; these larval structures persist well into the adult stage, though they are absent in mature females. The maximum length of the integument, with the animal retracted, is approximately 10 mm.

The body of *Hydromyles* is divided into a distinct, but small, head and a trunk. The wings are long and narrow for most of their length; they are fused medially on the ventral surface. The wings further differ from those of other gymnosomes by being expanded distally into thin, broad, deeply lobed, membranous tips. Figure 70b illustrates the positions of the wings and tentacles of a swimming individual. Rang and Souleyet (1852) described the swimming posture as similar to that of thecosomes, with the ventral surface directed obliquely upward in the water column.

The ventral foot is composed of three lobes, with the large, paired lateral lobes lying anteriorly to the base of the wings. Tesch (1950) accurately noted that the small, unpaired, median lobe is separated from the other two foot lobes, and is attached directly to the posterior border of the central area of the wings.

The head bears a conspicuous pair of tentacles, which are referred to here as primary tentacles. These are much longer than the anterior tentacles of other gymnosomes, and they also differ in structure. The surfaces of these tentacles of *Hydromyles* are covered with cilia arranged in parallel longitudinal rows, and each primary tentacle bears a smaller, secondary tentacle on the inner side of its base. There has been disagreement about whether the large primary tentacles are homologous with the anterior tentacles of other gymnosomes. Pelseneer (1887) regarded the primary tentacles as buccal appendages equivalent in origin and function to the buccal cones or sucker-bearing appendages, and the secondary tentacles as homologues of the anterior tentacles of other gymnosomes; Meisenheimer (1905b) regarded the primary and secondary tentacles as bifurcated anterior tentacles. Another set of small tentacles is located laterally on the posterodorsal surface of the head; these resemble in both position and structure the posterior head tentacles of other gymnosomes.

The usual gymnosome organs that are employed for prey capture—buccal cones, sucker-bearing arms, proboscis, and hook sacs—are lacking in this species. The mouth is bounded by two folds and opens directly into the buccal cavity. The radula is unusual, and resembles that of thecosomes in having only three teeth in each row (1-1-1), but the median tooth is sickle-shaped, as in some clionids. The lateral teeth have a broad base and a curved, single "hook." There is no dentition on any of the teeth. The reduced jaws consist of rows of small, square, chitinous plates and, unlike the jaws in some gymnosomes, are not spinous.

Two salivary glands, which have a simpler morphology than those of other gymnosomes, discharge into the buccal cavity and run posteriorly along the esophagus. The remainder of the gut resembles that of gymnosomes, in that the stomach is fused with the digestive gland. There is no gizzard and there are no gizzard plates. The intestine, however, is longer than in other species, and it emerges from the dorsal side of the stomach, not from the ventral side. The anus lies on the right ventral side near the genital openings.

Food capture by this species has not been described, but it has been suggested that *Hydromyles* (= *Anopsia*) is capable of absorbing dissolved organic material (Vyshkvartsev and Sorokin, 1978). Still, gizzard plates of unidentified thecosomes have been found in the stomachs of preserved specimens (Lalli, pers. obs.), which suggests that the diet of *Hydromyles* conforms with that of other gymnosomes.

One organ of *Hydromyles* has puzzled previous workers, and with good reason, since it is not present in any other planktonic gastropod. This structure lies on the right side of the viscera and is highly visible in living specimens because of its bright-orange color. In preserved specimens, the structure is white and filled with hardened fluid. MacDonald (1858) took this structure to be a spermatheca; Pelseneer (1888) believed it to be a gland of the reproductive tract; Meisenheimer (1905b), Tesch (1950), and Minichev (1963) referred to it as a blind appendage of the hindgut or a rectal sac, and offered no suggestion about its function. In fact, this very conspicuous organ is a gland that releases a brownish-colored fluid through the anal opening and into the surrounding water when the animals are disturbed (Harbison, pers. comm.) (Fig. 70c). This discovery, coupled with the animal's other anatomical differences from gymnosomes, continues to add doubt to the inclusion of *Hydromyles* in the Order Gymnosomata.

Hydromyles gaudichaudii is also the only gymnosome reported to be ovoviviparous. The animals are protandric hermaphrodites, and the reproductive anatomy of this species apparently does not differ in any essential feature from that of other gymnosomes (Meisenheimer, 1905b). Mature females, however, utilize the large space situated between the visceral mass and the integument as a brood chamber. According to Martoja (1965), but contrary to Meisenheimer's observations, the brood pouch does not arise from the reproductive tract, and it opens externally via a pore near the anus. Fertilized eggs enclosed in individual egg capsules enter the chamber through the common genital opening. The subsequent embryonic development has been described by MacDonald (1858). The embryonic integument expands into a globose shape and develops an anterior opening through which a bilobed velum can protrude; the young eventually swim freely within the brood chamber. The external surface of the integument gradually develops two transverse ciliated rings. All organs, except those of the reproductive tract, become well developed during the brooding period. The source of nourishment needed to accomplish advanced larval development is not known, but as larval development progresses, the brood sac expands greatly and the reproductive system and other organs of the incubating female undergo degeneration. The young emerge from the opening of the brood pouch as fully developed juveniles. Martoja (1965) noted the presence of an unusual mass of tissue, possibly glandular, that surrounds the esophageal ganglia of adult female *Hydromyles*. This tissue reaches its maximal development before the onset of incubation and diminishes greatly during brooding, which suggests that it may have an endocrine function.

Only one other species is included in the Suborder Gymnoptera, and there is little that can be said about *Laginiopsis trilobata*, since the species is based on a single specimen collected from deep water off the Azores in 1905. It was described (Pruvot-Fol, 1926) as having the typical gymnosome features of a distinct head with paired anterior and posterior tentacles, long paired wings, and three foot lobes. The species is set apart from other gymnosomes, however, by the possession of a long (ca. 10 mm), nonretractile proboscis terminating in three large lobes, and by the absence of the usual gymnosome buccal mass (the prehensile tentacles, jaws, radula, hook sacs, and salivary glands are absent). Differences also exist in the anatomy of the central nervous system and gut. Although the single specimen was captured alive, no behavioral observations were recorded.

Evolution

The gymnosomes, like the heteropods, remain an isolated group of mollusks that have become highly adapted to a pelagic existence. Their origins and phylogenetic history are obscure. Although the tiny larval shells of gymnosomes can be found in sediments, it is unlikely that they will provide clues about the time of origin of this group. But because gymnosomes are ecologically dependent on thecosomatous pteropods as prey, it is certain that the shell-less pteropods are also relatively recent arrivals to the planktonic realm. That is to say, they have evolved within the last 50 to 60 million years.

Numerous theories have been advanced and debated concerning the ancestral forms of gymnosomatous pteropods, and the main arguments have centered on whether gymnosomes and thecosomes arose monophyletically (i.e. had a common opisthobranch ancestor), or whether they derived biphyletically from different groups. Some of the early taxonomic work that placed gymnosomes and thecosomes in the same class (Pteropoda) and considered them as a monophyletic group has been discussed in Chapter 4. Boas (1886b) was the first to suggest separate origins for the two pteropod groups; he believed that the gymnosomes derived from a group of tectibranchs, whereas the thecosomes arose from bulloids (cephalaspideans). His ideas concerning a biphyletic origin were generally accepted by later workers, although debate continued on the specific ancestral group that gave rise to the gymnosomes. Although attempts have been made to derive gymnosomes from swimming aplysiomorph opisthobranchs (e.g. Pelseneer, 1888; Meisenheimer, 1905b), this link with herbivorous ancestors is not now generally accepted. Morton (1958), for example, has pointed out that there are no aplysioid gut features (crop, double gizzard, caecum) in the gymnosomes, nor in the aplysioids are there any structures corresponding to the gymnosome buccal cones or sucker-bearing tentacles and hook sacs. Further, the radulae are very different in these groups, and the nervous and reproductive systems are comparable only at a very general level.

On the other hand, Ghiselin (1965) has suggested that there is some basis for considering a monophyletic derivation of gymnosomes and thecosomes on the basis of reproductive-tract anatomy and spermatozoon morphology. He theorized that both pteropod groups should be considered members of the same clade as the Anaspidea, but not derived directly from them. He viewed the obvious differences between the shelled and shell-less pteropods as divergences related to different feeding mechanisms. Our own view is that a phylogeny based on reproductive-tract similarities is premature in light of the new observations presented in this book about reproductive biology in both groups and an obvious need for more detailed examination of reproductive anatomy in more species of both pteropod types.

Theories on phylogenetic relationships within the Gymnosomata are equally divided on whether the group itself is of monophyletic, biphyletic, or even polyphyletic origin. Part of the debate centers around the placement of *Hydromyles*, which is anatomically distinct in many respects from other gymnosomes, thus giving credence to a biphyletic theory of origin (Meisenheimer, 1905b). On the other hand, van der Spoel's (1967) hypothesis of polyphyletic origin, which derives gymnosomes from perhaps "six, only slightly related, ancestors," does not seem admissible. Nor has there been general agreement on the question of whether *Hydromyles* should be regarded as a specialized (Pelseneer, 1887) or primitive (Minichev, 1963) member of the group. Overall, and with the exception of *Hydromyles*, the gymnosomes are basically a homogeneous taxon.

Unlike the heteropods and thecosomes, which show a clear evolutionary progression from species that are totally encased in spiral shells to larger forms that lack external calcareous shells, none of the gymnosomes has a shell as an adult. Nor are there many other striking anatomical differences between the species, except in the extreme morphological diversity of structures used for prey capture and feeding. This is undoubtedly a reflection of the feeding specialization found in this group. Although diet is known for only a dozen species, all of these gymnosomes feed on thecosomes, and the shape of the prey shell probably has been a selective factor in the evolution of the gymnosome buccal armature and feeding mechanism. Adaptations for prey capture include tentacles with stalked suckers, tentacles with adhesive papillae (buccal cones), and chitinous hooks. Multiple adaptations have also evolved to permit prey extraction. This function can be performed by a proboscis or by long hooks, or by a combination of both. Gymnosomes that feed on spirally coiled thecosomes, such as *Clione* and *Paedoclione* feeding on *Limacina*, can and do extract the prey with their hooks. On the other hand, a long and flexible proboscis is ideally suited for extracting thecosomes of the type having long, straight, conical shells, as exemplified by *Pneumodermopsis paucidens* feeding on species of *Creseis*.

Despite limited behavioral observations and diet analyses, the present data and anatomical evidence allow us to assume that feeding specialization is a major feature of gymnosome biology. The disadvantages of such specialization are immediately obvious. In the short term, local populations of the specialist must rely on the availability of abundant prey within the same area at the same time or face local extinction through starvation. From an evolutionary point of view, the specialist's ultimate success as a species is irrevocably linked to the evolutionary success of its prey species. There are, however, a number of ways in which the risks of food specialization may be reduced, several of which have been discussed by Lalli (1970b) and Conover and Lalli (1972) as operating in the *Clione–Limacina* relationship. Any feeding specialist that faces the problem of exposure to periods with no food should be able to survive for a relatively long time without feeding. In fact, *Clione* can be maintained in the laboratory for as long as one month without food; more generally, only a small percentage of preserved gymnosomes ever contain food remains in the gut, which also indicates sporadic feeding. It might be expected too that feeding specialists would utilize as much of their prey as possible whenever it is available, and indeed the gymnosomes ingest everything but the shells of their prey. Further, because thecosomes are primarily microphagous, they are a more abundant food source than prey belonging to higher trophic levels.

There are also several mechanisms that may enforce the coupling of the prey and predator in space and time. *Clione*, for example, is a physiological generalist that is capable of living in waters of broader temperature and salinity ranges than are tolerated by either of its prey species individually. Thus *Clione* feeds on *Limacina helicina* in Arctic and Antarctic seas and on *L. retroversa* in more temperate regions. The disadvantages of specialization are thus lessened by the ability of *Clione* to feed on either of these closely related species, wherever and whenever they are encountered. This mechanism may not apply to subtropical and tropical gymnosomes, which tend to have more restricted temperature and salinity tolerances and narrower geographic ranges, but food specialization in these species may not be so limited. The number of species of both gymnosomes and thecosomes is higher in warm areas, and *Pneumodermopsis paucidens*, which is able to feed on several genera and species of thecosomes, may be more typical of warm-water gymnosomes. Similar reproductive patterns may also serve to maintain the association of prey and predator. Most gymnosomes and thecosomes lay eggs in floating masses, and the young hatch as veligers that filter-feed on phytoplankton. Thus the spawning periods of both the prey (*Limacina* spp.) and the predator (*Clione*) in polar and temperate areas tend to coincide with phytoplankton blooms (Mileikovsky, 1970; Conover and Lalli, 1972). Simultaneous spawning of the prey and predator is also important in maintaining the optimal size ratio. *Clione* veligers metamorphose within a few days of hatching, and the resultant polytrochous larvae begin carnivorous feeding on veliger or recently metamorphosed *Limacina*. The growth of both pteropods keeps pace in the following months, so that larger and larger prey become available for the maturing *Clione*; this is a critical factor governing the growth of the gymnosome, since growth rates and ultimate size are determined more by the size of the food than by the abundance of the prey. In tropical species, the spawning of prey and predators may take place over most of the year, but even there the young of thecosomes and gymnosomes tend to hatch within the same general depth range, are thus subjected to the same water movements, and so remain coupled in space. Although different mechanisms of prey–predator coupling may operate in different latitudes and habitats and with different species, all can lessen the risks of feeding specialization.

The advantages conferred by food specialization are more abstruse than the disadvantages. Conover and Lalli (1972) suggested that increased ecological efficiency may be a major beneficial result of specialization. When diet is limited to one or to a few, very similar, prey species, all aspects of feeding can be refined to maximize energy yield in respect to energy spent in such activities. For example, sense receptors may become highly attuned to recognize a particular prey species among other abundant, but unsuitable, organisms. Although very little is known about how gymnosomes locate prey from a distance, it has been demonstrated that *Clione* is capable of using its buccal cones to tactilely recognize a *Limacina* shell. The capture and extraction of different thecosome species from morphologically diverse shells also involves specialized functions that are reflected in the extreme anatomical diversity of gymnosome feeding structures, as discussed above. Further, gymnosomes do not waste energy in the capture, sorting, and rejection of unsuitable food, as do filter-feeding zooplankton. This also applies to the sorting of indigestible material in the gut, since only the soft parts of thecosomes are ingested; accordingly, the gut is anatomically simple and lacks a functional stomach or sorting area (Morton, 1958). Digestion and assimilation in specialists should also be extremely efficient, since the repertoire of enzymes needed to digest food can be restricted compared with that of food generalists; this has been confirmed in *Clione limacina*, which assimilates carbon from its prey with greater than 90-percent efficiency and with nearly 100-percent efficiency for nitrogen (Conover and Lalli, 1974). Finally, feeding specialization may also reduce interspecific competition, as has been suggested by Mayr (1964) and Mullin (1967). Though there is no direct evidence to support the idea that specialization in gymnosomes may have evolved as the result of intense interspecific competition for food, no other zooplankton are known to feed predominantly on thecosomes, although certain fish and whales may consume large numbers. Still, whereas food specialization may lessen interspecific competition, intraspecific competition for food is intensified. The aggressive behavior of starved *Clione* toward each other when presented with only small numbers of prey (Lalli, 1970b) is, perhaps, a behavioral reflection of the extremely limited diet of this gymnosome.

List of Recognized Species

(Synonymy is given in van der Spoel, 1976)

Suborder Gymnosomata
Family Pneumodermatidae
Pneumoderma atlanticum (Oken, 1815)
P. peroni (Cuvier, 1817)
P. mediterraneum (van Beneden, 1838)
?*P. meisenheimeri* Pruvot-Fol, 1926 (a species of uncertain validity)
P. degraaffi van der Spoel and Pafort-van Iersel, 1982
Spongiobranchaea australis d'Orbigny, 1836
S. intermedia Pruvot-Fol, 1926
Pneumodermopsis ciliata (Gegenbaur, 1855)
P. paucidens (Boas, 1886)
P. polycotyla (Boas, 1886)
P. simplex (Boas, 1886)
P. minuta (Pelseneer, 1887)
?*P. oligocotyla* Massy, 1917 (a species of uncertain validity)
P. canephora Pruvot-Fol, 1924
P. pupula Pruvot-Fol, 1926
?*P. macrocotyla* Zhang, 1964 (a species of uncertain validity)
P. teschi van der Spoel, 1973
Schizobrachium polycotylum Meisenheimer, 1903
Crucibranchaea macrochira (Meisenheimer, 1905)
C. michaelsarsi Bonnevie, 1913
Abranchaea chinensis Zhang, 1964
Platybrachium antarcticum Minichev, 1976
Family Notobranchaeidae
Notobranchaea macdonaldi Pelseneer, 1886
N. inopinata Pelseneer, 1887
N. valdiviae Meisenheimer, 1905
N. hjorti (Bonnevie, 1913)
N. longicollis (Bonnevie, 1913)
N. tetrabranchiata Bonnevie, 1913
N. grandis Pruvot-Fol, 1942
N. bleekerae van der Spoel and Pafort-van Iersel, 1985
Family Cliopsidae
Cliopsis krohni Troschel, 1854

Pruvotella pellucida (Quoy and Gaimard, 1832)
P. danae Pruvot-Fol, 1942
Family Clionidae
Subfamily Thliptodontinae
Thliptodon gegenbauri Boas, 1886
T. diaphanus (Meisenheimer, 1903)
?*T. antarcticus* Meisenheimer, 1906 (a species of uncertain validity)
T. schmidti Pruvot-Fol, 1942
?*T. akatukai* Tokioka, 1950 (a species of uncertain validity)
Cephalobrachia macrochaeta Bonnevie, 1913
C. bonnevii Massy, 1917
Massya longicirrata (Massy, 1917)
Subfamily Clioninae
Clione limacina (Phipps, 1774)
Paraclione pelseneeri Tesch, 1903
P. longicaudata (Souleyet, 1852)
P. flavescens (Gegenbaur, 1855)
Fowlerina zetesios Pelseneer, 1906
F. punctata (Tesch, 1903)
Paedoclione doliiformis Danforth, 1907
Thalassopterus zancleus Kwietniewski, 1910
Suborder Gymnoptera
Family Hydromylidae
Hydromyles gaudichaudii (Souleyet, 1852)
Family Laginiopsidae
Laginiopsis trilobata Pruvot-Fol, 1922

References Cited

Works with recommended keys or aids for the identification of species are indicated by an asterisk.

Boas, J. E. V. 1886a. Spolia Atlantica. Bidrag til Pteropodernes. Morfologi og Systematik samt til Kundskaben om deres geografiski Udbredelse. *K. danske Vidensk. Selsk. Skr.*, 6 Raekke, naturv. mat. Afd., 4: 1–231.

———. 1886b. Zur Systematik und Biologie der Pteropoden. *Zool. Jb.* 1: 311–40.

Conover, R. J., and C. M. Lalli. 1972. Feeding and growth in *Clione limacina* (Phipps), a pteropod mollusc. *J. expl mar. Biol. Ecol.* 9: 279–302.

———. 1974. Feeding and growth in *Clione limacina* (Phipps), a pteropod mollusc. II. Assimilation, metabolism, and growth efficiency. *J. expl mar. Biol. Ecol.* *16*: 131–54.

Cooper, G. A., and D. C. T. Forsyth. 1963. Continuous plankton records: Contribution towards a plankton atlas of the North Atlantic and the North Sea. VII. The seasonal and annual distributions of the pteropod *Pneumodermopsis* Keferstein. *Bull. mar. Ecol.* *6*: 31–38.

Dexter, R. W. 1962. Further studies on the marine mollusks of Cape Ann, Massachusetts. *Nautilus* *76*: 63–70.

Ghiselin, M. T. 1965. Reproductive function and the phylogeny of opisthobranch gastropods. *Malacologia* *3*: 327–78.

Glover, R. S. 1957. An ecological survey of the drift-net herring fishery off the north-east coast of Scotland. II. The planktonic environment of the herring. *Bull. mar. Ecol.* *5*: 1–43.

Hamner, W. M., L. P. Madin, A. L. Alldredge, R. W. Gilmer, and P. P. Hamner. 1975. Underwater observations of gelatinous zooplankton: Sampling problems, feeding biology, and behavior. *Limnol. Oceanogr.* *20*: 907–17.

Ito, J. 1964. Food and feeding habit of Pacific salmon (genus *Oncorhynchus*) in their oceanic life. *Bull. Hokkaido reg. Fish. Res. Lab.* *29*: 85–97.

Knipowitsch, N. 1891. Zur Entwicklungsgeschichte von *Clione limacina*. *Biol. Zbl.* *11*: 300–303.

Lalli, C. M. 1970a. Morphology of *Crucibranchaea macrochira* (Meisenheimer), a gymnosomatous pteropod. *Proc. malac. Soc. Lond.* *39*: 1–14.

———. 1970b. Structure and function of the buccal apparatus of *Clione limacina* (Phipps) with a review of feeding in gymnosomatous pteropods. *J. expl mar. Biol. Ecol.* 4: 101–18.

———. 1972. Food and feeding of *Paedoclione doliiformis* Danforth, a neotenous gymnosomatous pteropod. *Biol. Bull. mar. biol. Lab., Woods Hole* *143*: 392–402.

Lalli, C. M., and R. J. Conover. 1973. Reproduction and development of *Paedoclione doliiformis*, and a comparison with *Clione limacina* (Opisthobranchia: Gymnosomata). *Mar. Biol.* *19*: 13–22.

———. 1976. Microstructure of the veliger shells of gymnosomatous pteropods (Gastropoda: Opisthobranchia). *Veliger* *18*: 237–40.

Lebour, M. V. 1931. *Clione limacina* in Plymouth waters. *J. mar. biol. Ass. U.K.* *17*: 785–95.

MacDonald, J. D. 1858. On the anatomy of *Eurybia Gaudichaudi*, as bearing upon its position amongst the Pteropoda. *Trans. Linn. Soc. Lond.* 22: 245–49.

Mackie, G. O. 1985. Midwater macroplankton of British Columbia studied by submersible PISCES IV. *J. Plankton Res. 7*: 753–77.

Mackie, G. O., C. L. Singla, and C. Thiriot-Quiévreux. 1976. Nervous control of ciliary activity in gastropod larvae. *Biol. Bull. mar. biol. Lab., Woods Hole 151*: 182–99.

Manteufel, B. P. 1937. On the biology of *Clione limacina* Phipps. *Bull. Soc. Nat. Moscow, Sect. Biol., N.S. 46*: 25–35.

Martoja, M. 1965. Sur l'incubation et l'existence possible d'une glande endocrine, chez *Hydromyles globulosa* Rang (*Halopsyche Gaudichaudi* Keferstein), Gastéropode Gymnosome. *C. r. hebd. Séanc. Acad. Sci., Paris 260*: 2907–9.

Massy, A. L. 1917. The Gymnosomatous Pteropoda of the coasts of Ireland. *Scient. Proc. R. Dubl. Soc. 15*: 223–44.

———. 1932. Mollusca: Gastropoda Thecosomata and Gymnosomata. *"Discovery" Rep. 3*: 267–96.

Mayr, E. 1964. *Systematics and the Origin of Species*. New York: Dover. 334 pp.

McGowan, J. A. 1968. The Thecosomata and Gymnosomata of California. *Veliger 3*(Suppl.): 103–29.

Meisenheimer, J. 1905a. Die arktischen Pteropoden. *Fauna arct.* 4: 408–30.

———. 1905b. Pteropoda. *Wiss. Ergebn. dt. Tiefsee-Exped. "Valdivia" 9*: 1–314.

Mileikovsky, S. A. 1962. [Pelagic larvae of Gastropoda from the region of the White Sea Biological Station of Moscow State University.] In: *The Biology of the White Sea*. L. A. Zenkevitch, ed. Moscow: Moscow State Univ. Press. Pp. 171–200. [In Russian.]

———. 1970. Breeding and larval distribution of the pteropod *Clione limacina* in the North Atlantic, Subarctic and North Pacific Oceans. *Mar. Biol. 6*: 317–34.

Minichev, Yu. S. 1963. Anatomy of *Anopsia gaudichaudii* (Souleyet) and systematic position of Gymnosomata (Opisthobranchia). *Zool. Zh.* 42: 1317–28. [In Russian.]

———. 1976. On the morphology of the pelagic molluscs Pneumodermatidae (Opisthobranchia, Gymnosomata) from the Antarctic waters. *Explorations of the Fauna of the Seas. 18*: 102–06. [In Russian.]

Morton, J. E. 1958. Observations on the gymnosomatous pteropod *Clione limacina* (Phipps). *J. mar. biol. Ass. U.K. 37*: 287–97.

Mullin, M. M. 1967. On the feeding behaviour of planktonic marine copepods and the separation of their ecological niches. In: *Proceedings of a Symposium on Crustacea, Ernakulum, Jan. 12–15, 1965*, 2: 955–64.

Pelseneer, P. 1887. Report on the Pteropoda collected by H.M.S. Challenger during the years 1873–76. I. The Gymnosomata. *Scient. Rep. "Challenger," Zoology 19*: 1–74.

———. 1888. Report on the Pteropoda collected by H.M.S. Challenger during the years 1873–76. III. Anatomy. *Scient. Rep. "Challenger," Zoology* 23: 1–97.

Pratt, H. W. 1982. The sea butterfly. *Sea Frontiers 28*: 229–30.

Pruvot-Fol, A. 1924. Étude de quelques gymnosomes Méditerranéens des pêches de l'"Orvet" en 1921 et 1922. *Archs. Zool. exp. gén. 62*: 345–400.

———. 1926. Mollusques ptéropodes gymnosomes provenant des campagnes du Prince Albert 1 de Monaco. *Résult. Camp. scient. Prince Albert 1 Monaco 70*: 1–60.

———. 1938. Sur les apparences trompeuses de quelques échantillons de gymnosomes a l'état conservé. *J. Conch., Paris 82*: 256–58.

———. 1942. Les Gymnosomes I. *Dana Rep.* No. *20*: 1–54.

———. 1954. Mollusques opisthobranches. *Faune Fr.* No. *58*. 460 pp.

Rang, P. C. A. L., and L. F. A. Souleyet. 1852. *Histoire naturelle des mollusques ptéropodes*. Paris: J.-B. Baillière. 86 pp.

Rosewater, J. 1959. Intertidal stranding of *Clione limacina* in Massachusetts. *Nautilus* 73: 76–77.

Russell, H. D. 1960. Heteropods and pteropods as food of the fish genera, *Thunnus* and *Alepisaurus*. *Nautilus* 74: 46–56.

Schiemenz, P. 1906. Die Pteropoden der Plankton-Expedition. *Ergebn. Plankton-Exped. der Humboldt-Stiftung* 2: 1–38.

Sentz-Braconnot, E. 1965. Sur la capture des proies par le ptéropode gymnosome *Pneumodermopsis paucidens* (Boas). *Cah. Biol. mar. 6*: 191–94.

Sheader, M., and F. Evans. 1975. Feeding and gut structure of *Parathemisto gaudichaudi* (Guerin) (Amphipoda, Hyperiidea). *J. mar. biol. Ass. U.K. 55*: 641–56.

van der Spoel, S. 1967. *Euthecosomata, a Group with Remarkable Developmental Stages (Gastropoda, Pteropoda)*. Gorinchem: J. Noorduijn. 375 pp.

*———. 1976. *Pseudothecosomata, Gymnosomata and Heteropoda (Gastropoda)*. Utrecht: Bohn, Scheltema, and Holkema. 484 pp.

van der Spoel, S., and L. Diester-Haass. 1976. First records of fossil gymnosomatous protoconchae (Pteropoda, Gastropoda). *Bull. zool. Mus. Univ. Amst. 5*: 85–87.

van der Spoel, S., and T. Pafort-van Iersel. 1982. *Pneumoderma degraaffi* n. sp., a pteropod new to science (Mollusca, Opisthobranchia). *Bull. zool. Mus. Univ. Amst. 9*: 17–20.

———. 1985. Note on the taxonomy of the Family Notobranchaeidae and description of *Notobranchaea bleekerae* n. sp., a species new to science (Gastropoda, Pteropoda). *Basteria* 49: 29–36.

Stock, J. H. 1971. *Micrallecto uncinata* n. gen., n. sp., a parasitic copepod from a remarkable host, the pteropod *Pneumoderma*. *Bull. zool. Mus. Univ. Amst.* 2: 77–81.

———. 1973. *Nannallecto fusii* n. gen., n. sp., a copepod parasitic on the pteropod, *Pneumodermopsis*. *Bull. zool. Mus. Univ. Amst.* 3: 21–24.

Stock, J. H., and S. van der Spoel. 1976. *Pteroxena papillifera* n. gen., n. sp., an endoparasitic organism (Copepoda?) from the gymnosomatous pteropod, *Notobranchaea*. *Bull. zool. Mus. Univ. Amst.* 5: 177–80.

Synkova, A. I. 1951. Food of Pacific salmon in Kamchatka waters. In: *Pacific Salmon: Selected Articles from Soviet Periodicals* (1961). Jerusalem: Israel Program for Scientific Translations. Pp. 216–35.

*Tesch, J. J. 1950. The Gymnosomata II. *Dana Rep.* No. 36. 55 pp.

Vyshkvartsev, D. I., and Yu. I. Sorokin. 1978. [On the intensity of feeding of some marine invertebrates on dissolved organic material.] *Nauch. Soobshch. Inst. Biol. Morya, Vladivostok.* 3: 27–31. [In Russian.]

Zhang, Fu-sui. 1964. The pelagic molluscs off the China coast. I. A systematic study of Pteropoda (Opisthobranchia), Heteropoda (Prosobranchia) and Janthinidae (Ptenoglossa, Prosobranchia). *Studia mar. Sinica* 5: 125–226. [In Chinese.]

6

The Planktonic Nudibranchs

Swimming Sea Slugs

Class Gastropoda
  Subclass Opisthobranchia
    Order Nudibranchia
      Suborder Dendronotacea
        Family Phylliroidae
      Suborder Aeolidacea
        Family Glaucidae
        Family Fionidae

Nudibranchs, or sea slugs, form the largest and most diverse order of opisthobranchs. All of the sea slugs lack a shell and mantle cavity, and the body is generally elongate and flexible. The great majority of nudibranchs are benthic animals that crawl slowly on their well-developed foot. Loss of the shell and flexibility of the body, however, have resulted in the ability of some nudibranchs to swim. Swimming usually is accomplished by dorsoventral or lateral flexions of the whole body, resulting in an inefficient movement that cannot be sustained for long periods. Other nudibranchs are propelled by rowing movements of the dorsal cerata or by undulations of an expanded mantle edge (Farmer, 1970; Thompson, 1976). Swimming often seems to be an escape response to benthic predators, although in some cases it may enable an animal to search for new habitats with more abundant prey, or it may be related to breeding behavior and the search for a mate. In any event, swimming is always a temporary behavior in these bottom-dwelling animals.

Only a few nudibranchs are truly holoplanktonic or pelagic, spending their complete life cycles in the oceanic environment. *Cephalopyge trema-*

toides and at least two species of *Phylliroë*, which form a separate family of dendronotaceans, are highly modified for life in warm oceanic waters. The remaining species discussed here are aeolid nudibranchs: the Family Glaucidae contains two species, both exhibiting a unique buoyancy mechanism and forming part of the circumtropical pleustonic community; the Family Fionidae contains only one species, one that shows little modification for a pelagic existence and is usually found crawling over the surface of its prey or other floating objects.

The inclusion of *Fiona pinnata* with planktonic species is somewhat arbitrary, since this nudibranch is not capable of swimming, lacks special buoyancy mechanisms, and is dependent on attachment to floating organisms or materials. This species has nonetheless been included, because it is a common inhabitant of warm-water, oceanic, pleustonic communities and because considerable knowledge has been amassed concerning its behavior and biology. At the same time, we have omitted discussion of several nudibranch genera, such as *Scyllaea* and *Hancockia*, that might also have been included, since they are found at sea attached to *Sargassum* or other floating plants. Both genera are dendronotaceans belonging to the Family Scyllaeidae, and they are sometimes regarded as pelagic or semi-pelagic. In truth, their omission here has been dictated by the fact that very little is known about their biology. *Scyllaea pelagica* Linnaeus, 1758, is reported to be capable of temporary swimming, using two pairs of lateral expansions of the dorsum (Thompson, 1976), and to feed on attached hydroids (Pruvot-Fol, 1954; Thompson and Brown, 1981). These references include descriptions of the anatomy of this species and a discussion of taxonomy.

Family Phylliroidae

This family includes at least two species of *Phylliroë* and *Cephalopyge trematoides*, all of which are highly specialized for a planktonic existence in circumtropical waters. These nudibranchs are relatively small (<55 mm in length), elongate, and highly transparent. They are capable of swimming by undulating movements of the whole body, and in keeping with their pelagic life styles, they have a very reduced foot. They are carnivorous and feed mostly upon planktonic coelenterates. Although *Phylliroë* and *Cephalopyge* share many anatomical features, their unusual life cycles differ in several important respects, and thus the species are treated separately below.

Phylliroë

Phylliroë bucephala (Color Fig. 14; Fig. 71a) is the well-known Mediterranean species that also occurs in the Atlantic Ocean. *P. atlantica*, described from the Atlantic and Pacific oceans, shows only slight anatomical differ-

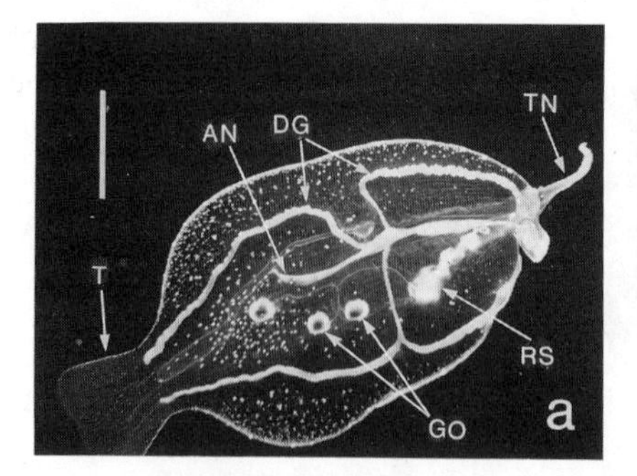

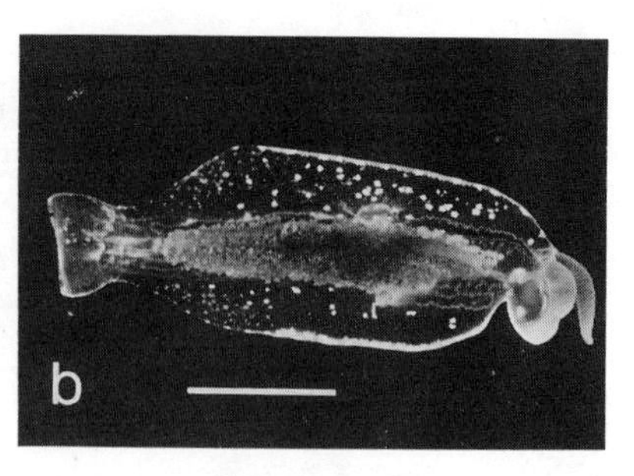

FIG. 71. *a,* An adult *Phylliroë bucephala* from the Gulf Stream, with somewhat contracted tentacles; *b,* an unidentified species of *Phylliroë* from Australian waters (6°31′ S, 150°21′ E) (by M. Jones). AN, anus; DG, lobes of digestive gland; GO, lobes of the ovotestis; RS, reproductive ducts and accessory glands; T, tail; TN, tentacle (rhinophore). Scale lines represent 10 mm.

ences from *P. bucephala* (Pruvot-Fol, 1954), and it is possible that these names apply only to one species (van der Spoel, 1970). The unidentified species of *Phylliroë* in Figure 71b, from Australian waters, appears to be a different species. A taxonomic revision of the genus is warranted.

Phylliroë is characterized by its elongate, laterally compressed, leaflike shape (*phyll-* = leaf). The transparent body is smooth and streamlined and drawn out to a slightly forked and expanded tail. Maximum body length is 55 mm. A single pair of long, smooth tentacles (rhinophores) are the only projections from the body surface; gills and cephalic tentacles are lacking, and the foot is reduced to a glandular remnant (pedal gland) embedded in the smooth body outline. The head extends forward from the rhinophores as a downwardly directed snout bearing the terminal mouth. The anus is situated on the right lateral side, approximately at the center of the body.

One of the most remarkable features of *Phylliroë* is that the nudibranch is highly bioluminescent. The light originates from glandular cells located in the integument and scattered over the body suface. In addition, the bright golden lobes of the digestive gland and other variously pigmented internal organs, which can be seen through the transparent integument, contribute to the colorful beauty of this animal. The adaptive significance of this striking coloration is not known.

Adult *Phylliroë* are capable of swimming by lateral undulations of the body. Waves of muscle contractions pass from the head posteriorly along the body and propel the animal forward at speeds approaching 15 cm/sec (Gilmer, pers. obs.).

Biological observations of living animals have been restricted largely to

Phylliroë bucephala from the Mediterranean. The following descriptions of feeding, reproduction, and development apply to that species only.

For many years, it was thought that *Phylliroë bucephala* was parasitized by a medusa often found attached to the pedal gland of the nudibranch. This misconception remained until Ankel (1952) suggested that *Phylliroë* is the parasite of the association and not the host, a fact confirmed by later work of Martin and Brinckmann (1963). Juvenile *Phylliroë* are found living within the hydromedusa *Zanclea costata* (formerly called *Mnestra parasites*), where they attach to the inner bell surface by their rudimental foot (Figs. 72a, b). They feed on the ring and radial canals and the manubrium of the coelenterate by employing a sucking action of the muscular pharynx. The growth of the nudibranch is rapid: a young *Phylliroë* may increase its length from 1.6 to 11.0 mm in 10 days, during which time the medusa diminishes in size and becomes deformed. When a young *Phylliroë* attains a size larger than that of the medusa and begins to swim actively, it consumes the tentacles and remaining parts of the manubrium of the *Zanclea*. Feeding at this time presumably involves biting actions of the paired denticulate jaws and the small radula; the latter is composed of a single, central, denticulate tooth bordered on each side by three to five lateral, needlelike teeth in each of several rows. The diminished bell of the medusa remains attached for some time to the foot gland of the nudibranch (Fig. 72c); it is this connection of a deformed and reduced medusa attached to *Phylliroë* that first gave rise to the idea that the medusa is a parasite of the mollusk. Eventually, the remaining medusa tissues may become detached, and *Phylliroë* larger than about 12 mm may no longer be physically associated with *Zanclea*. Martin and Brinckmann (1963) suggest, however, that more than 50 percent of freshly collected, living adults will still have remnants of medusae attached to the foot. Medusae almost always become detached from *Phylliroë* during preservation, giving a false ratio in oceanic collections (Ankel, 1952).

Although the association with a medusa seems to be species-specific and obligatory for young *Phylliroë bucephala*, the natural prey of the free-living adults are not well known. In the laboratory, adults will seize and eat *Zanclea* medusae; they use their rhinophores to locate and capture the prey, seizing and eating the prey tentacles one by one before ingesting the remainder of the medusa (Martin and Brinckmann, 1963). Adults can also be fed on siphonophore tentacles, indicating that the *Zanclea–Phylliroë* association is not obligatory in the adults, and that nematocysts of other coelenterates are not a deterrent to predation by the nudibranch. Martin (1966) was further able to induce adult *Phylliroë* to swallow *Zanclea* polyps detached from stolons, but only when the food was directed repeatedly to the

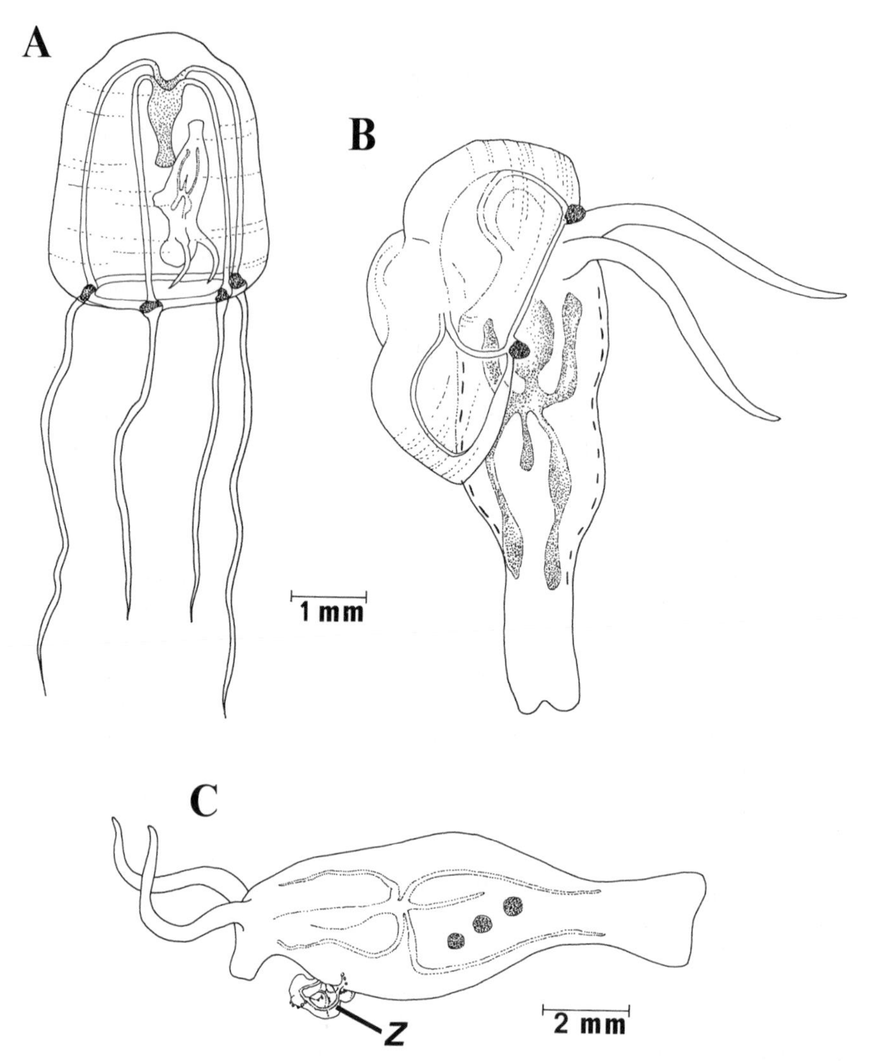

FIG. 72. Life cycle of *Phylliroë bucephala* (*A* and *B* modified and redrawn from Martin and Brinckmann, 1963; *C* modified and redrawn from Ankel, 1952): *A*, a young specimen attached to the inner bell surface of the medusa *Zanclea costata*; *B*, an older specimen that has ingested part of the medusa; *C*, a free-swimming adult with the reduced bell of *Zanclea* (Z) still attached to its pedal gland.

mouth of the nudibranch; this is not, however, presumed to be indicative of natural food.

On several occasions, divers have seen adult *Phylliroë* feeding on dense surface swarms of the larvacean, *Oikopleura albicans* (Gilmer, pers. obs.). Each of the eight observed nudibranchs approached individual, free-swimming larvaceans from below and used its jaws to quickly seize the prey. Ingestion took less than 30 seconds, but no single *Phylliroë* captured more than one prey animal during the periods of observation. Adults also have been observed to feed on the medusa *Aequorea.* These data confirm that adult *Phylliroë* do not retain the food specificity exhibited by their young.

Although young *Phylliroë* have generally been regarded as ectoparasites of *Zanclea*, it can reasonably be argued that the nudibranch is not a true parasite, but rather a carnivorous feeding specialist, during its early development. Unlike successful parasites, young individuals of *Phylliroë* eventually consume all, or most, of their initial prey. Further, adult *Phylliroë* are free-living animals that can repeatedly capture and completely ingest *Zanclea* medusae in the laboratory and feed on other zooplankton in nature. Nor are the anatomical modifications of *Phylliroë* suggestive of a parasitic life style, but rather of adaptation to a pelagic, free-living existence.

The term parasitism has been employed by earlier authors to describe the relatively long time interval in which young *Phylliroë* are attached to *Zanclea* medusae. In fact, this results from the unusual, inverted, size relationship between predator and prey. A small *Phylliroë* feeding upon a relatively large medusa requires a longer time in which to consume its prey than is usual in the typical association of a larger predator with its smaller prey. This nudibranch–medusa food link illustrates the biological similarities and differences that exist between parasitism and extreme food specialization, as well as the difficulties that can arise in attempting to separate these types of feeding. Both parasites and food specialists depend on specific food items—either a host, in the case of parasites, or a specific prey. Parasites generally live for long periods of time in association with a specific host and, though the host is usually weakened, it is not usually killed by the parasite. On the other hand, encounters between a carnivorous food specialist and its specific prey are generally of very brief duration, and many prey individuals are eaten in the course of the predator's life. By these definitions, the *Phylliroë–Zanclea* association is clearly one of predator and prey, with the long residence time of a young *Phylliroë* on its initial prey resulting from the inability of the small nudibranch to quickly consume prey considerably larger than itself.

The anatomy of the reproductive system of this nudibranch has not been completely described. The hermaphroditic gonad is divided into three to five round lobules (Fig. 71a). Ducts leading from the lobules unite to form a hermaphrodite duct, which then branches into male and female channels. The channels are provided with various, but undescribed, accessory glands. The penis is unusual in having an enlarged, distal branch covered with numerous small papillae; this feature has led to the penis being referred to as "armed." The genital openings are close together, on the right ventral side and posterior to the pedal gland.

Divers have observed single mating pairs on several occasions, and fertilization appears to be a reciprocal process. Mating individuals unite head to tail, with the long penes of each coiled together; they swim actively in a spiral pattern and remain paired for at least 6 minutes. In the two cases where mating specimens were collected, the body lengths of partners did not vary by more than 5 mm; the smallest mating pair measured 32 and 34 mm when alive (Gilmer, pers. obs.).

In the Mediterranean, *Phylliroë bucephala* is known to spawn from November to June. Laboratory animals successively lay about 20 uncoiled egg strings, each about 6 mm long and containing approximately 12 eggs measuring 0.11 mm in diameter (Martin and Brinckmann, 1963). Thus fecundity seems to be remarkably low, with each nudibranch producing only about 240 eggs. At 13° C, veligers hatch within 7 to 9 days (Martin, 1966). The veligers are small (0.18 mm in largest diameter) and are active swimmers. Attempts to feed them in the laboratory have been unsuccessful. Martin (1966), however, succeeded in maintaining veligers for 2 weeks in the laboratory, during which time he tested their reactions to living *Zanclea* hydroid polyps and medusae. Contact with *Zanclea* polyps elicited no perceptible change in the behavior of the veligers, and did not appear to cause discharge of the nematocysts of the coelenterate. Although it is assumed that *Phylliroë* veligers metamorphose upon finding *Zanclea* medusae, veligers did not respond to young *Zanclea* medusae (two-tentacle stage) in the laboratory. When placed directly into the bells of medusae, veligers did not attach to the prey. Martin concluded from these trials that the free-swimming larval stage must be longer than 2 weeks. It is also possible that unsuitable laboratory conditions and lack of natural food for the veligers prohibited sufficient development and growth for the attainment of metamorphosis. It should further be considered that metamorphosis, with the loss of the veliger shell and velum, may occur before the young actively seek out *Zanclea*. Since the fecundity of *Phylliroë* appears to be low, and the completion of the life cycle is dependent on a specific prey species, it is not surprising that *Phylliroë* is one of the rarer planktonic mollusks.

Zooxanthellae have been found in the branches of the digestive gland of *Phylliroë* (Zirpolo, 1923), but the nature of this symbiosis has not been well studied, nor is it known whether all species and all individuals of *Phylliroë* harbor these symbiotic algae. It is possible that the zooxanthellae are obtained from the nudibranch's coelenterate prey. The only reports of parasites are of trematodes, which have been found on the surface and interior of the body of some specimens (Pruvot-Fol, 1954).

Cephalopyge trematoides

In 1956, Steinberg observed a large number of living animals of this species and concluded that there is considerable individual morphological variation. As a result of her work, she reviewed the taxonomic status of several animals originally described as separate species and synonymized the genera *Ctilopsis*, *Dactylopus*, *Nectophyllirhoe*, *Boopsis*, and *Bonneviia* with *Cephalopyge*. She also concluded that five other species of *Cephalopyge*—*C. mediterranea*, *C. orientalis*, *C. picteti*, *C. michaelsarsi*, and *C. arabica*—are synonymous with *C. trematoides*. Thus the genus is monotypic.

Cephalopyge (Fig. 73), like *Phylliroë*, has an elongate, streamlined body. The specific name, *trematoides*, describes the wormlike form. The nudibranch is laterally compressed, but less so than *Phylliroë*. The body extends to a maximum length of 25 mm and terminates in a truncate tail. The head bears only one pair of tentacles, which are moderately long, smooth, contractile rhinophores. The head extends forward as a small rostrum (or cephalic disk) that overlies the terminal mouth. The foot is not so reduced as in *Phylliroë* and is distinctive as a small ventral projection, just behind the rhinophores. The foot is capable of considerable extension and contraction. A glandular groove runs from the lower edge of the mouth to the foot. The descriptive generic name is taken from the position of the anus (*-pyge*) on the middorsal surface immediately behind the head (*cephalo-*). The animal is very transparent, but bioluminescence has not been reported in this nudibranch, as it has in *Phylliroë*.

Adult *Cephalopyge* swim in a manner similar to that of *Phylliroë*, but the lateral undulations appear to be faster. The effector stroke begins anteriorly and moves posteriorly in a growing wave from side to side. By moving at rates as fast as seven undulations per second, *Cephalopyge* can propel itself through the water at speeds approaching 12 cm/second (Gilmer, pers. obs.).

The buccal armature of *Cephalopyge* is very similar to that of *Phylliroë*. There is a muscular pharynx, covered at the anterior end by paired chitinous jaws. The very small radula has often been overlooked by earlier authors (e.g. Stubbings, 1937), but it consists of three teeth in each of about twelve rows. The central tooth has a long pointed cusp with denticles at the base; the cusps of the denticulate lateral teeth are curved inwardly as hooks.

Cephalopyge has been found both swimming freely and attached to the siphonophore *Nanomia bijuga* (Figs. 73b, c). Feeding observations of cap-

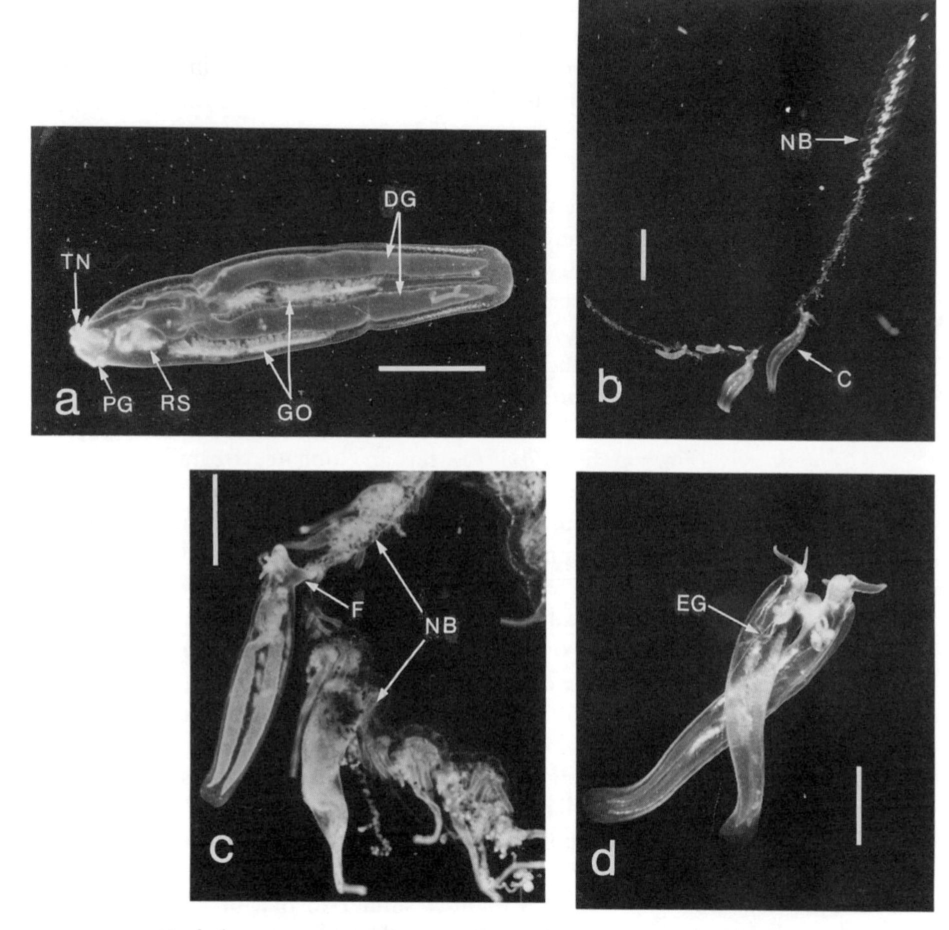

FIG. 73. *Cephalopyge trematoides*: *a,* a free-swimming specimen from the western North Atlantic, with slightly contracted tentacles; *b,* five specimens of *Cephalopyge* attached to the siphonophore *Nanomia bijuga*; *c,* a *Cephalopyge* feeding on *Nanomia* (the nudibranch is attached by the pedal gland of its foot to the prey); *d,* mating specimens, with connected penes and an egg string issuing from the nearest individual. C, *Cephalopyge* (the largest of five specimens in *b*); DG, lobes of digestive gland; EG, egg string; F, pedal gland on the foot of *Cephalopyge*; GO, lobes of ovotestis; NB, *Nanomia bijuga*; PG, pedal gland; RS, reproductive ducts and accessory glands; TN, tentacles (rhinophores). Scale lines represent 5 mm in *a,* *c,* and *d,* and 10 mm in *b.*

tive animals have been made by Sentz-Braconnot and Carré (1966). *Cephalopyge* usually expands its foot and uses its pedal gland to attach to the stolon of a *Nanomia* colony, then proceeds to eat the tentacles, zooids, and eventually the entire colony. Depending on the relative sizes of predator and prey, the predator may detach from *Nanomia* several times before ingesting the entire colony, or the nudibranch may ingest the entire colony at one time. If the *Nanomia* is relatively small, the predator may not attach directly to the stolon, but will remain swimmming while capturing and ingesting the siphonophore.

Cephalopyge seems to be monospecific in its choice of prey; laboratory specimens never attached to or ate other offered siphonophores (*Abylopsis*, *Rhizophysa*, *Apolemia*, *Sulculeolaria*) or hydromedusae (*Liriope*, *Cunina*). This nudibranch, like *Phylliroë*, has been termed a parasite, but it too is more properly regarded as a carnivorous feeding specialist that attaches only temporarily to relatively large-sized prey.

The reproductive tract of *Cephalopyge* has been described by Stubbings (1937) and Steinberg (1956). The ovotestis is divided into several elongate lobules (Fig. 73a), the total number increasing with the age of the animal. Some of the lobules extend between the body wall and ventral parts of the alimentary canal, and others are positioned dorsally to these in the midsection of the body. The separate ducts from each lobule fuse to form the common hermaphroditic duct. Male and female channels separate from this duct, with the vas deferens differentiating into a prostatic region before entering the penis. The penis is not papillate (or "armed" as in *Phylliroë*) but, when everted, bears a proximal, wing-shaped branch. The female channel contains a seminal receptacle and two accessory glands. The male and female channels join near the common genital pore, which is located on the right side just behind the foot.

The mating behavior of *Cephalopyge* is notably different from that of *Phylliroë*. On five of the six occasions when mating has been observed by divers in the field, the animals were in small swarms of up to 30 animals within an area of 30 cm to 1 meter. The animals swam rapidly about, mating and then breaking apart, and in some instances mating again with different individuals. The pairing lasted for only about 1 minute. Individuals pair head to head (Fig. 73d) and fertilization appears to be reciprocal.

Sentz-Braconnot and Carré (1966) described development from eggs laid in the laboratory during May and December. The eggs are deposited over several days in floating, cylindrical, transparent filaments (Fig. 73d). The total egg mass per individual measures several tens of centimeters and contains at least 3,000 eggs. Free-swimming veligers have a two-lobed velum, a shell that is first ovoid and later spiraled, a small foot, an operculum, and two statocysts at the base of the velar lobes. Veligers placed in the presence of young colonies of *Nanomia* did not attach or metamorphose.

The growth of juvenile *Cephalopyge* can be rapid under laboratory conditions in which they are supplied with unlimited food. One individual increased its length from about 7 to 11 mm in 2 days; a larger specimen grew from 10 to 14 mm in 5 days (Sentz-Braconnot and Carré, 1966).

Family Glaucidae

The Family Glaucidae contains only two genera and two species, the more common being *Glaucus atlanticus*, which is found throughout the circumtropical belt of the Atlantic, Pacific, and Indian oceans. *Glaucilla marginata* is confined to warm waters of the Pacific. Both species are pleustonic and are found in association with the coelenterates *Velella*, *Porpita*, and *Physalia*, and with other pleustonic residents such as the snail *Janthina* (see Chapter 2). Like other surface dwellers, glaucids are subject to drift by wind, and occasionally are stranded on beaches, where they may be sighted by vigilant mollusk collectors. Despite the wide distribution and abundance of *Glaucus*, few observations have been made of living animals, and fewer still of the less common *Glaucilla*.

Adaptation for a pelagic existence has not involved anatomical modification so much as behavioral change, and the glaucids remain easily recognizable to malacologists as nudibranchs that belong to the Suborder Aeolidacea. *Glaucus atlanticus* (Color Fig. 16; Fig. 74) retains many of the features of its benthic relatives. The shell-less body is flattened and elongate, with a length of up to 30 mm. The short, blunt head bears a pair of small oral tentacles near the mouth and a pair of short, chemosensory rhinophores situated farther back on the dorsal surface. Three or four clusters of long, flattened papillae (referred to as "cerata"; singular, "ceras") project from lobes on each side of the body. The cerata are arranged in single rows in each cluster, and the total number may be as many as 84 (Thompson and Bennett, 1970). The flat, ventral foot is long and slender, and the metapodium projects beyond the body as a tail. The body coloration is striking, with the ventral (uppermost) surface being deep bluish purple and the dorsal (lower) surface a silvery white. The coloration and countershading are common features of tropical, pleustonic animals. Three pores are evident on the body: the genital opening is on the ventral surface, to the right of the foot; the excretory opening is on the right dorsolateral surface between the first and second clusters of cerata; and the anal papilla is situated dorsolaterally on the right side between the second and third ceratal clusters.

Glaucilla marginata (Fig. 75) is a somewhat smaller animal of up to 12 mm in length, and the coloration differs in that the ventral surface is

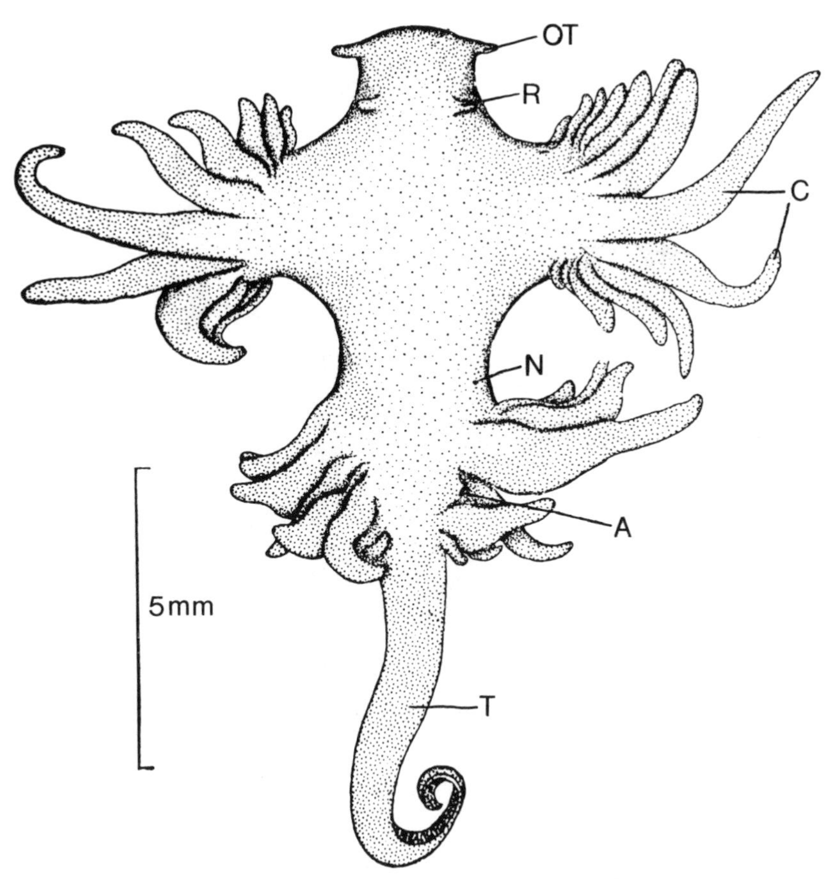

FIG. 74. *Glaucus atlanticus*, dorsal view (from Thompson and McFarlane, 1967). A, anus; C, cerata; N, renal opening; OT, oral tentacle; R, rhinophore; T, tail (metapodium).

brown and the dorsal silver-blue (Thompson and Bennett, 1970). This species also differs from *Glaucus* in typically having four clusters of cerata, arranged in more than one row in each cluster; by having more cerata (up to 139 per individual); by the more posterior location of the excretory pore, which is immediately above the anal papilla, between the second and third ceratal clusters; and by a much shorter metapodium. Further differences in internal anatomy between the genera are noted below.

Neither *Glaucus* nor *Glaucilla* are active swimmers. Instead, they are usually found floating upside down at the sea surface. They are capable of slow gliding movements on the underside of the surface film, but any long-

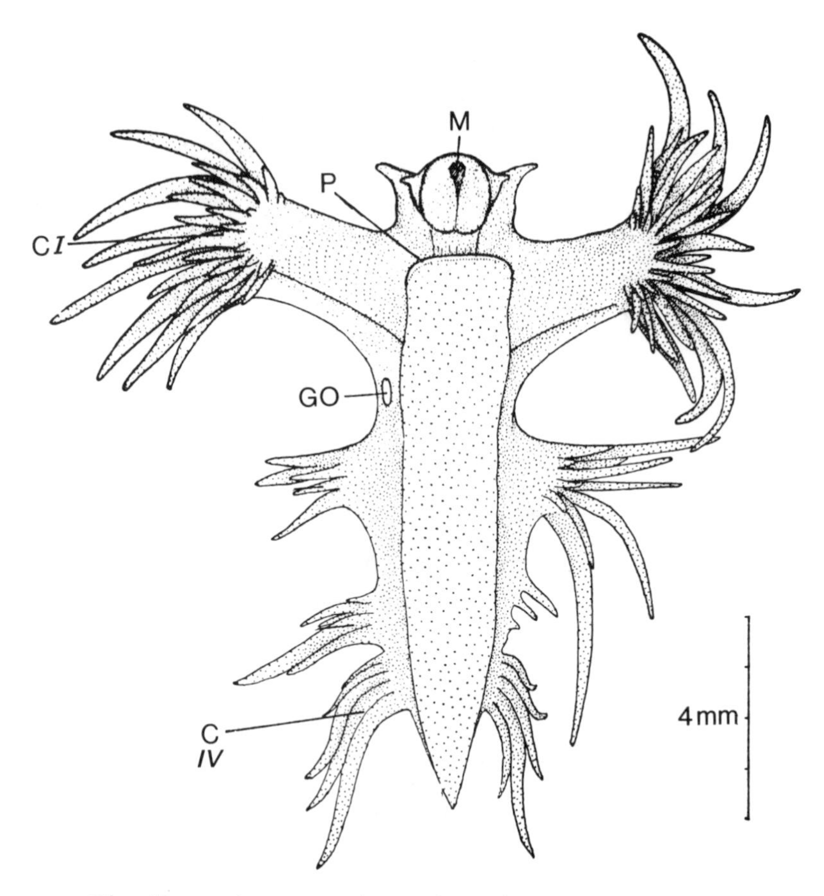

FIG. 75. *Glaucilla marginata,* ventral view (from Thompson and Bennett, 1970). CI, first ceratal cluster; CIV, fourth ceratal cluster; GO, genital opening; M, mouth; P, propodium of foot.

distance transport is achieved passively through the actions of wind and currents. Buoyancy is achieved in both species by swallowing air that is then stored in the spacious and usually distended gastric cavity. Thompson and McFarlane (1967) have noted the presence of strong sphincter muscles between the stomach and the hindgut, and have suggested that these play a role in the maintenance of the hydrostatic mechanism. If living animals are squeezed to expel gas, they sink to the bottom of a laboratory container, then attempt to regain the surface to ingest more air (Thompson, 1976). This dependence on access to air recalls the pleustonic janthinids, which are also passive drifters but which achieve buoyancy through trapping air bubbles by the foot rather than through ingestion.

Glaucus and *Glaucilla* are carnivores that prey on pleustonic coelenterates. The buccal organs consist of strong, chitinous, hinged jaws and a radula with a single tooth in each row (Fig. 76). Each radular tooth has a projecting central cusp (proportionately longer in *Glaucilla*) that is bordered by a variable number of denticles.

These nudibranchs feed chiefly, if not exclusively, on species of the chondrophores *Velella* and *Porpita*, and on the siphonophore *Physalia* (Bennett, 1836; Bieri, 1966; Thompson and Bennett, 1969). They can also be canni-

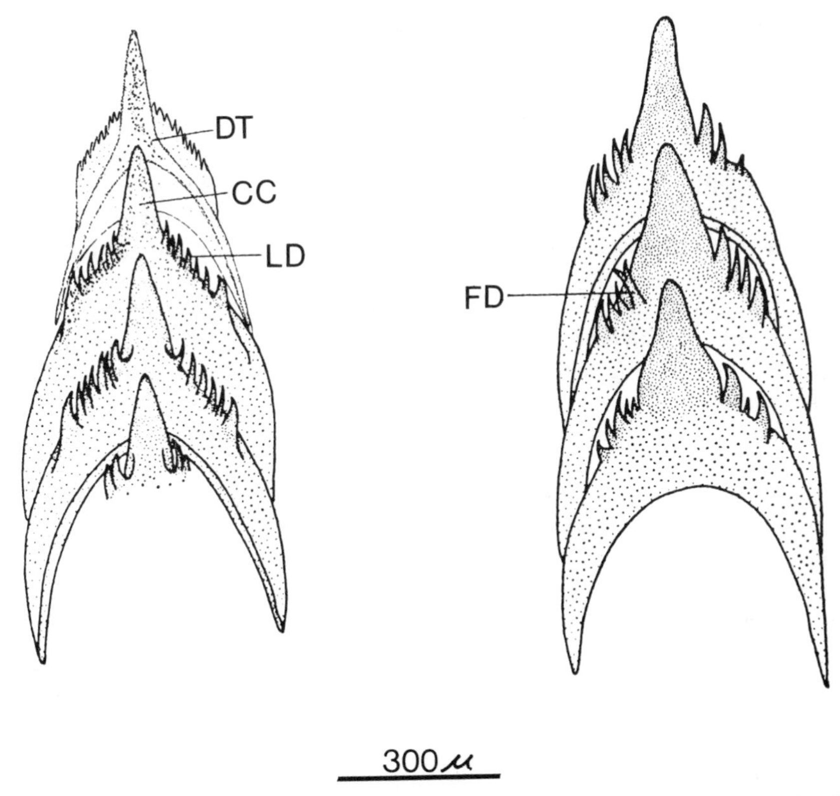

FIG. 76. The radular teeth of *Glaucus atlanticus,* taken from the proximal ends of the radulae of two 12-mm-long individuals (from Thompson and McFarlane, 1967). Radular formula = 0-1-0. CC, central cusp; DT, incompletely formed tooth; FD, forked denticle; LD, lateral denticles.

balistic (Bieri, 1966); the smaller (15 mm body length) of the animals shown in Color Figure 16 was later eaten by its mate (18 mm). It can be assumed that location of prey is by chance encounter, since glaucids are sightless, passive drifters. The jaws are used to seize the prey. There are no detailed accounts of prey ingestion, but the structure of the radular teeth suggests that they can be used to penetrate the tough integument of the prey, and the hinged jaws, with denticulate edges, are well equipped for biting and tearing prey tissue. All soft tissues, including the tentacles of the coelenterate prey, are eaten (Bennett, 1836).

The nematocysts of the prey, both fired and undischarged, are ingested along with other tissues. Not only do *Glaucus* and *Glaucilla* derive nourishment from their prey, but certain of the ingested and undischarged nematocysts become the predators' main defensive mechanisms (Thompson and McFarlane, 1967; Thompson and Bennett, 1970), just as they do in benthic eolid nudibranchs that feed on benthic coelenterates (Edmunds, 1966; Thompson, 1976). Nematocysts of the prey are transported along ciliated branches of the digestive gland that lead into each ceras. Nematocysts are stored in special sacs (cnidosacs), which are located at the tip of each ceras and open to the exterior via a pore. Pressure on the cerata causes the expulsion of nematocysts through the cnidopores, and the nematocysts discharge on contact with sea water. Thompson and Bennett (1970) usually found the larger nematocysts of *Physalia* in the cnidosacs of glaucids; nematocysts obtained from *Velella* or *Porpita* were rarely present. This led them to suggest that the nudibranchs selectively store the more virulent nematocysts of *Physalia* and digest those of *Velella* and *Porpita*. Day and Harris (1978) have pointed out, however, that nematocysts stored by eolids are continually being replaced, perhaps because they remain functional for limited periods. Therefore, the type of nematocysts present in glaucid cnidosacs may not be independent of diet and indicative of selection, but may simply reflect the most recent type of prey consumed. Although no experiments have been done to test the effectiveness of nematocyst discharge by glaucids on potential predators such as small fish, there is no reason to believe that it would be less than for the coelenterates themselves. Further, Thompson and Bennett (1969) were able to identify *Glaucus atlanticus* as being responsible for the stings reported by bathers in Port Stephens, Australia, in 1968. The stings were unpleasant enough (mild pain for 1 to 2 hours) that both *Glaucus* and *Glaucilla* are considered as potentially harmful to man and should be treated with the same caution as their prey.

Glaucids, like other nudibranchs, are hermaphroditic. The anatomy and histology of the reproductive tract of *Glaucus atlanticus* have been studied by Thompson and McFarlane (1967), and the following account is taken from their description.

The ovotestis is located dorsally in the posterior part of the body. A hermaphrodite duct, which functions to store sperm, runs forward from the gonad before separating into male and female channels. The male duct continues as the vas deferens to the base of the penis. The penis of *Glaucus* is tipped with a chitinous spine; that of *Glaucilla* is unarmed. The female channel branches to a seminal receptacle, where sperm received during copulation are stored, and then the channel elaborates into the albumen and mucous glands. There is a single, external, genital opening (Fig. 75) for both the efferent female duct (vagina) and the opening to the penial sheath.

Animals pair ventrally or laterally during copulation (Color Fig. 16). Short (20 mm), uncoiled egg strings containing 10 to 36 small (63 μm) eggs each are shed into the sea or are attached to prey (Bennett, 1836; Bebbington, 1986). Laursen (1953) also found *Glaucus* egg ribbons on shells of *Janthina*. Veliger larvae hatch within 48 hours at 25° C and have been illustrated by Bebbington (1986), but subsequent development has not been described. It would be of interest to obtain further information, not only on development and metamorphosis, but also on such questions as growth rates; size at sexual maturity; breeding seasons, if any; and whether self-fertilization is possible. There is no information on reproduction and development in *Glaucilla marginata*.

Family Fionidae

This eolid family consists of a single species, *Fiona pinnata*, which closely resembles the glaucids in many respects. Like *Glaucus atlanticus*, *Fiona* is a member of the tropical pleustonic community and is oceanic and circumtropical in distribution. But *Fiona* is not as specialized for pelagic life as the glaucids, since it lacks both swimming ability and buoyancy mechanisms; it also has a wider diet. Despite the wide distribution of *Fiona*, many aspects of its biology remain unknown.

Fiona (Color Fig. 15) does not differ in any obvious anatomical trait from benthic eolid nudibranchs. The elongate body reaches 50 to 60 mm in length and bears numerous, scattered cerata projecting from the sides of the dorsal surface. Each ceras bears an immobile, undulating membrane along the length of the mesial surface; these sail-shaped cerata are unique to this genus. The rounded head bears a pair of long, oral tentacles and a pair of long, smooth rhinophores. The margins of the flattened foot extend slightly beyond the sides of the body. The anal opening is located in a laterodorsal position, slightly anterior to the middle of the body. The coloration of individual *Fiona* varies, according to diet, from bluish purple or brown on the dorsal surface to white or pale yellow on the ventral surface.

Fiona shows the least adaptation to a pelagic existence of all the gastropods considered here. It is not capable of swimming, and no buoyancy mechanisms have been described for the species. Although it has been reported to creep inverted on the underside of the water surface (Kropp, 1931), it is most often found clinging to the surface of *Velella* or to gooseneck barnacles that are attached to driftwood or other floating objects. It has also been found on floating *Sargassum* and kelp. Thus, *Fiona* seems to require a floating substrate for attachment. Its inclusion in the tropical pleustonic community marks the animal as pelagic, but it cannot properly be considered as a planktonic species.

Fiona pinnata is carnivorous, preying on both pleustonic coelenterates and stalked barnacles. The nudibranch is equipped with feeding structures similar to those of the glaucids. There are strong paired jaws with serrated cutting edges and a uniserial (0-1-0) radula. The single radular tooth, in each of up to approximately 40 rows, is very similar to that of *Glaucus* and *Glaucilla*; it is wishbone-shaped, with a large, pointed, central cusp bounded by six to ten smaller denticles on each side.

Fiona frequently is found crawling on the upper surface of *Velella*, one of its prey species (Color Fig. 15). The nudibranch employs its jaws and radula to remove tissue from the float and sail of the prey. Bayer (1963) noted that a detached *Fiona* made active searching movements to locate a *Velella* placed near its head; once contact was made, the predator immediately crawled onto the upper surface of the prey and began feeding. However, he also reported that *Fiona* exhibited avoidance reactions to the lower surface and tentacles of *Velella*, and consistently refused to cling to the lower surface of the prey. This suggested that the nudibranch was not immune to the prey's nematocysts, but Bayer found many, mostly undischarged, nematocysts in the feces of *Fiona*, indicating that zooids were consumed. More recent observations (by Gilmer) have shown that a large *Fiona* (26 mm long) can ingest all of the various polyps, including the larger dactylozooids from the lower margin of a *Velella* float (60 mm in diameter) in 2 hours. There are no reports to indicate that *Fiona* feeds on either *Porpita* or *Physalia*, although these coelenterates are common inhabitants of the tropical pleuston community. In the laboratory, *Fiona* responds to *Physalia* with avoidance reactions or mucus emission, and it cannot be induced to feed on or even cling to any part of this siphonophore (Bayer, 1963). Unlike the glaucids, *Fiona* does not store and utilize ingested nematocysts of prey for defense (Alder and Hancock, 1845–55).

Fiona also feeds on the stalked barnacle *Lepas anatifera* (Bayer, 1963; Bieri, 1966; Holleman, 1972). The nudibranch chews into the stalk of the barnacle and completely consumes the inner tissues, leaving the tough outer skin of the stalk and the empty shell plates. A 15-mm–long *Fiona* is capable of eating nine small barnacles, measuring 5 mm in length, in approximately

12 hours (Bieri, 1966). Normally, the nudibranch is unable to penetrate the harder stalks of *Pollicipes polymerus* (Holleman, 1972).

The coloration of *Fiona* is dependent on its diet. Animals that have fed on *Velella* have a deep-blue color (indistinguishable from the prey color) on the head and dorsal cerata. This coloration fades within 3 days of starvation to a pale, milky blue, but can be restored within 24 hours of commencing to feed on the chondrophore prey (Kropp, 1931; Bayer, 1963). Similarly, *Fiona* that has fed on *Lepas* develops a pink to brown coloration on the dorsal surface.

The anatomy of the hermaphroditic reproductive tract has been described by MacFarland (1966). A hermaphrodite duct emerges from the ovotestis and then diverges into separate male and female channels. The male channel consists of a vas deferens, the walls of which differentiate into prostatic glandular tissue before the duct enters the penial sheath and attaches to the base of the penis. The retractile, unarmed penis is very long and filamentous; it can be extruded through the male pore, immediately behind the right tentacle. The female channel consists of an oviduct, mucous gland, and seminal receptacle. The female opening lies posterior to the male pore.

Copulation and egg laying have been described by Holleman (1972). Mating individuals pair head to tail, with the right sides of their bodies in contact. Copulation is reciprocal and lasts 5 to 15 minutes; the same individuals may mate twice within a short interval. White, spiraled egg masses (Color Fig. 15) are attached to the floats of *Velella* within 12.5 to 48 hours after copulation. In nature, *Fiona* egg masses can also be found on drifting bottles, cuttle bones, logs (Bennett, 1966), and empty *Spirula* shells (Thompson and Brown, 1984). Most individuals lay at least two egg masses on separate occasions. Each egg capsule contains one or two eggs. Eggs measure a remarkable 495 μm in diameter, much larger than reported for other opisthobranchs. Embryonic development from the first cleavage of the fertilized egg to the early veliger stage was described in considerable detail by Casteel in 1904. Free-swimming veligers hatch within 46 hours (Bayer, 1963), 5 days (Holleman, 1972), or 10 to 11 days (Thompson and Brown, 1984); the variable developmental time may be temperature-dependent. The veliger shell is inflated and egg-shaped, corresponding to Type 2 in the classification of opisthobranch veligers (Thompson, 1976). The length of the planktonic stage is unknown. Metamorphosis probably depends on the ability of the veliger to locate its adult food.

The growth of young metamorphosed *Fiona* is remarkably rapid. Holleman (1972) followed growth for 24 days in 12 individuals fed on *Lepas*. The animals increased in length from an average of 3.37 mm to 29.97 mm, thus increasing size by 1.11 mm/day or attaining a 900-percent increase in length over 24 days. Average wet weight increased from 1.3 mg to 448.8 mg over the same interval, or by 18.64 mg/day. Bayer (1963) also reported extremely rapid growth and early attainment of sexual maturity. In 5 days, four specimens feeding on *Velella* increased from 8 mm to about 32 mm, at which size they began copulating. Egg masses were laid 2 days later, and within the ensuing 3 weeks the four individuals had produced a total of 89 egg masses. Thus the life cycle of *Fiona* can be completed in approximately 1 month.

Evolution

It is clear that adaptation to a holopelagic existence has evolved independently in the three nudibranch families considered here and, in two of the families (Glaucidae and Fionidae), has not depended on the attainment of swimming ability. *Fiona pinnata* is the least specialized species and remains dependent on attachment to floating substrates to remain in the pelagic realm. The glaucids also are not active swimmers, but their habit of storing ingested air permits them to float independently at the sea surface. Only *Phylliroë* and *Cephalopyge* are active swimmers capable of existing below the sea surface.

It is of particular interest that all of these pelagic nudibranchs feed on pelagic coelenterates. Only *Fiona* and adult *Phylliroë* are known also to include other prey types in their diets. Diet specialization is common in all nudibranchs and appears to have evolved independently in the pelagic families (Thompson, 1976). The evolution of pelagic life styles and the success of these nudibranchs probably has much to do with the fact that pelagic coelenterates are abundant and have few other predators; most animals avoid coelenterate prey because of the toxic nematocysts. As far as is presently known, the major feeding competitors with *Glaucus*, *Glaucilla*, and *Fiona* are janthinid mollusks and sea turtles (see Chapter 2). Competitors for *Zanclea* and *Nanomia*, the planktonic coelenterate prey of *Phylliroë* and *Cephaloypge*, respectively, are not known.

As with the janthinids, feeding specialization on coelenterate prey was probably not a new adaptation in the transition from a benthic to a pelagic habitat. Many benthic dendronotaceans and most eolids feed on attached coelenterates, such as hydroids, soft corals, and sea anemones. Preadaptation to such prey was probably a major step in permitting expansion into the warm areas of the open ocean where siphonophores and chondrophores are abundant at the sea surface and accessible to rafted nudibranchs.

Feeding on coelenterates also has permitted *Glaucus* and *Glaucilla* to deter potential predators by storing (in their cerata) the undischarged nematocysts obtained from prey. *Fiona*, *Phylliroë*, and *Cephalopyge* do not accumulate nematocysts from their prey, but their prolonged physical contact with coelenterate prey probably confers temporary protection from predatory animals that avoid medusae and siphonophores. In this regard, the development of juvenile *Phylliroë* within *Zanclea* bells may offset the apparent low numbers of young produced. Protective coloration also is exhibited by all of the species. *Glaucus* and *Glaucilla* are countershaded, which presumably decreases mortality from both aerial and pelagic visual predators. *Phylliroë* and *Cephalopyge* are transparent, a common camouflage among zooplankton exposed to visual mid-water predators. And *Fiona* quickly acquires the coloration of its prey, becoming blue when feeding on *Velella* or brown when preying on *Lepas*.

All of the pelagic nudibranchs have wide distributions throughout the circumtropical waters, with the exception of *Glaucilla marginata*, which is confined to the warm areas of the Pacific Ocean. As expected, their distributions coincide with those of their prey, which also are essentially circumtropical. Since *Glaucilla* is thought to feed on the same prey items as *Glaucus*, its restriction to a single ocean must be due to factors other than prey distribution.

There seem to be no major adaptive changes in the reproduction and development of these nudibranchs, compared with their benthic relatives. All are hermaphrodites, which would be a major advantage to animals of limited mobility living in the vast spaces of the open ocean if self-fertilization were possible. This mode of reproduction, however, has not yet been demonstrated. All of the species lay egg masses and have a free-swimming veliger stage. It is presumed, but not yet demonstrated, that metamorphosis involves contact with suitable prey. It would be surprising if this were not the case, since veliger larvae of benthic nudibranchs generally demonstrate an extraordinary ability to recognize, and metamorphose on, the specific adult food (Thompson, 1976).

The one striking developmental feature common to *Phylliroë*, *Cephalopyge*, and *Fiona* is the rapid growth rate of metamorphosed juveniles. This suggests that sexual maturity is attained within a very short time, and that life cycles are rapid (about 1 month for *Fiona*). This may be related to opportunistic behavior: when large or abundant prey are available, the animals remain attached for some period of time ingesting much of the prey and expending little or no energy on locomotion. A large proportion of ingested energy could thus be channeled into production (i.e. growth and reproduction). When a prey organism or prey patch is exhausted, the veligers have already been produced that are then dispersed in the water column to locate new food.

The advantages conferred by the adoption of a pelagic life style include abundant prey, few feeding competitors, expansion of range, unlimited space, and possibly fewer predators. In view of the swimming ability of so many nudibranchs and the pre-adaptation of shell loss in this group, it is perhaps surprising that not more species have become pelagic inhabitants.

List of Recognized Species

Suborder Dendronotacea
Family Phylliroidae
Phylliroë bucephala Péron and Lesueur, 1810
?*P. atlantica* Bergh, 1871 (a species of uncertain validity)
Cephalopyge trematoides (Chun, 1889)
Suborder Aeolidacea
Family Glaucidae
(Synonymy is given in Thompson and McFarlane, 1967, and Thompson and Bennett, 1970.)
Glaucus atlanticus Forster, 1777
Glaucilla marginata Bergh, 1868
Family Fionidae
Fiona pinnata (Eschscholtz, 1831)

References Cited

Works with recommended keys or aids for the identification of species are indicated by an asterisk.

Alder, J., and A. Hancock. 1845–55. *A Monograph of the British Nudibranchiate Mollusca*. London: Ray Society.

Ankel, W. E. 1952. *Phyllirrhoe bucephala* Per. & Les. und die Meduse *Mnestra parasites* Krohn. *Pubbl. Staz. zool. Napoli* 23: 91–140.

Bayer, F. M. 1963. Observations on pelagic mollusks associated with the siphonophores *Velella* and *Physalia*. *Bull. mar. Sci. Gulf Caribb.* *13*: 454–66.

Bebbington, A. 1986. Observations on a collection of *Glaucus atlanticus* (Gastropoda, Opisthobranchia). *Haliotis* *15*: 73–81.

Bennett, G. 1836. Observations on a species of *Glaucus*, referred to the *Glaucus hexapterygius*, Cuvier. *Proc. zool. Soc. Lond.* *1836*: 113–19.

Bennett, I. 1966. Some pelagic molluscs and associated animals in south-eastern Australian waters. *J. malac. Soc. Aust.* 9: 40–51.

Bieri, R. 1966. Feeding preferences and rates of the snail *Ianthina prolongata*, the barnacle *Lepas anserifera*, the nudibranchs *Glaucus atlanticus* and *Fiona pinnata*, and the food web in the marine neuston. *Publs Seto mar. biol. Lab.* *14*: 161–70.

Casteel, D. B. 1904. The cell-lineage and early larval development of *Fiona marina*, a nudibranch mollusk. *Proc. Acad. nat. Sci. Philadelphia 56*: 325–405.

Day, R. M., and L. G. Harris. 1978. Selection and turnover of coelenterate nematocysts in some aeolid nudibranchs. *Veliger 21*: 104–9.

Edmunds, M. 1966. Protective mechanisms in the Eolidacea (Mollusca Nudibranchia). *J. Linn. Soc.* (*Zool.*) *46*: 27–71.

Farmer, W. M. 1970. Swimming gastropods (Opisthobranchia and Prosobranchia). *Veliger 13*: 73–89.

Holleman, J. J. 1972. Observations on growth, feeding, reproduction, and development in the opisthobranch, *Fiona pinnata* (Eschscholtz). *Veliger 15*: 142–46.

Kropp, B. 1931. The pigment of *Velella spirans* and *Fiona marina*. *Biol. Bull. mar. biol. Lab., Woods Hole 60*: 120–23.

Laursen, D. 1953. The genus *Ianthina*. *Dana Rep.* No. *38*. 40 pp.

MacFarland, F. M. 1966. Studies of opisthobranchiate mollusks of the Pacific coast of North America. *Mem. Calif. Acad. Sci.*, vol. 6. San Francisco. 546 pp.

Martin, R. 1966. An attempt to infect *in vitro* medusae of *Zanclea costata* (Anthomedusae) with the veliger of *Phyllirrhoe bucephala* (Opisthobranchia). *Pubbl. Staz. zool. Napoli 35*: 130–31.

Martin, R., and A. Brinckmann. 1963. Zum Brutparasitismus von *Phyllirrhoe bucephala* Per. & Les. (Gastropoda, Nudibranchia) auf der Meduse *Zanclea costata* Gegenb. (Hydrozoa, Anthomedusae). *Pubbl. Staz. zool. Napoli 33*: 206–23.

Pruvot-Fol, A. 1954. Mollusques opisthobranches. *Faune Fr.* No. *58*. 460 pp.

Sentz-Braconnot, E., and C. Carré. 1966. Sur la biologie du nudibranche pélagique *Cephalopyge trematoides*. Parasitisme sur le Siphonophore *Nanomia bijuga*, nutrition, développement. *Cah. Biol. mar. 7*: 31–38.

van der Spoel, S. . 1970. The pelagic Mollusca from the "Atlantide" and "Galathea" Expeditions collected in the East Atlantic. *Atlantide Rept. 11*: 99–139.

Steinberg, J. E. 1956. The pelagic nudibranch *Cephalopyge trematoides* (Chun, 1889) in New South Wales, with a note on other species in this genus. *Proc. Linn. Soc. N.S.W. 81*: 184–92.

Stubbings, H. G. 1937. Phyllirhoidae. *Scient. Rep. John Murray Exped. 5*: 1–14.

Thompson, T. E. 1976. *Biology of Opisthobranch Molluscs*, vol. I. London: Ray Society. 207 pp.

Thompson, T. E., and I. Bennett. 1969. *Physalia* nematocysts: Utilized by mollusks for defense. *Science 166*: 1532–33.

*———. 1970. Observations on Australian Glaucidae (Mollusca: Opisthobranchia). *J. Linn. Soc. (Zool.)* 49: 187–97.

Thompson, T. E., and G. H. Brown. 1981. Biology and relationships of the nudibranch mollusc *Notobryon wardi* in South Africa, with a review of the Scyllaeidae. *J. Zool., Lond.* 194: 437–44.

———. 1984. *Biology of Opisthobranch Molluscs*, vol. II. London: Ray Society. 229 pp.

*Thompson, T. E., and I. D. McFarlane. 1967. Observations on a collection of *Glaucus* from the Gulf of Aden with a critical review of published records of Glaucidae (Gastropoda, Opisthobranchia). *Proc. Linn. Soc. Lond.* 178: 107–23.

Zirpolo, G. 1923. Ricerche sulla simbiosi fra zooxantelle e *Phyllirhoe bucephala* Péron et Lesueur. *Boll. Soc. Nat. Napoli* 35: 129–38.

Glossary

Glossary

Amphipods Laterally compressed, planktonic or benthonic crustaceans; *Themisto, Hyperia.*
Aphallic In males, lacking a copulatory organ.
Appendicularia See *Larvacea.*
Aragonite A form of calcium carbonate present in shells of thecosomatous pteropods and heteropods.
Assimilated food That portion of ingested food that is absorbed and utilized by an animal, the remainder being discarded as feces.
Assimilation efficiency The percentage of ingested food that is assimilated by an animal.

Bathypelagic zone The water column below about 1,000 m depth.
Benthic Pertaining to the sea-floor environment.
Benthos Plants or animals that live on or in the sea floor.
Bioluminescence The production of light by living organisms.
Biomass The number of individual organisms (in some area or volume or region) multiplied by the average weight of the individuals.
Bipolar species Those species that live in both Arctic and Antarctic waters, but are not present in mid-latitudes.
Buccal cavity The oral cavity of mollusks.
Buccal cones Eversible, prehensile tentacles of certain gymnosomes that are used in prey capture.
Buccal organs Mouthparts used in feeding.
Bulimoid shells Smooth gastropod shells with a tall conical shape.

Calcification The formation of calcium carbonate skeletal structures.
Calcite A form of calcium carbonate present in the shells of Foraminifera and most benthic mollusks.

Carnivores Animals that feed exclusively or primarily on other animals.

Cerata External papillae arising from the dorsal surface of eolid nudibranchs that function as respiratory organs and, in some species, as defensive structures (singular, *ceras*).

Cestodes A class of parasitic flatworms of the Phylum Platyhelminthes.

Chaetognaths Holoplanktonic, unsegmented "arrow worms"; *Sagitta.*

Chalimus A life stage of parasitic copepods intermediate between the copepodite stages and the adult.

Chitin A horny substance forming the hard part of crustacean exoskeletons, and the radular teeth, gizzard plates, hooks, and opercula of mollusks.

Chondrophores Polymorphic, colonial coelenterates forming part of the pleuston community; *Velella, Porpita.*

Chromatophores Epidermal sacs of pigment under muscular control that cause color changes in cephalopods, some gymnosomes, and some thecosomes.

Cladocera Planktonic crustacea with a bivalved carapace; *Podon, Evadne.*

Cnidaria See *Coelenterates.*

Cnidosacs Internal sacs, located at the tips of the cerata of some eolid nudibranchs, that contain nematocysts obtained from coelenterate prey.

Coccolithophorids Small, flagellate, unicellular phytoplankton with calcareous plates (coccoliths) embedded in their cell walls.

Coelenterates Animals of the Phylum Cnidaria (= Coelenterata), including pelagic jellyfish and benthic sea anemones and corals.

Cohort A group of organisms produced at the same time; one generation.

Columella The central axis of a gastropod shell.

Common genital pore In hermaphrodites, a single external opening used for the emission of both male and female gametes.

Copepodite A life stage of copepods intermediate between the naupliar stages and the adult.

Copepods Small, holoplanktonic crustacea that usually form the numerically dominant group of zooplankton in most marine areas; *Calanus, Metridia.*

Cosmopolitan species Those species with a very broad geographical distribution; present in extensive areas of the Atlantic, Pacific, and Indian oceans.

Countershading Color difference in the dorsal and ventral surfaces of an animal; a protective mechanism against visual predators.

Crustacea A class of primarily aquatic arthropods characterized by a segmented body and chitinous exoskeleton; including copepods, ostracods, amphipods, euphausiids.

Ctenidium The molluscan gill; used for respiration and, in some species, for feeding.
Ctenophores Gelatinous zooplankton characterized by eight longitudinal rows of fused cilia ("ctenes") used in swimming; *Beroë*, *Pleurobrachia* ("sea goose-berries").

Dactylozooid A type of polyp bearing nematocysts and specialized for prey capture and defense in hydrozoan colonies.
Denticle A small toothlike projection.
Detorsion A process in which torsion is partially or wholly reversed, with the mantle cavity and visceral mass moving back to the right side or to the posterior end of the body relative to the fixed head and foot; characteristic of pulmonates and most opisthobranchs.
Dextral coiling Gastropod shells coiled in a clockwise direction.
Diatoms Unicellular phytoplankton with siliceous cell walls.
Digeneans Flatworms belonging to an order (Digenea) of the Phylum Platyhelminthes.
Dinoflagellates Unicellular phytoplankton having two flagella and a cellulose test.
Dioecious Having separate sexes (male and female) in a species.
Direct development Development in which there is no free larval stage; the young hatch or are born as immature adults.
Dissolution The breakdown of calcareous shells to dissolved calcium and carbonates.
Doliolids Barrel-shaped gelatinous zooplankton of the Phylum Chordata, Subphylum Urochordata; *Doliolum*.

Ecological efficiency The efficiency by which energy is transferred from one trophic level to another.
Ectoparasite A parasite that lives on the external surface of its host.
Embryonic shell That portion of the veliger shell that is formed before hatching.
Endogenous sperm Sperm produced by an animal.
Endoparasite A parasite that lives in the internal organs of its host.
Eolids Nudibranchs of the Suborder Aeolidacea that are characterized by having paired cephalic tentacles, unsheathed rhinophores, and, usually, cerata; *Glaucus*, *Glaucilla*, *Fiona*.
Epifauna Animals that live on, or attach to, a substrate.
Epipelagic zone The upper region of the sea from the surface to approximately 200 meters depth.
Epitoniids Marine, benthic "wentletrap" snails belonging to the same suborder as the janthinids; *Epitonium*.

Euphausiids Shrimplike, holoplanktonic crustaceans; *Euphausia*, *Meganyctiphanes* ("krill").

Euphotic zone The surface waters of the oceans that receive sufficient light to support photosynthesis.

Eupyrene spermatozoa Sperm used to fertilize eggs.

Exogenous sperm Sperm received by an animal during fertilization.

Fecundity The rate of production of eggs or young.

Foraminifera Planktonic or benthonic protozoa characterized by a calcareous exoskeleton and pseudopodia; *Globigerina*.

Gill A respiratory organ; the molluscan ctenidium.

Gizzard An organ of the thecosome alimentary tract which is lined with chitinous plates and is used to crush mineralized parts of food.

Globose shells Gastropod shells with a low spire and an inflated, rounded, body whorl.

Growth efficiency The amount of growth attained per unit of ingested or assimilated food.

Hemocoel A blood-filled body cavity present in animals with an open circulatory system.

Herbivores Animals that feed exclusively or primarily on plants.

Hermaphrodite An animal that produces both male and female gametes.

Holoplankton Planktonic organisms that complete their life cycles in the water column; permanent residents of the plankton community.

Hook sacs Paired sacs of gymnosomes that contain eversible, chitinous hooks used to capture and ingest prey.

Hydroid A life stage of attached polyps of the Phylum Cnidaria (Coelenterata), Class Hydrozoa.

Hypobranchial gland A mucoid gland present in the mantle lining of many mollusks.

Indirect development Development of young involving a free-swimming larval stage.

Ingestion The act of swallowing food.

Iridocytes (= iridophores) Epidermal pigment cells that effect color changes by the reflection of light.

Keel The thin, transparent extension of shell from the outer shell whorl of atlantid heteropods.

Larvacea Zooplankton of the Subphylum Urochordata that build mucous houses and filter-feed on nanoplankton; *Oikopleura*, *Fritillaria*.

Macrophagy Feeding on large-sized particles.
Mantle An extension of the body wall of mollusks that secretes the shell and lines the largest shell whorl.
Mantle cavity The space between the mantle lining and the rest of the mollusk body.
Manubrium In medusae, a tubelike projection bearing the mouth at its terminus.
Medusae Zooplankton of the Phylum Cnidaria (Coelenterata); "jellyfish."
Melanophores Cells containing melanin, a dark pigment.
Meroplankton Temporary residents of plankton communities, usually larvae of benthonic or nektonic adults; larval clams, larval fish.
Mesogastropods An order of prosobranchs that includes janthinids and heteropods as well as many marine benthic snails.
Mesopelagic zone The water column from the bottom of the epipelagic zone (ca. 200 m) to about 1,000 m depth.
Mesopodium The central area of the gastropod foot.
Metamorphosis The transformation in form from one distinctive stage to another in an animal's life history.
Metapodium See *Postpodium*.
Microphagy Feeding on small-sized particles.
Monoecious Hermaphroditic.
Monophagy Feeding on only one or several, closely related, prey species.
Monotypic genus A genus comprising a single species.

Nanoplankton Planktonic organisms in the size range of 2 to 20 μm, including small phytoplankton and bacteria.
Nauplius A free-swimming larval stage of crustacea.
Nekton Pelagic animals capable of swimming against a current; squid, fish, marine mammals.
Nematocysts Stinging cells on the tentacles of coelenterates.
Neoteny The retention of external larval characteristics in sexually mature animals.
Neritic Referring to inshore waters of less than 200 m depth that overlie continental shelves.

Oceanic Referring to offshore waters in areas deeper than 200 m.
Oligopyrene spermatozoa Atypical sperm used to transport normal (eupyrene) sperm to females.

Oligotrophic area A region of low biological productivity.

Operculum A chitinous or (rarely) calcareous structure on the foot of mollusks that is used to seal the shell aperture.

Opisthobranchia A subclass of gastropods that includes pteropods, nudibranchs, and sea slugs.

Osphradium A sensory organ in mollusks used to sample water entering the mantle cavity.

Oviparity The release of eggs that hatch outside the parent's body and develop without parental care.

Ovotestis The gonad of a hermaphroditic animal in which both male and female gametes are produced.

Ovoviviparity The retention of fertilized eggs within the parent until hatching.

Pallial gland See *Hypobranchial gland.*

Parapodia The paired swimming wings of thecosomes and gymnosomes.

Pelagic Referring to the oceanic water column and the organisms living therein.

Periostracum The outermost layer of the molluscan shell, composed of conchiolin.

Phytoplankton Microscopic planktonic plants; diatoms, dinoflagellates.

Plankton Plants or animals that live in the water column and are incapable of swimming against a current.

Pleuston Organisms that float passively at the sea-air interface.

Polychaetes Marine, segmented worms belonging to the Phylum Annelida.

Polyp The sessile stage, or form, in the life history of many coelenterates.

Polytrochous larva The second larval stage of gymnosomes characterized by three ciliary bands encircling the shell-less body.

Postembryonic shell That portion of the veliger shell that is formed after hatching.

Postpodium The posterior portion of the gastropod foot.

Proboscis An extensible feeding organ that bears the mouth terminally.

Propodium The anterior portion of the gastropod foot.

Prosobranchs A subclass of gastropods that includes janthinids, heteropods, and most benthic marine snails.

Protandrous hermaphrodite An animal that begins life as a male, then becomes a female.

Protoconch The larval shell of mollusks.

Pseudoconch A thin, internal, noncalcareous, "false" shell present in some euthecosomes and pseudothecosomes.

Pseudofeces Particles captured by ciliary-mucus feeders that are sorted and rejected before ingestion.

Ptenoglossate radula The type of radula present in janthinids and epitoniids, consisting of an indefinite number of lateral hook-shaped teeth but no median tooth (n-0-n).

Pteropod ooze Sea-floor sediments composed of more than 30 percent $CaCO_3$ from pteropod or other pelagic mollusk shells.

Radiolaria Planktonic protozoa with a siliceous skeleton and pseudopodia; *Acanthometron*.

Radula A feeding organ unique to mollusks, consisting of rows of chitinous teeth (absent in bivalves).

Radular formula Designation of the number and type of teeth in one row of the radula; e.g. 2-1-1-1-2 denotes a total of four marginal teeth, two lateral teeth, and one median tooth per row. (The designation "n" = an indefinite number of particular teeth.)

Reciprocal fertilization In hermaphrodites, the fertilization of eggs in both members of a mating pair.

Rhinophores Sensory, cephalic tentacles present in many nudibranchs; located posteriorly to the oral tentacles, if present.

Salps Barrel-shaped gelatinous zooplankton of the Phylum Chordata, Subphylum Urochordata; *Salpa, Thalia*.

Scyphozoa A class of jellyfish of the Phylum Cnidaria (Coelenterata); *Aurelia*.

Sexual dimorphism Morphological difference between sexes of the same species.

Sibling species Closely related species, usually in the same genus.

Silicoflagellates Small, flagellate, unicellular phytoplankton with an internal siliceous skeleton.

Sinistral coiling The counterclockwise coiling of gastropod shells.

Siphonophora Pelagic coelenterates composed of colonies of individuals; *Physalia, Nanomia*.

Siphonophore bracts A type of zooid forming a transparent, protective cover over certain siphonophore polyps.

Spermatophore A capsule in which numerous sperm are packaged and transferred during fertilization.

Spermatozeugma A single oligopyrene spermatozoon with attached eupyrene sperm.

Spiral cleavage A type of embryonic development in which successive cell divisions result in the blastomeres having a spiral arrangement around the polar axis.

Statocysts Sense organs used for balance and orientation.

Stolon A stemlike attachment connecting individuals of colonial coelenterates.

Symbiosis A close physiological association between two species, often for mutual benefit.

Taenioglossate radula A type of mesogastropod radula with seven teeth per row (radular formula = 2-1-1-1-2).

Tintinnids Shelled, planktonic, ciliate protozoa.

Torsion A process in which the gastropod mantle cavity and visceral mass rotate 180 degrees counterclockwise relative to the fixed head and foot; characteristic of prosobranchs and some opisthobranchs.

Trematodes A class of parasitic flatworms of the Phylum Platyhelminthes.

Trochoid shells Gastropod shells with the shape of a broad inverted "V."

Trophic level The functional position occupied by an organism in a food web; e.g. primary producers (plants); primary consumers (herbivores); secondary consumers (carnivores), etc.

Vas deferens A sperm duct.

Veliger A larval stage of mollusks.

Velum A membranous structure edged with cilia and used by veliger larvae for swimming and food collection.

Viviparity Giving birth to live young that are nourished by a transfer of organic food from the parent throughout development.

Water mass A large volume of seawater having a common origin and a distinctive combination of temperature, salinity, and density characteristics.

Wingplate The locomotory organ of pseudothecosomes formed by fusion and expansion of paired wings.

Zooid An individual member of a colony of animals.

Zooplankton Planktonic animals.

Zooxanthellae Dinoflagellates that live symbiotically in the tissues of organisms such as corals and mollusks.

Indexes

Systematic Index

Numbers in italics refer to pages with text-figures

Abranchaea, 174, 208
Abylopsis, 223
Actaeon, 146
Aeolidacea, 214, 224, 232, 234
Aequorea, 219
Alepisaurus brevirostris, 42
  ferox, 42
Anomia tridentata, 59
Anopsia, 202
Apolemia, 223
Architectonicacea, 146
Architeuthis, 3
Argonauta boettgeri, 98
Atlanta, 29, 35, 39–45 *passim*, 49–52 *passim*, 92, 142
  fusca, 47
  gaudichaudi, 40
  helicinoides, 92
  inclinata, 39
  inflata, 29, 39, 46
  lesueuri, 37, 38, 46
  peroni, 39, 42, 47, 50, Color Fig. 2
Atlantidae, 27, 29–30, 34–35, 39–40, 51f

Bonneviia, 221
Boopsis, 221
Brachyscelus rapacoides, 130

Calyptraeidae, 23
Campaniclava cleodorae, 127ff
  clionis, 127f
Capulidae, 23
Cardiapoda, 30–36 *passim*, 40f, 53
  placenta, 30, 31, 32, 36, 40–45 *passim*, 49
  richardi, 30, 32, Color Fig. 5
  sublaevis, 43
Cardiodectes, 21
  medusaeus, 49, 130
Carinaria, 30–35 *passim*, 41–53 *passim*
  cristata, 28, 30, 38, 40ff, 49, 130
  japonica, 49
  lamarcki, 31, 41f, 46, 47, 49f, Color Fig. 3
  mediterranea, 49
Carinariidae, 27–42 *passim*, 53
Cavolinia, 63, 65, 77f, 85–86, 99, 116, 145–51 *passim*
  gibbosa, 114, 122
  globulosa, 95
  inflexa, 84, 95, 137, 186
  longirostris, 75, 84–87 *passim*, 94–98 *passim*, 109, 113–16 *passim*, 130, 143, 186–87, 188
  tridentata, 59, 75, 77, 83–87 *passim*, 98, 113–19 *passim*, 128ff, 138, 188, Color Fig. 7
  uncinata, 65, 66, 75, 82–87 *passim*, 95, 109–*11*, 113, *115*
Cavoliniidae, 58, 65, 75–79 *passim*, 85–87, 109–23, 145–51 *passim*
Cavoliniinae, 151
Centrobranchus nigroocellatus, 42
Cephalobrachia, 174, 209
Cephalopyge, 221, 232ff
  arabica, 221
  mediterranea, 221
  michaelsarsi, 221

orientalis, 221
picteti, 221
trematoides, 214f, 221–24, 222
Cerithiacea, 23
Clio, 63, 77f, 85, 87, 129, 145, 150f
antarctica, 132, *133*
balantium, *see Clio recurva*
campylura, 116
chaptali, 116
cuspidata, 95, *127f*, 130
polita, 87, 110, 149
pyramidata, 64, 78–87 *passim*, *81*, 94f, 99, 110–*13*, *118–22 passim*, 129f, 136, 149, 188
pyramidata convexa, 119
pyramidata lanceolata, 119
recurva, 63, 116, *127f*, 130
sulcata, 87, 132, *133*
Clioinae, 151
Clione, 174, 177, 205, 209
limacina, 59, 168, *169*, *170f*, *176–91 passim*, 179, *180*, 195–99, *198*, 206–7, Color Fig. 12
Clionidae, 167, 209
Clioninae, 209
Cliopsidae, 167, 175, 208
Cliopsis, 170, 174, 208
krohni, *169*, *176*, *187f*
Clytia linearis, 127
striata, 127f
Conularia, 145
Corolla, 68–79 *passim*, 88, 92, 98f, 123, 130, 152, *187f*
calceola, 70, 92, 123, *124*, *126*, Color Fig. 10
ovata, 75
spectabilis, 75, 92, 98, 100, 130
Coryphaena, 39, 41
Creseis, 39, 61, 63, 77, 85, 87, 95, 129, 142–47 *passim*, 151, 205
acicula, 64, 75–79 *passim*, 92, 101–13 *passim*, 119–20, 136f, 143, 186, 188
virgula, 39, 77f, 85, 116, 122, 132, 137, 186, 188
virgula conica, 109–10, 119–20, 131, 136
virgula virgula, 84, 136
Crucibranchaea, *173f*, *177*, 208
macrochira, *191*
Cryptobia carinariae, 49
Ctilopsis, 221
Cunina, 223
Cuvierina, 63, 77f, 85, 151
columnella, 64, 79, *83–87 passim*, 95–96, *114*, 118–22 *passim*, *121*, 127–30 *passim*, 139–43 *passim*
Cuvierininae, 151
Cymbulia, 60, *68–78 passim*, 88, 130, 152
peroni, 92, 123–24
valdiviae, 92
Cymbuliidae, 58, 145, 149, 152

Dactylopus, 221
Dendronotacea, 214f, 232, 234
Desmopteridae, 58, 145, 152
Desmopterus, 72, 76, 78, 88, 93, 147–52 *passim*
papilio, 72, *73*, 75f, 93
Diacria, 65, 77, 85, 127, 129, 145, 151
major, 64, 129
quadridentata, 75–79 *passim*, 84–87 *passim*, 129
trispinosa, 78, 84, 87, 95f, 109–*11*, 120, 122, 128ff, 186–*87*, 188, Color Fig. 8
trispinosa minor, 129
Doliolum, 40

Epitoniidae, 9, 12, 23
Epitonium, 18
Euribia gaudichaudii, 199n
globulosa, 199n

Fiona, 14f, 234
pinnata, 215, 229–32, 233, Color Fig. 15
Fionidae, 214f, 229–32, 234
Firoloida, 32–36 *passim*, 44–49 *passim*, 53
desmaresti, 28, 32, 33, 36, 40f, 45, 46, 47
Fowlerina, 174, 177, 209

Glaucidae, 214f, 224–34 *passim*
Glaucilla, 15, 230, 232ff
marginata, 224–29, 226, 233
Glaucus, 15, 230, 232ff
atlanticus, 21, 224–29, *225*, 227, Color Fig. 16
Gleba, 68–72 *passim*, 76, 88–92 *passim*, 123, *125*, 130, 149, 152
cordata, 59, *71*, 75–79 *passim*, *89*, 90, 91f, 98, 123f, 130
Globerigerinoides, 87
Globigerina, 87, 92
Gonyaulax excavata, 100
Gymnoptera, 167f, 175, 199–203, 209

Gymnosomata (Suborder), 167, 168–99, 208

Halobates, 13
Hancockia, 215
Hyalocylis, 77f, 85, 151
striata, 39, 78, 84, 116
Hydrobia, 22
Hydromyles, 174, 205, 209
gaudichaudii, 188, 199–203, *200*, Color Fig. 13
Hydromylidae, 167, 209
Hyolithes, 145
Hyperia, 41

Ianthina, 23
Isochrysis, 194

Janthina, 8–23 *passim*, 224, 229
exigua, 12f, 20f
globosa, 21, 24
janthina, 9–21 *passim*
pallida, 12–20 *passim*, *16*, *17*
prolongata, 10f, *13*, 20f, Color Fig. 1
umbilicata, 9, *13*, 19f
Janthinidae, 4, 8–23 *passim*, 232

Kinetocodium danae, 128–29, Color Fig. 8

Laginiopsidae, 167, 209
Laginiopsis, 174, 209
trilobata, 175, 203
Laomedea, 21
geniculata, 21
striata, *see Clytia striata*
Lecithocladium, 21
Lepas, 21
anatifera, 99, 230f, 233
Lernaeoceridae, 21, 129–30
Limacina, 6, 62–69 *passim*, *68*, 78–84 *passim*, 99–107 *passim*, *101*, 142–50 *passim*, 205–7
bulimoides, 101, 103, 108–9, 136, 143, 186, 188
helicina, 59f, 74, *81–85 passim*, *83*, 94, 99–108 *passim*, *102*, *106*, 132, 143, 178–88 *passim*, 206, Color Fig. 6
helicina antarctica, 63, 75, 95, 132–*33*
helicoides, 84f, 95–96, 103, 104–5
inflata, 6, 79, 88, 92, 94, 103–9 *passim*, *107*, 131–38 *passim*, 143, 145, 149f, 188
lesueuri, 101, 103, 143
retroversa, 62, 73–85 *passim*, *74*, 94, 99–108 *passim*, *105*, 131–38 *passim*, *133*, 178–88 *passim*, 206
trochiformis, 63, 101, 103, 108–9, 131–32, 136, 143
Limacinidae, 58, 65, 101–9, *120*, 147–50 *passim*
Liriope, 223
Lycaea, 130

Massya, 174, 209
Micrallecto uncinata, 199
Minyas, 15
Mnestra parasites, 217
Muggiaea, 40

Nannallecto fusii, 199
Nanomia, 40
bijuga, 221–24, *222*, 232
Nautilus, 3, 147
Nectophyllirhoe, 221
Notobranchaea, 171, 174, 208
macdonaldi, 199
Notobranchaeidae, 167, 208

Octopus, 3
Oikopleura albicans, 219
Opalinopsis carinariae, 49
Oxycephalus, 99
Oxygyrus, 29, 34f, 45
keraudreni, 29f, 34f, 39–42 *passim*, 51

Paedoclione, 174, 205, 209
doliiformis, *172*, *185–96 passim*, *192*, *194*, *195*
Pandea conica, 127ff
Paraclione, 174, 188, 209
longicaudata, 190
Parathemisto guadichaudi, 189
Pelagea noctiluca, 42
Pennellidae, 21, 129–30
Peraclididae, 58, 67, *120*, 145, 149, 152
Peraclis, 67–70 *passim*, *68*, 75f, 80, 88, 90, 98, 145–52 *passim*
apicifulva, 88, 92
bispinosa, *83*, 123
moluccensis, 67
recticulata, *67–71 passim*, 75, 78, 88, *89*, 92, 123, Color Fig. 9
Perigonella, 127
Perigonimus sulphureus, 127f
Phylliroë, 40, 215–21, *216*, 223, 232ff
atlantica, 215–16

bucephala, 215, 216–21, 218, Color Fig. 14
Phylliroidae, 214, 234
Physalia, 14f, 22, 224–28 *passim*
physalis, 12
Platybrachium, 208
Pneumoderma, 174, 188, 208
atlanticum, 170, 176, 188f, 199
Pneumodermatidae, 167, 175, 177, 208
Pneumodermopsis, 171, 174, 208
canephora, 169, 186–87, 188f, Color Fig. 11
ciliata, 188
paucidens, 186, 187f, 199, 205f
Pollicipes polymerus, 231
Porpita, 15, 22, 224–30 *passim*
porpita, 13
Protatlanta, 29, 34
Pruvotella, 174, 209
Psyche globulosa, 199n
Ptenoglossa, 4, 8f, 12, 22f
Pterosoma, 30, 32, 41, 45–49 *passim*, 53
planum, 31, Color Fig. 4
Pterotrachea, 32–36 *passim*, 41–53 *passim*, 46
coronata, 32–41 *passim*, 45, 47, 49
hippocampus, 33, 41–45 *passim*
minuta, 38
scutata, 33, 37
Pterotracheidae, 27, 32–53 *passim*

Recluzia, 9–15 *passim*, 10, 21, 24
rollandiana, 12, 15, 19
Rhizophysa, 223
Rosacea cymbiformis, 42

Salpa cylindrica, 40
Sargassum, 22, 215, 230
Schizobrachium, 174, 208
polycotylum, 168
Scolex pleuronectis, 21
Scyllaea, 215
pelagica, 215
Scyllaeidae, 215
Spirialis recurvirostra, 69
Spirula, 3, 231
Splanchnotrophidae, 199
Spongiobranchaea, 174, 208
australis, 171, 188
Styliola, 77f, 85, 151, 186
subula, 84, 120, 129, 136
Sulculeolaria, 223
Syringopharynx pterotracheae, 49

Thalassopterus, 174, 209
Thalia, 40
Thliptodon, 169f, 174, 209
diaphanus, 176
Thliptodontinae, 209
Thunnus albacares, 41
Tiedemannia, 92
neapolitana, 124
Tomopteris, 40
Tonnacea, 146

Velella, 13ff, 22, 224–33 *passim*
velella, 10, 12

Zanclea, 232f
costata, 217–20, 218

Subject Index

Numbers in italics refer to pages with text-figures

Aberrant individuals, 112–*13*
Abundance, *see* Population densities
Accessory copulatory organ, *see* Penis
Aggression, 179, 207
Albumen, 19
Albumen gland, 103f, 106, 190
Ammonium, 42–43, 95, 98
Amphipods, 41, 49, 99, 130, 139, 189
Anaspidea, 204
Anatomical artifacts, *see* Morphological artifacts
Antarctic Convergence, 132
Antarctic Ocean, 60, 62, 108, 132f, 168, 178, 206
Anus, 65, *172*, *173*, *192*, *196*, 202, *216*, 221, *225*, 229
Aphallism, 17, 23, 105
Aplacophora, 1f
Aplysiomorphs, 204
Appendicularia, *see* Larvacea
Aragonite, in shells, 6, 9, 27, 29, 50, 60, 86, 118, 139, 199; sedimentation of, 6, 139–45 *passim*
Aragonite compensation depth, 139f, 144–45
Arctic Ocean, 60, 62, 99, 108, 168, 178, 206
Argentina, 132
Arthropods, 1
Asexual reproduction, 112–13, 149–50
Assimilation, 97, 184, 207
Atlantic Ocean, 62, 99, 126, 134f, 140, 144, 168, 215, 224
Australia, 28, 200f, 216

Bacteria, 84, 90, 100
Balancer, 65
Barbados, 21, 108, 122, 135
Barnacles, 20–21, 99f, 230–31
Bathypelagic species, 60, 62, 84, 95f, 104, 116, 150, 168, 203
Benthos, 2, 99
Bermuda, 122, 132, 135
Bib, *33*, 41
Bicarbonate ions, 6, 139f
Bioluminescence, 37, 216, 221
Bipolar species, 108, 168, 178, 206
Birds, 9, 15, 98f, 189
Bismarck Sea, 44
Bivalves, 1f, 12, 87
Black gut, 6, 99–100
Breeding seasons, 44f, 107–8, 113, 192–99 *passim*, 206, 220
Brine shrimp, 41
British Columbia, 8
Brood chamber, 150, 201, 203
Brood protection, 106, 116, 150
Buccal cavity, 12, 14, 175, *177*, 202
Buccal cones, *172–81* *passim*, *179*, *185*, 202, 205
Buccal mass, *33*, 72, 88, *106*, *173–80* *passim*, *177*, 203
Bulloids, 146, 204
Buoyancy regulation, 3f, 35ff, 51, 65, 74–81 *passim*, 149, 171, 226f. *See also* Float; Ionic regulation
By-the-wind sailor, *see Velella*

Calcite, 9, 50, 86, 139–45 *passim*
Calcite compensation depth, 144–45

Calcium carbonate, 6, 9, 50, 134, 139–45 *passim*
Calcium carbonate budget, 140–43
California Current, 44, 49, 137
Camouflage, 9, 42
Cannibalism, 13, 39ff, 227–28
Cape Hatteras, 131–32
Carbon dioxide cycle, 6, 61, 139–44
Caribbean Sea, 21, 108, 116, 122, 135, 144
Cenozoic, 22
Cephalaspideans, 146, 204
Cephaloconi, *see* Buccal cones
Cephalopods, 1ff, 27, 35, 45, 98, 175, 177
Cerata, 214, 224–32 *passim*, *225*, *226*
Cestodes, 21, 130
Chaetognaths, 23, 39ff, 92, 98, 131, 133f, 189
Challenger Expedition, 59f, 134
Chemical composition, 93–95. *See also* Aragonite; Calcite; Conchiolin
Chitons, 1
Chondrophores, 12, 224–32 *passim*. *See also Porpita; Velella*
Chromatophores, 72, 124, *173*, 175
Chromosomes, 17, 123
Cilia, 64, 80–90 *passim*, 149, 201
Ciliary bands, in gymnosomes, *169*f, *171*, *172*, *173*, 195–203 *passim*, *196*
Ciliates, 49, 130
Circumtropical species, 8, 12, 28, 60, 73, 215, 224, 229, 233
Cladocera, 100
Clasper, *see* Tail filaments
Climatic change, 134–38
Cnidosacs, 228
Coccolithophorids, 85ff, 92f, 97, 139ff
Cod, 99
Coelenterates, *see* Chondrophores; Hydroids; Medusae; Siphonophores
Collection methods, 5–6, 28, 43–44, 60–62, 79, 122, 167, 201. *See also* Scuba diving; Submersibles
Coloration, 9, 11, 31–32, 33–34, 72, 216, 224–25, 229–33 *passim*. *See also* Camouflage; Chromatophores; Countershading
Columella, 67
Columellar muscle, *106*, 117
Competition, 51, 186, 207, 232
Conchiolin, 29, 34, 51
Convergent evolution, 51, 146, 175
Copepods, 13, 39ff, 45, 87, 92, 100, 131, 142; parasitic species, 21, 49, 129–30, 199
Copulation, *10*, 18, 44–45, *102–3*, 109–*11*, 123, 190–*91*, 220, 222f, 229, 231f
Corals, 12, 232
Corner glands, 65
Countershading, 9, 224–25, 233
Crabs, 21, 92
Cretaceous, 145
Cross-fertilization, 18, 44, 103, 110, 190–92, 220, 223, 231
Crustacea, 39, 41, 86–92 *passim*
Crystalline style, 86, 88
Ctenidium, *see* Gills
Ctenophores, 23, 41, 59, 98, 189

Dana Expedition, 28, 200
Deep scattering layer, 28
Defensive mechanisms, 9, 33, 42, 79, 214, 228–33 *passim*. *See also* Coloration
Deglacial periods, 144
Detorsion, 65, 147f
Detritus, 84, 90
Development, *see* Embryology; Polytroch larvae; Veliger larvae
Dextral coiling, 8, 29, 146
Diatoms, 85, 87, 92f, 97
Diet, *see* Food
Digeneans, 49, 130. *See also* Trematodes
Digestion, 39ff, 86, 93, 97, 207
Digestive gland: of adults, *16*, 30, *107*, *114*, *173*, *179*, *185*, 202, *216*, 220, 222; of larvae, *106*, *125*, *194*, *196*
Dinoflagellates, 85, 87, 92f, 100
Dinoflagellate toxins, 6, 100, 138
Dissolution, 50, 138–45 *passim*, 199
Dissolved organic material, 202
Distribution, *see* Geographic distribution; Vertical distribution
Diurnal vertical migration, 28, 48, 60, 79, 88, 96, 119, 132–37 *passim*, 168
Doliolids, 40
Dolphin fish, 39, 41, 99
Downwelling, 132, 134

Egg capsules, of *Janthina*, 18, *19–20*
Egg filament, of *Firoloida*, *33*, 45, 48
Egg masses, 45, *102*f, 113, 123–24, *192–97 passim*, 220, 222ff, 229, 231f
Eggs, 45, 103f, 106, 123, 192–97 *passim*, *194*, 203, 220, 229, 231
Eggs, numbers of, 20, 45, 103f, 113, 123, 193, 197, 220, 222, 229. *See also* Fecundity

Embryology, 18, 45, 106, 113, 123, 193, 197, 203, 231. *See also* Polytroch larvae; Veliger larvae
Embryos, 18, 20, *107*, *115*f
Endocrines, 203
Energy budgets, 96–98, 182–84
England, 8
English Channel, 103, 108, 197f
Enzymes, 86, 88, 207
Eocene, 51, 145, 148
Epifauna, 20–21, 49–50, 126–29, 130, 199
Epipelagic species, 28, 60, 134, 137, 168
Escape behavior, *see* Defensive mechanisms
Euphausiids, 40, 45, 134
Euthecosomes, 60–65 *passim*, 73–150; as prey, 39, 87f, 92, 178–89 *passim*
Evolution, of pelagic gastropods, 4, 22–23, 51–52, 145–50, 175, 204–7, 232–34
Excretion, 42–43, 95, 98
Exogastric form, 147
Eyes, 9, 22, 73; of veligers, 20, 104, *107*; of heteropods, 27–37 *passim*, 29, *31*, *33*, *36*. *See also* Light perception; Vision

Fecal pellets, 15, 86, 88, 92–93, 97, 100
Fecundity, 20, 103f, 106, 193, 197, 220. *See also* Eggs, numbers of
Feeding mechanisms, 12–14, 37–41, 52, 80–93, 178–89, 205, 217–23 *passim*, 228, 230–31
Feeding rates, 14, 183–84, 230–31. *See also* Filtration rate
Feeding specialization, 182–89 *passim*, 205–7, 219, 223, 232
Feeding webs, 60, 80–93 *passim*, 100, 148ff; of euthecosomes, 78, *81*, *82*, 186–*87*; of pseudothecosomes, *89*, *90*, *91*, 187–88
Fertilization, *see* Cross-fertilization; Self-fertilization
Filtration rate, 96f, 148
Fin, of heteropods, *29*–*36* *passim*, *31*, *33*, 48
Fin sucker, 29f, *31*f, 37, 44–45, 48
Fish, 21, 27, 35, 49, 51, 130, 137; as predators, 15, 39, 41–42, 98ff, 189
Fish eggs, as prey, 41
Fish larvae, as prey, 40f, 189
Fish mortality, 100
Flagellates, 49, 84, 87, 92. *See also* Dinoflagellates
Float, of janthinids, 8, *10*, 11–12, 18–23 *passim*
Flotation, *see* Buoyancy regulation
Food, of pelagic gastropods, 10, 12–15, 39–41, 80–93, 178–89, 202, 217–23 *passim*, 227, 230–31
Foot, 4; of janthinids, *10*–*11*, 22; of heteropods, 27, 30, 48; of thecosomes, 58, 146; of gymnosomes, 168, 172; of nudibranchs, 215, 221, 224, 229. *See also* Fin; Footlobes
Footlobes: of thecosomes, *63*f, 67, 69, *71*f, 88, *89*, 148f; of gymnosomes, *170*, *171*, *172*, *173*, 192, *196*, 201, 203
Foraminifera, 39, 84–88 *passim*, 92f, 131, 133, 138ff, 141
Fossils, 20, 48, 137, 199. *See also* Geological records
Fungi, 139

Gastropoda, 1ff
Geographic distribution, 12, 28, 60, 62, 73, 131–34, 168, 200, 233
Geological records, 22, 51, 61, 134–39, 145, 204
Gills: of janthinids, 11; of heteropods, 30, *31*f, *33*; of thecosomes, 65, 69, 147–48; of gymnosomes, *169*, *171*, *173*–*75*
Gizzard, 84–93 *passim*, *107*, 149, 202
Gizzard plates, 39, 146, 202
Glacial periods, 136, 144f
Gonad, *see* Reproductive anatomy
Greenland, 81, 191
Growth rates, 48–49, 108–9, 117–22, 126, 184, 194, 217, 224, 231–32, 233. *See also* Shell growth
Gulf of Maine, 107, 186
Gulf of Thailand, 134
Gulf Stream, 14, 108, 131f, 216
Gullet bladder, *169*
Gymnosomes, 58f, 72, 98–99, 167–209

Head, 9, 30f, 63f, 72, 171, 201, 203, 216, 221, 224
Heart, 30, *114*
Hemocoel, 178
Hermaphroditism: of janthinids, 15, 23; in thecosomes, 100–101, 104, 109, 123; in gymnosomes, 190f, 203; in nudibranchs, 219–20, 228–33 *passim*
Herring, 99f, 189
Heteropods, 4, 27–53, 59, 98, 130, 139, 142; as prey, 87, 92, 189

Historical accounts, 4–5, 9, 19, 27–28, 59–60, 199n
Holocene, 48
Holoplankton, 3, 214
Hooks, 175–81 *passim*, *176*, *177*, *179*, *180*, *185*
Hook sacs, *172–80 passim*, *177*, *179*, *185*f, 202
Hydroids, 21, 49–50, 126–29, *127*, 215, 232
Hydromedusae, *see* Medusae
Hypobranchial gland, 14, 16

Indian Ocean, 12, 62, 73, 137, 140, 200, 224
Indicator species, 131–39
Indo-Pacific, 44, 200
Ink gland, 202–3
Integument, 173, 175, 201
Intestine, 30, *125*, *173*, *194*, 202
Ionic regulation, 37, 77f, 171
Ireland, 8
Iridiocytes, 72
Iridophores, 33
Irminger Current, 186
Isopods, 139

Janthinids, 4, 8–24, 27, 226–27, 232
Jaws: of janthinids, 14; of thecosomes, 72, 81, 88; of gymnosomes, 174f, *177*, 186, 202; of nudibranchs, 217, 221, 227f, 230
Jellyfish, *see* Medusae
Jurassic, 22

Keel, 29, 34, 51
Kelp, 230
Kidney, 30

Labrador Current, 131
Labrador Sea, 191, 197
Lancetfish, 42, 99, 189
Lanternfish, *see* Myctophids
Larvacea, 23, 219
Larvae, *see* Polytroch larvae; Veliger larvae
Larval shell, *see* Protoconch; Veliger larvae
Length of life, *see* Longevity
Light perception, 9, 65, 73, 172
Limpets, 23
Lipids, *see* Oil droplets
Longevity, 108–9, 122, 232f
Luminescence, *see* Bioluminescence

Mackerel, 99, 189
Malaysian Archipelago, 200
Mantle, 30, 168; of thecosomes, 65, 67, *68*, 77f, 81, *89*, 106, 116f, 150
Mantle appendages, 65, *66*, 77, 82, 85–86, *111*, 129, 150
Mantle cavity, 11, 30, 65, 69, 72, 106, 116, 150, 168, 214
Mantle gland, *see* Hypobranchial gland; Pallial gland
Mantle lobes, *see* Mantle appendages
Marlin, 15
Mating, *see* Copulation
Mediterranean Sea, 41, 45, 136f, 215–20 *passim*
Medusae, 23, 41f, 98, 127, 129f, 189, 217–23 *passim*
Melanophores, 72
Meroplankton, 2
Mesopelagic species, 60, 62, 116, 137, 150, 168
Mesopodium, 10, 19, 48
Metabolism, 42–43, 96, 98
Metamorphosis: in janthinids, 12, 19f; in heteropods, 48; in thecosomes, *105–9 passim*, *107*, *115–16*, 124–*25*; in gymnosomes, 194–98 *passim*, *196*; in nudibranchs, 220, 231, 233
Metapodium, 48, 224, *225*. *See also* Postpodium
Migration, *see* Diurnal vertical migration
Minute developmental forms, 117
Miocene, 51, 145, 148
Monoplacophora, 1f
Morphological artifacts, 5, 28, 61, 65, 67, 112–*14*, 117, 147, 167–68, 190
Mouth: of heteropods, 29–33 *passim*, *31*; of thecosomes, *63*f, *68–73 passim*, 80–90 *passim*, *105*; of gymnosomes, 172, *173*, *175*; of nudibranchs, 216, *226*
Mucous glands: in *Janthina*, 11, 20; in thecosomes, *70*, *71*, 88, 103f, 106; in gymnosomes, 190. *See also* Pallial gland
Mucous strands, for flotation, 12, 35, *36*, *78*
Mucous webs, *see* Feeding webs
Myctophids, 21, 42, 49, 99, 130

Nanoplankton, 90
Needle whelks, 23
Nekton, 2
Nematocysts, 14f, 23, 41, 128f, 217, 220, 228–33 *passim*

Nematodes, 49
Neoteny, 146, *172*, 185f, 190, 195
New Zealand, 8
Nudibranchs: pelagic species, 15, 214–34; benthic species, 199, 214, 224, 228f, 232
Numbers of species, 2ff, 9, 27–32 *passim*, 60–67 *passim*, 167, 215, 224, 229

Oil droplets, in integument, 171, *172*, 201
Oligotrophic environments, 90, 148
Ooze, *see* Pteropod ooze
Operculum: of janthinids, 10–11, 20, 22; of heteropods, 29f, 48, 50; of thecosomes, 63, 67, *68*, 104, *106*, 124, *125*, 146
Orientation axes, 34, *68*f, 147
Osphradium, 11, 30, 65, 73, 172
Oviparity, 18, 19–20
Ovotestis, *see* Reproductive anatomy
Ovoviviparity, 104–5, 116, 150, 203. *See also* Brood protection

Pacific Ocean, 62, 99, 103, 133–44 *passim*, 200, 215, 224, 233
Palaeoecologic change, 134–39
Palaeozoic, 145
Pallial gland, *64*, *68*, *70*, *71*, 80–88 *passim*, 93, *114*, 149f. *See also* Hypobranchial gland
Papillon de mer, 60
Parapodia, *see* Wings
Parasitism, 21, 49, 128, 129–30, 199, 217–23 *passim*
Pedal gland, of nudibranchs, 216f, *218*, 222
Penis: of heteropods, 29, 44f; of thecosomes, *101*ff, 105, 109f, *111*, 123; of gymnosomes, 190; of nudibranchs, 220, 223, 229, 231
Periostracum, 11
Phylogeny, 23, 51f, 145–49, 204–5
Phytoplankton, 81–93 *passim*, 108, 194f, 206. *See also* Coccolithophorids; Diatoms; Dinoflagellates; Flagellates
Pigmentation, *see* Coloration
Plankton collection, *see* Collection methods
Pleistocene, 48, 134–38 *passim*, 145
Pleuston, 8, 15, 22, 215, 224, 229
Pollution, 137–43 *passim*
Polychaetes, 40f
Polyplacophora, 1f
Polytroch larvae, 170f, 195–99 *passim*, *196*
Population densities: of thecosomes, 6, 60ff, 73, 135–39 *passim*, 150; of janthinids, 15; of heteropods, 28, 51; of gymnosomes, 168, 199f; of nudibranchs, 220
Portuguese man-of-war, *see Physalia*
Postglacial periods, 136f, 144
Postpodium, 10. *See also* Metapodium
Predators: of janthinids, 9, 15; of heteropods, 41–42; of thecosomes, 98–100; of gymnosomes, 168, 189
Preservation peak, 144f
Preservatives, chemical, 5, 61, 112–17 *passim*, 167. *See also* Morphological artifacts
Prey location, 13, 37, 40, 217, 228, 230
Prey recognition, 179–81
Prey selection, 12, 40f, 181–82
Primary specimen, 112f
Proboscis: of janthinids, *10*, 14; of heteropods, 29f, *31*, *32*, *33*; of pseudothecosomes, 61, 67, 69, *70*, *71*f, 88, *90*, 149; of gymnosomes, *173*ff, 186, *187*, 202f
Proboscis sac, 177
Propodium, 10f, 48, 226
Prostate gland, 45, *101–5 passim*, 109, 190
Protandry, *see* Hermaphroditism
Protoconch, 116–17. *See also* Veliger larvae
Protozoa, 84–92 *passim*. *See also* Ciliates; Flagellates; Foraminifera; Radiolaria
Pseudoconch: of pseudothecosomes, 61, *68*f, *70*, *71*, 77f, 124, 126, 147, 149f; of cavoliniids, *66*, 77–78, 129, 148ff
Pseudocopulation, 18
Pseudofeces, *82–89 passim*, 100
Pseudothecosomes, 61–73 *passim*, 88, 92, 98, 187–88
Pteropoda, 4, 58f, 204
Pteropod ooze, 6, 28, 50, 134, *135*, 140, 144
Pulmonates, 1f

Radiolaria, 84f, 87, 92
Radula: of ptenoglossans, 12, *13*f; of heteropods, 30f, 37–41 *passim*, *38*, 52; of thecosomes, 81, *83*, 88, 146; of gymnosomes, *172–80 passim*, *176*, *177*, *185*f, 202; of nudibranchs, 217, 221, 227f, 230

Raft, of janthinids, *see* Float
Redfish, 99
Red Sea, 136, 139
Red tide, 138
Reproductive anatomy: of janthinids, 15–17, *16*, 23; of heteropods, 44; of thecosomes, *101–9 passim*, *107*, 123, 149; of gymnosomes, 189–90, 204; of nudibranchs, *216–23 passim*, *222*, 228–29, 231. *See also* Albumen gland; Penis; Sperm groove
Respiration, 11, 42–43, 95–98
Rhinophores, *216*f, 221–29 *passim*, *222*, *225*

Salivary glands, 14, 72, 88, *177*, *179*, *180*, *185*, 202
Salmon, 99, 189
Salps, 23, 40f
Santa Barbara Basin, 137
Sargasso Sea, 108, 119, 131f
Scaphopoda, 1f
Scattering layer, 28
Schizogamy, 112–13, 149
Scuba diving, 5, 35f, 40, 44, 61, 65, 80f, 93, 109, 112, 167, 190, 201
Scyphozoa, *see* Medusae
Sea anemones, 12, 15, 232
Sea angel, 178
Sea butterfly, 60
Sea level, changes in, 137–38
Seals, 99
Sea slugs, 1, 214
Sediments, 50–51. *See also* Pteropod ooze
Sediment traps, 50, 141–42
Self-fertilization, 15, 18, 23, 103, 112f, 193, 229, 233
Sex change, 15, 100–101, 123, 190. *See also* Hermaphroditism
Sexes, proportion of, 44
Shell growth, 50, 106–7, 117–22, *121*, 194–95, 197
Shell microstructure, 50, 63, 117–18, *120*, 148
Shell reduction and loss, 3, 27–34 *passim*, 51, 63, 145, 168
Shell thickness, 29, 50, 63, 118, 122
Sinistral coiling, 62f, 67, 146f
Sinking rates, 35, 74–78 *passim*, 142f
Siphonophores, 12, 14f, 40ff, 98, 217, 221–32 *passim*. *See also Nanomia*; *Physalia*
Sipunculids, 139
Size: of cephalopods, 3; of janthinids, 9, 11; of heteropods, 29f, 32; of thecosomes, 63, 67, 70, 73; of gymnosomes, 168, 184, 201; of nudibranchs, 215f, 221, 224, 229. *See also* Weight
Skinny developmental forms, 117
Slope Water, 131f
Sole (fish), 99
South China Sea, 134, 136, 144
Spawning, 45, 113, 123, *124*, *192–93*, 195, *222*
Species, numbers of, *see* Numbers of species
Spermatophores, *31*, 45, 52, *102–11 passim*, *107*, 149, 191
Spermatozeugmata, *17*–18, 23
Spermatozoa, 17–18
Sperm groove, 45, *101*f, 109, 190
Sporozoa, 49
Squid, 2f, 51
Statocysts: in veligers, 20, 103–4, 115, 124, 197, 224; in adults, 22, 37, 65, 73
Stomach, 30, *106*, *125*, *173*, 202
Strobilation, 112, 149
Style sac, 86, 88, 149
Stylets, 14, 109
Subantarctic, 132f
Subarctic, 103, 108, 168. *See also* Labrador Current; Labrador Sea
Submersibles, 6, 190
Subtropical species, *see* Circumtropical species
Sucker-bearing arms, *173–77 passim*, 186, 202, 205
Suckers, 174, 177, 186
Swimmers' complaints, 61, 228
Swimming, 34–36, 73–79 *passim*, 147, 170–71, 201, 215f, 221, 232
Swimming rates, 35f, 74ff, 79, 170–71, 216, 221
Symbiosis, 129, 220–21

Taenioglossate radula, 38
Tail filaments, 31–32, *33–34*, 42
Tectibranchs, 204
Teloconch, 117
Tentacles: of janthinids, 9; of heteropods, 30ff; of thecosomes, 65, 69, 72f; of gymnosomes, 171–72, *200–203 passim*; of nudibranchs, 224, *225*, 229. *See also* Rhinophores; Wing tentacles
Tintinnids, 85, 87, 92, 100
Toxins, 14. *See also* Dinoflagellate toxins
Transparency, 27–34 *passim*, 42, 72, 173, 215f, 221, 233

Trematodes, 21, 130, 221. *See also* Digeneans
Tropical species, *see* Circumtropical species
Tuna, 41–42, 99
Turtles, 15, 42, 232

Veliger larvae: of janthinids, 12, 18–20, *21*; of heteropods, 46, 47f; of thecosomes, 101–7 *passim*, *105*, *106*, 113, *115*–16, 123–*25*; of gymnosomes, 193, *194*, *195*, *196*f, *198*; of nudibranchs, 220, 223, 229, 231
Veligers, as prey, 39, 86, 88, 92
Velum, *see* Veliger larvae
Vertical distribution, 28, 37, 60, 62, 132–34, 137, 168, 198–99, 203
Vertical migration, *see* Diurnal vertical migration
Vision, 37, 40, 42. *See also* Eyes; Light perception
Viviparity, 18

Water mass identification, 131–34
Webs, *see* Feeding webs
Weight, 50, 94, 97, 118f, 182ff, 232
Wentletrap snails, 12, 22
Whales, 99, 168, 189
White Sea, 197
Wingplate, of pseudothecosomes, *67*–*78* *passim*, *68*, *70*, *71*, *73*, 88, *89*, *90*, *124*, *125*f
Wings: of euthecosomes, 58, *62*–*68 passim*, *63*, *64*, *66*, *73*–*74*, *81*, 85f, 104, *107*, *111*, *114*; of gymnosomes, 58, 168–*70*, *169*, *179*, *192*–203 *passim*, *200*
Wing tentacles, *62*, *72*, *73*, 78

Yolk, 18

Zooxanthellae, 220–21

Library of Congress Cataloging-in-Publication Data

Lalli, Carol M.
Pelagic snails : the biology of holoplanktonic gastropod mollusks
Carol M. Lalli and Ronald W: Gilmer.
p. cm.
Bibliography: p.
Includes index.
ISBN 0-8047-1490-8 (alk. paper) :
1. Gastropoda. I. Gilmer, Ronald W., 1950– II. Title.
III. Title: Holoplanktonic gastropod mollusks.
QL430.4.L28 1989
594'.3—dc19 88-20116
CIP